CLIMATE, HISTORY AND THE MODERN WORLD

We live in a world that is increasingly vulnerable to climatic shocks – affecting agriculture and industry, government and international trade, not to mention human health and happiness. Serious anxieties have been aroused by respected scientists warning of dire perils that could result from upsets of the climatic regime.

In this internationally acclaimed book, Hubert Lamb explores what we know about climate, how the past record of climate can be reconstructed, the causes of climatic variation, and its impact on human affairs now and in the historical and prehistoric past.

This second edition incorporates important new material on: recent advances in weather forecasting, global warming, the ozone layer, pollution, and population growth. Providing a valuable introduction to the problems and results of the most recent research activity, this book extends our understanding of the interactions between climate and history, and discusses implications for future climatic fluctuations and forecasting.

H. H. Lamb is Emeritus Professor in the School of Environmental Sciences and was the Founder and first Director of the Climatic Research Unit at the University of East Anglia.

To Moira, my wife,
for her loving constant support

CLIMATE, HISTORY AND THE MODERN WORLD

Second edition

H. H. Lamb

London and New York

First published 1982
Reprinted 1985
Second edition 1995
by Routledge
11 New Fetter Lane, London EC4P 4EE

Simultaneously published in the USA and Canada
by Routledge
29 West 35th Street, New York, NY 10001

© 1982, 1995 H.H. Lamb

Typeset in Garamond 10/12 by
Florencetype Ltd, Stoodleigh, Devon

Printed and bound in Great Britain by Biddles Ltd, Guildford, Surrey

British Library Cataloguing in Publication Data
A catalogue record for this book is available from the British Library

Library of Congress Cataloguing in Publication Data
A catalogue record for this book has been requested

ISBN 0–415–12734–3
0–415–12735–1 (pbk)

CONTENTS

CONTENTS

CONTENTS

CONTENTS

Part III Climate in the modern world and questions over the future

ILLUSTRATIONS

TABLES

ACKNOWLEDGEMENTS

My first thanks are due to my wife and family for their loyalty, tolerance and understanding of my domestic shortcomings and nights shortened by late hours and early rising.

My debt to Shell International, to the Nuffield and Wolfson Foundations, and to the University of East Anglia, as well as to individual members of those bodies, must also be acknowledged for making possible the setting up of the Climatic Research Unit at the university in Norwich in 1972 and ensuring its survival through the very difficult first years. I also wish to express my great gratitude to the Rockefeller Foundation for its vision and its trust in sponsoring, and sustaining over many years, the Historical Weather Mapping Project in the Unit. My thanks are due to the same foundation and to the Ford Foundation for their financial support of the First International Conference on Climate and History, which was held at Norwich in 1979.

These are the things which made the writing of this book possible.

At this point I also express my sincere thanks to Mrs Margaret Gibson, who once again has produced a book typescript for me with remarkable speed and accuracy.

I acknowledge with happy memory many other debts: to my old colleagues and friends, now dead, my former chief, Sir Graham Sutton, FRS, Director General of the United Kingdom Meteorological Office until 1965, and to Professor Gordon Manley, founder of the School of Environmental Sciences at the University of Lancaster, for their inspiration and constant encouragement; also to colleagues too numerous to mention individually, both here in the university and far away in many countries, for valued discussions and permission to quote their work, but especially to Professor A. Berger and Dr G. Woillard of the Catholic University of Louvain, Professor T. O'Riordan of the University of East Anglia and Dr C. J. E. Schuurmans of the Koninklijk Nederlands Meteorologisch Instituut, De Bilt, Holland, who kindly read and made suggestions on various chapters for me. Nevertheless, any errors of fact or interpretation are my own. I also wish to thank for valuable discussions, guidance and basic information the leaders of the Nordic Abandoned Farms Project in all the Scandinavian countries and most specifically Professor A. Holmsen of Oslo, Professor J. Sandnes and Dr H. Salvesen of Trondheim, Professor Sv. Gissel of Copenhagen, Eva Österberg of Lund and Björn Teitsson of Reykjavik, as well as the research staff of the Arkeologisk Museum in Stavanger. I am particularly indebted for basic information also to Dr A. Bourke, formerly Director of the Irish Meteorological Service and President of the World Meteorological Organization's Commission for Agricultural Meteorology, Dr P. Brimblecombe of the University of East Anglia, Professor R. A. Bryson of the University of Wisconsin, Professor H. Flohn of the University of Bonn, Dr K. Frydendahl of the Danish Meteorological Institute, Dr J. Maley of Montpellier, Mr V. Morgan of the School of English and American Studies, University of East Anglia, Dr E. J. Moynahan of Guys Hospital, London, Dr M. L. Parry of the University of Birmingham, and, perhaps most of all, to the young historian Dr Chr. Pfister of the University of Bern, Switzerland. I think with happiness also of the kindness of Hr Øystein Bottolfsen of Stokmarknes in the Lofoten Islands, Dr H. Rohde of Hamburg, Professor R. S. Scorer of Imperial College, South Kensington, and Dr P. A. Tallantire for their successful efforts in helping me to obtain the material for figs. 81, 69, 115, 116 and 89. I also wish to thank particularly Dr J. Murray Mitchell of the United States Environmental Data Service, Washington, DC, for information and encouragement.

Many individuals and organizations seem to have taken pleasure in supplying their own diagrams, maps and pictures for use in this book, and I am indebted also for their leave to use their copyright. In particular, my thanks go to Arkeologisk Museum, Stavanger for fig. 49; Dr Keith Barber for fig. 73; Mr Guy Beresford of Rolvenden, Kent for fig. 63; Professor A. Berger for fig. 118; Hr Øystein Bottolfsen and Ivar Toften of Vesterålen for figs. 81a and b; Dr Humberto Bravo of the Centre for Atmospheric Sciences in the University of Mexico for fig. 116; the Trustees of the British Museum for fig. 57a; Mr Bruce Dale (photographer) and the National Geographic Magazine for fig. 58; Professor W. Dansgaard of Copenhagen for figs. 35, 36 and 112; Mr P. E. Baylis and the Department of Electronics and Electrical Engineering, University of Dundee for fig. 13; General Fea of the Servizio Aeronautica Militare Italiano, Rome for figs. 68a, b; Professor H. Flohn for figs. 33b and 96; Dr C. U. Hammer, Copenhagen for fig. 112; the Controller of Her Majesty's Stationery Office for figs. 4 and 6 (Crown copyright); Dr D. V. Hoyt of the National Center for Atmospheric Research, Boulder, Colorado for fig. 110; the Icelandic Weather Bureau, Reykjavik for fig. 97; Mr J. A. Kington of the Climatic Research Unit for fig. 27; the Kunsthistorisches Museum, Vienna for fig. 84; Professor V. C. La Marche of the Laboratory of Tree Ring Research, Tucson, Arizona for figs. 52, 53; Professor Leona M. Libby for fig. 33a; Library of Congress,

Washington, DC for fig. 109; the Museum of London for fig. 83a; Macdonald, Dettwiler & Associates Ltd, Richmond, BC, Canada for fig. 2; Dr J. D. McQuigg, Columbia, Missouri for fig. 105; Dr V. Markgraf of Tucson for figs. 43 and 53; Messrs Methuen of London for use of the copyright of figs. 8, 9, 15, 64 and 65; Professor J. K. St Joseph of Cambridge and the Ministry of Defence, London, Air Photography Branch (Crown copyright) for fig. 74; the National Climatic Center, Washington, DC for figs. 3a and b; the Trustees of the National Gallery, London for figs. 85a, 88a and 88b; the Oskar Reinhart Collection, Winterthur, Switzerland for fig. 85b; Oxfam for fig. 107; Dr M. L. Parry for figs. 102, 103; Professor R. E. Peterson of Texas Tech University and Mr Carl Holland of Plainsview, Texas (photographer) for fig. 11; Dr Chr. Pfister for fig. 76; Mr I. J. W. Pothecary of the Meteorological Office for figs. 80a, b; the Public Records Office, London for fig. 32; Dr F. Röthlisberger of Zürich for fig. 61; the Royal Society, London for figs. 46 and 47 (which originally appeared in the Philosophical Transactions); Drs K. R. Saha and D. A. Mooley of the Indian Institute of Tropical Meteorology, Pune for fig. 100; Scandia Photopress AB and Sydsvenska Dagbladet, Malmö for figs. 89a, b; Fraulein L, Schensky of Schleswig for kindly supplying her late father's, F. A. Schensky's, photographs used in figs. 69a, b; Dr W. Schneebeli of Zürich for figs. 94a, b; Dr G. Singh of the Australian National University. Canberra for fig. 54; Professor Trygve Solhaug of Bergen for fig. 81c; the late Professor A. Thom of Dunlop, Ayrshire, for fig. 47; Professor Sigurdur Thorarinsson of Reykjavik for fig. 62; the United States Geological Survey, Reston, Virginia for fig. 111; Verlag Brüder Rosenbaum, Vienna for figs. 38a, b; Drs T. M. L. Wigley and T. Atkinson for the data used to construct fig. 104; and Dr T. Williamson of the Science Museum, South Kensington for fig. 29.

Many maps and diagrams drawn by Mr David Mew, and photocopied by Mr Peter Scott, of the School of Environmental Sciences in the University of East Anglia have been incorporated in this work, and the contribution of their skills and advice is gratefully acknowledged.

Finally, I would like to thank Mr Peter Wait and Miss Janice Price of Methuen & Co. for the original invitation many years ago and for their continual encouragement, which have led to the publication of this book.

PREFACE

We live in a world that is increasingly vulnerable to climatic shocks. After some decades in which it seemed that technological advance had conferred on mankind a considerable degree of immunity to the harvest failures and famines that afflicted our forefathers, population pressure and some other features of the modern world have changed the situation. In the years since about 1960, moreover, the climate has behaved less obligingly than we had become used to earlier in the century. And there is alarm about how man's activities might inadvertently upset the familiar climatic regime and therefore disrupt the food production which is geared to it. This concern has in recent years largely replaced the debate which had begun earlier about the possibilities of deliberate action to change world climate so as to increase the total cultivable area. Serious anxieties have been aroused by respected scientists, acknowledged as experts in the field, warning of dire perils: that the next ice age may be now due to begin, and could come upon us very quickly, or that the side-effects of man's activities and their ever-growing scale may soon tip the balance of world climate the other way and for a few centuries produce a climate warm enough to melt the Greenland and Antarctic ice-caps, raising the sea level and drowning most of the world's great cities.

This book examines what we know about climate, and its impact on human affairs now and in the historical and prehistoric past, and how we may better understand the problem of climatic fluctuations and changes. Climatic forecasting in the strict sense may be far off, though premature claims are made from many sides. But much has been learnt about the laws which govern the behaviour of climate. We are already in a much better position than previous generations to understand the past and assess our present situation, so as to make more rational provision for the future than our forefathers could.

Many parts of the world have experienced more extremes of weather of various kinds in the last fifteen to twenty-five years than for a long time past and have suffered losses, which have affected political decisions and managerial decisions in industry and land-use. Energy problems are also involved.

In these and other ways climate and our understanding of it are very much part of the problems of the modern world. The writer hopes that this book may serve as a guide to the present state of knowledge and the potential capacity of science in these matters, and also that it may provide some helpful insights now to those on whom the burden of weighty decisions falls – affecting practical matters in agriculture and industry, government and international trade, not to mention human health and happiness.

H. H. Lamb
September 1981

PREFACE TO
THE SECOND EDITION

Since this book was published in 1982 its subject has been continually in the limelight and research has been active. Also, as is by no means unusual, further noteworthy weather events have been in the news. Some additional reports, remarks and comments have therefore become desirable, yet the main body of past historical work is still not well known. It has therefore been decided to issue this revised text which incorporates notices of much new, important, material, thus making our knowledge of the past – particularly the interactions between climate and history – more accessible and providing a handy introduction to some of the problems and results of ongoing research.

Some of the climatic problems affecting humanity arise perhaps more fundamentally from the pressures of the burgeoning human population of the world than from climate. But climate has been the trigger – repeatedly in recent years – for natural disasters such as famines in Africa and typhoon floods killing large numbers in Bangladesh. The wars in Iran and Iraq as well as some of the outbreaks of violence in Korea, China, Vietnam and Cambodia a few years earlier may usefully be considered in relation to climate as well as man-made stresses.

Anxieties about the possibility of drastic warming of world climates resulting from the continual build-up of carbon dioxide (and other intrusions) in the atmosphere due to human activities have been forced upon the notice of politicians and industrial managements. Even more urgently the discovery of serious damage to the protective ozone layer in the stratosphere, exposing us all to lethal amounts of the sun's ultra-violet radiation, demands attention, including some reversal of widely popular human habits.

In these years there has also been a succession of very great volcanic eruptions that have loaded the atmosphere with debris and, perhaps more importantly, with gases and vapours that veil the sun's radiation and may be interrupting or even reversing the tendencies towards warming of world climates. One eruption, that of Mount St Helens in 1980 in the western USA, has forced us to note how limited must be the usefulness of applying statistically based rules of thumb connecting measures of the magnitude of any eruption with the climatic effect: for the main thrust of ejected material on that occasion was nearly horizontally, very much less being directed to the stratosphere.

There have been very notable advances in these years in weather forecasting by mathematical models, enormously improving the forecasting for up to five to seven days ahead. But much of the gain is jeopardized by modern tendencies to use sloppy and inappropriate language in forecasts. Thus, it is now fashionable to speak of 'best temperatures' in forecasts rather than 'highest' or 'lowest' whichever may really be best for the activities in prospect. And forecasters in southern England seem to like to assume that summer temperatures in England are much the same as in the Mediterranean, or if they are not, they *should* be and it is a bad year.

The idea of climatic change has at last taken on with the public, after generations which assumed that climate could be taken as constant. But it is easy to notice the common assumption that Man's science and modern industry and technology are now so powerful that any change of climate or the environment must be due to us. It is good for us to be more alert and responsible in our treatment of the environment, but not to have a distorted view of our own importance. Above all, we need more knowledge, education and understanding in these matters.

Hubert Lamb
Holt, Norfolk
December 1994

1

INTRODUCTION

Most generations of mankind in most parts of the world have regarded climate as an unreliable, shifting, fluctuating thing, sometimes offering briefly unforeseen opportunities but at other times bringing disaster by famine, flood, drought or disease – not to mention frost, snow and icy winds. Before the days of records and reference books with figures for past years, there could hardly be any clear perception of trends. In old writings, including those by fine observers such as John Evelyn and Samuel Pepys, we come across too frequent references to such items as 'the severest winter that any man alive had known in England', 'so deep a snow that the oldest man living could not remember the like', and so on. Yet here and there we do find recognition of long-lasting changes. There was no mistaking this when the glaciers in the Alps, in Iceland and in Norway, during the seventeenth century and thereabouts, were advancing over farms and farmland. Doubtless, the nomadic peoples of the past or present in every continent have been aware of such changes at times when their pastures were drying up. It must have been equally clear, at least to some, when in various countries in the late Middle Ages traditional crops and croplands had to be given up and taxes 'permanently' reduced. When, on the other hand, the climate becomes warmer or more convenient for human activities, it tends to be taken for granted and the change may for a long time pass unnoticed. A probably rare awareness of a change of this sort occurs in a passage in the ancient Roman horticultural work *De Re Rustica* (Book I) by Columella, citing a statement by 'the trustworthy writer Saserna' in the early part of the first century BC that 'regions [in Italy] which previously on account of the regular severity of the weather could give no protection to any vine or olive stock planted there, now that the former cold has abated . . . produce olive crops and vintages in the greatest abundance'. In another situation much later in history we may detect a slowly dawning awareness, possibly in very vague form, of a climatic change when the vineyards of medieval England, some of them cultivated for hundreds of years, were given up after many years of dismal failure.

Yet for eighty years or more, down to about 1960, it was generally assumed that for all practical purposes and decisions climate could be considered constant. This view seemed at the time to be soundly based in science; the first long series of regular meteorological observations made with instruments in the cities of Europe and North America showed the climate of the late nineteenth century to be very similar to the period about a hundred years earlier when the observations had been instituted. Many working practices in applying climatology to forward planning are, for better or for worse, still based on this assumed constancy of climate. Everyone will agree that if climate were defined by the statistics of weather over a sufficiently long time, it would be effectively constant. But how long would this period have to be? If we were to take the conditions of the last million years as a basis, the repeated swings from ice age to warm inter-glacial conditions and back again would have to be regarded as part of the normal climate. Yet the changes would be sufficient to wreck the economy many times. The practical choice, a definition relevant alike to our indi-vidual concerns and to national and international affairs, is surely to consider that the climate has changed if the conditions over some large part of a human lifetime differ significantly from those prevailing over an earlier or later period of similar duration.

We live in a time of renewed perception of climatic and environmental change. For many people this arises from fears about the possibility that man's activities, and their increasing scale and variety, may have side-effects that disturb the climatic regime, just as they are visibly changing other aspects of the environment about us. Others may be interested in the possi-bility of using the increasing power of our technology deliberately to modify the climate: for instance, to increase the total cultivable area of the world or, sad to say, to change the pattern of climate as a possible strategy of war. In any case, many people now know that there have been significant shifts of climate during the twentieth century: at first, a more or less global warming to about 1950, then some cooling. More recently, a notable increase in the incidence of extremes of various kinds in almost all parts of the world has hit agriculture and created difficulties for planning in many fields.

The former assumption of constancy of climate is thus widely felt to be unsatisfactory today. And, after many decades in which there was little or no inquiry about climatic development and change, the leading institutes of meteorology and climatology are now pressed for advice on future climate. The position is doubly unfortunate in that the forecast opinions ventured by the 'experts' have often increased the confusion, the views of the theor-eticians sometimes contradicting those whose study has been concentrated on reconstructing the actual past behaviour of the (natural) climate.

The assumptions that were common until recently among knowledge-able people outside the sciences of meteorology and climatology are well illustrated by Jacquetta Hawkes and Sir Leonard Woolley writing in volume 1 of the UNESCO *History of Mankind* (London, Allen & Unwin, 1963). After recounting the drastic changes of the ice ages, interglacials and early postglacial times, they stated

> by about 5000 BC when the first agricultural communities were already extending in Asia, the climate, the distribution of vegetation and all the related factors had settled to approximately their present condition. When true civilization at last began, not only was Homo Sapiens and the agricultural basis of his existence firmly established, but the natural environment which was to form the background of all subsequent history had already assumed the form which we ourselves have inherited.

Most archaeologists today realize that climate and environment have a more interesting history than that.

In reality, during our lifetime and that of the structures which we build, the climate is always changing to a greater or less degree. And the landscape that goes with it, the ranges of vegetation and of the animal species, birds and insects that inhabit its provinces, change too – mostly rather slowly but sometimes more quickly. The changes in these realms are on the whole more gradual than the swings of weather and climate which instigate them, but they also undergo their disasters and depopulations, recoveries and advances. This book, besides introducing the evidence on which past climate can be reconstructed, presents the story of this continual ebb and flow and of the more lasting shifts of climate. We shall see in outline how these changes happen continually and how the fortunes of the flora, fauna and human populations are forever being affected. Just how some of the impacts work will be examined in more detail in chapter 15.

It is true that, as Jacquetta Hawkes and Leonard Woolley put it, for many thousands of years the zones we know have been present and identifiable somewhere on this planet, 'the jungle has been there for the pygmy, the grassland for the nomad or the cultivator, and the ice-floes for the Eskimo', but the movements of their margins have caused much trouble from generation to generation and continue to do so. In looking for evidence of climatic impact in the course of history, it is sensible to look most at the marginal areas near the poleward and arid limits of human settlement and activity, for it is there that vulnerability is likely to be greatest. In regions like the lowlands of western and southern Europe most of the effects of climatic changes are liable to be obscured by successful competition of the societies living there with the inhabitants of regions more adversely affected.

Often people think about history (and some historians have written about it) as if it were basically a tale of the deeds of great men and women. These heroes and heroines, the causes which they led, and the crises and battles which resulted, are commonly thought of as having determined the structure of society in the times that followed. Of course, economic crises arose from time to time and had some influence on the course of events. But many aspects of the economy, and the landscape which developed with it, have been largely seen as products of great decisions and decisive battles. Alternatively, from the Marxist view of history it is all a question of the development of man's technology and the tools which at any given time were at his disposal to conquer and exploit the world about him. The assumption that the climate, the opportunities which it offers and the constraints it places upon man and the environment are effectively constant generally underlies all these views.

Some readers may at this point decide that the interpretation of history in these pages is but a resurrection of climatic determinism, an over-simplified view which they rejected long ago. Such labelling only tends to restrict freedom of thought. Who can deny that there are cases when a desert or a marsh, an ice-cap or a glacier, or indeed the sea, has advanced over land that had been settled and used for agriculture; and in these extreme cases there is no doubt that a climatic change, or the accumulating consequences of some tendency of the climate over previous years, has dictated human action. Most situations are, of course, far more complex and allow the human populations some choices. But, even in many of these, to write history without reference to the record of climate is to make matters more obscure than they need be and may amount to making nonsense of the story.

Progress towards understanding inevitably has its difficulties. Some historians of yesteryear who were interested in the possible impact of climate were not helped by inaccuracies that were probably unavoidable at the time in the first reconstructions of the climatic sequence during the centuries before meteorological instrument records began. But the last thirty years or more have seen great advances in the quantity and variety of evidence of past climate and in the methods available to interpret the evidence. Gradually, we are gaining a more reliable record of the climate, the main features of which have already been corroborated by independent data and methods.

When we compare this record with the course of human history and the still longer record revealed by archaeology, we cannot fail to be struck by the many coincidences of the more catastrophic events in both. This again raises the question: what exactly was the role of climatic disturbance in the human story in each case? It will certainly be difficult, and may be dangerous, to generalize. There is room for many detailed investigations to improve our understanding. But, in general, it seems helpful at this stage to think of climate as a catalyst or at the least a trigger of change: in the major breakdowns of societies and civilizations climatic shifts may often be found to have played the role of a trigger, rather like the recently

4

recognized trigger action of the variations of the tidal force in setting off earthquakes and volcanic eruptions.

Historians and others have also been confused and uncertain as to how far changes of climate and environment, some of which have had an impact on history, could have been caused by human activities. Many people in every part of the world today, including those who are generally well informed (and among them some meteorologists), are plainly predisposed to the opinion that *if* the climate is not as constant as we used to think it, this must be due in some way to the impact of man. The impact of human activities on other aspects of the environment is only too obvious and began at least as far back as the first clearance of forests for settled agriculture thousands of years ago. And, with the now rapidly growing scale and power of our technology, new possibilities of effects upon the climate – whether intentional or inadvertent – must in all reason be watched for.

Both human history and the history of climate are often thought of as cyclic, as if we are all caught in a wheel of fate whose turning is not only remorseless but knows no intelligible causes. In these pages an attempt is made to understand rather more about these variations. We shall see the workings of nature, and latterly some possible intrusions of man, in the continual development and fluctuations of climate. And we shall observe how these break in upon the course of human affairs.

By now, it is clear that the only approach that is likely to be profitable to the hints of cyclic recurrences in climate is to seek to identify the evolutions in the atmosphere, oceans and terrestrial or extraterrestrial environment which mark – and in some cases cause – their successive phases. In this way we may come not only to understand the physical processes which lead from one phase to the next, and what controls or varies the timing, but ultimately to discern the origins of the whole sequence. One example, illustrated in chapter 4, is the sequence which typically follows a great volcanic eruption and leads to the formation of a persistent dust veil in the stratosphere. It is equally clear that there are various cycles in human affairs, whose causes also need to be understood. Some of these are linked to cyclic phenomena in climate and the environment. Others certainly are not. The cycle of day and night is linked with variations in the death rate and in the incidence of criminal and other activities, some of which disturb the peace. Next in the scale, the seasonal round of the year and each year's seed-time and harvest mark out times when people's health and energy are commonly at their best, times when the stresses of dearth, undernourishment and starvation are most likely, times when travel is easiest and times favourable for military adventures. Operating over a longer time-scale covering a few years, we observe cycles of confidence in business activity, the trade cycle, and, similarly, the swings of the political pendulum which seem not to be wholly masked even in totalitarian states. Over longer periods, ranging from one generation to several centuries in length, we observe swings from strong or dictatorial rule to democracy, too often gradually degenerating into muddle and chaos, followed by dictatorship again. And in the realm of moral and social life and family discipline we also see oscillations as each generation, in establishing its independence, veers off from the ways of its predecessor, often thereby turning once more to some of the habits of earlier generations. And there are those like Arnold Toynbee who believe that the mere ageing (or 'wearing out') of human institutions is sufficient in the course of time to bring down civilizations. In all these cases, however, as with the cycles in climate, some external event may cut the cycle short and start a new train of events. Thus, climate and human history present not wholly independent but partly interactive systems. It should be worth while to trace cause and effect in the linkages and certainly to look for any regularities.

One of the least happy lessons of human history may be read between the lines of the late medieval decline in Europe from the genial climate of the high Middle Ages which coincided with the twelfth and thirteenth century climax of cultural development and energetic activity. When former croplands were failing and being abandoned in the north and on the uplands of Europe (and also, as we now know, in the Middle West of North America), when farms and villages were being deserted and fields enclosed for sheep, in the riots and revolts which followed blame for all the sufferings and troubles was fastened on those who (for whatever motives) were in fact turning land to new and more productive use. Is it too much to hope that with better understanding of the behaviour of climate, accompanied by some wise preparations and sympathetic explanation to the people affected, we may cope better with such tensions in the future? Is it always appropriate when things go wrong to ask whose fault it is?

Some advance in sympathy, which we like to think characteristic of our own century, was registered by a speaker in a BBC religious affairs broadcast ('Thought for Today', 21 July 1978) who said: 'Who is responsible for mass unemployment. . . . Who is responsible for the world recession? . . . the answer must be *us*.' Such speaking is suited well enough to awaken our moral responsibility for one another throughout the world community, but the case in reality demanded some allowance for extremes of weather in the 1970s, for extensive crop failures, as in 1972 (and 1975), and their effect on world grain stocks.

This book provides an introduction to the development of climate, the record of its vicissitudes and their impact on the affairs of mankind. Human history is not acted out in a vacuum but against the background of an environment in which many sorts of change are always going on: besides the changes imposed by man, a never-ending competition goes on among the species of the plant and animal worlds, whose fortunes, like those of the soil and of the physical landscape itself, are continually affected by the vagaries of the climate. Some of the changes are slow and gradual, others are sharp and register abrupt events. We shall see examples of all these things.

The next three chapters are necessarily concerned with the physical basis of climate and climatic changes, with just enough illustrations to provide an adequate picture of the behaviour of this changing background to human life. In the rest of the book the history and development of climate in the past and in our own day are presented interwoven with allusions to aspects of human affairs and to other changes in the environment where the effects of climatic vicissitudes are registered.

We shall see that, contrary to the thinking of a generation ago, mankind is by no means emancipated by science and the technological revolution from the effects of climatic changes and fluctuations. Vulnerability to the effects, which included great famines in the past, seems rather to be increasing once more after some decades when a degree of immunity had indeed been achieved. Exposure to risks attending climatic shifts is increased greatly by the population explosion and the difficulty of producing enough food. The situation is made worse by the demand for an ever-rising standard of living in all parts of the world. And the systematic exploitation of resources to the limit, especially in agriculture, maximizes the risk.

It is reported (e.g. by Professor R. W. Kates of Clark University, Worcester, Mass., at the World Climate Conference held in Geneva, 1979) that three-quarters of the estimated world total cost of $40 billion yearly from natural hazards is accounted for by the major climatic causes of disaster: e.g. floods 40 per cent, tropical cyclones/hurricanes/typhoons 20 per cent, drought 15 per cent. The national and international organization of our present civilization with its advanced technology undoubtedly enables us, as never before, to rush help and supplies to relief of the immediate distress caused by natural disasters. It may be doubted, however, whether this complex world-wide community, with its interlocking arrangements and finely adjusted balances, is any more able than its predecessors to absorb the effects of long-term shifts of climate – particularly if they come on rapidly – entailing significant geographical displacement of crop zones and areas suited to various kinds of food production or are accompanied by mass migration of people.

It is important therefore to seek better knowledge of the pace of climatic change, especially the more rapid and drastic events of climatic history, and to identify the early symptoms which may have signalled the changes. On the other side, study must be given to the flexibility needed in the organization of human society if we are to be able to adjust to such things.

2

THE CLIMATE PROBLEM

CLIMATE DEFINED

By climate we mean the total experience of the weather at any place over some specific period of time. By international convention the period to which climate statistics relate is now normally thirty years, e.g. at the time of writing 1941–70, although we shall see arguments for preferring different periods for different purposes, particularly somewhat longer periods, such as fifty or a hundred years, and for preferring (as our grandfathers did) the decades that correspond to our linguistic usage based on our system of numbers, e.g. 1940–9 . . . 1970–9, and so on.

Climate was sometimes wrongly defined in the past as just 'average weather': the statistics required to specify a climate comprise not only averages but the extremes and the frequencies of every occurrence that may be of interest. The Classical Greek word κλιμα originally referred to a zone of the Earth between two specific latitudes, being associated with the inclination of the sun; and hence it came to be associated with the warmth and weather conditions prevailing there. This association was still embodied in the word 'clime' when first used in English in the sixteenth century and for long after. It was commonly used to refer not only to the prevailing climate as we mean it but to the terrestrial environment, vegetation, etc., that goes with that.

Climate has been too much taken for granted in recent times. Since some time in the late nineteenth century it has been usual to suppose that for all practical decisions climate can be taken as constant, however obvious the year-to-year fluctuations may be. The latter seemed best treated as random in their occurrence, although a few shadowy cycles might play a part in them and perhaps be of some limited use in forecasting, e.g. to indicate which was likely to be the finest European summer in a decade or to predict the years of high or low level of the great east African Lake Victoria. Anyway, such forecasts often failed. It was known that ice ages had occurred in the distant, 'geological' past; but the climate in Roman

8

times seemed to be not too much different from now, and it was assumed that this must be true of all the centuries in between.

As we shall see in later chapters, those centuries in fact brought a succession of changes in Europe and elsewhere which included a long period of evidently genial warmth in the high Middle Ages followed by the development world-wide of a colder climate, especially in and around the seventeenth century, with probably the greatest spread of ice since the last major ice age. Such a sequence can hardly have been withstood by the primitive human economies of those times without effects on their history.

EARLY WRITING ABOUT CLIMATE AND HISTORY

By coincidence, many writers at the time of this Little Ice Age and some of the early scientists then living were much interested in what has become known as the 'climate theory' of the 'humours' and character tendencies of the various peoples of the world. This is a theory which goes back at least to Hippocrates and ancient Greece. Aristotle had described the 'natural character' of men respectively in the cold, warm and middle zones of the Earth, and so arrived at a basis for believing in the superior quality of the Greeks of his time (the Hellenes) which should fit them to rule the world if only they could be united among themselves. This was a dangerous theory, which each nation soon took up in whatever form was most flattering to its own ego, thus contributing to the heady growth of chauvinism in the seventeenth century. It lingered on into some of the fanatical nationalisms of the twentieth century. Theatre audiences in Shakespeare's time loved presentations of foreign parts with overdrawn stereotypes of the peoples who inhabited them, a taste no doubt engendered by the Age of Discoveries. Seventeenth-century English Protestant preaching had much to say about the immoral peoples of the Catholic south of Europe, whereas the northern nations were considered 'dull and lumpish': there was some concern therefore over the actual latitude of England! One English writer – harmlessly enough – attributed his country's vaunted sense of humour to the climate 'and our gross diet', whereas another justified the immorality of the Restoration theatre as needed to disperse the spleen and gloominess of mind to which 'the British climate, more than any other' made men liable!

Scientists were naturally concerned with the theory, and whatever grain of truth might be in it, only as a step towards understanding the truth about mankind. Clearly there were problems: how was one to understand the 'barbaric' state of contemporary Greece and the lapsed state of Italy since its days of classical order and power? And what of the disappearance of the free democratic ways of the early Germanic peoples in the Denmark,

Germany and indeed in the England of the seventeenth century? It is curious to note the static conception of the world that these questions implied, though that may have been a necessary stage in sorting out the new knowledge which was then beginning to grow rapidly. There seems to have been no thought that climate, and, for that matter, the racial mixture and biological inheritance of nations, could change in the course of the centuries. Or is it possible that the leading men of affairs, and perhaps some of the scientists, in London, Paris, Florence and particularly in Zurich and Copenhagen were after all aware of the growth of the Alpine glaciers which were then threatening the villages and swallowing up the pastures about Chamonix and Grindelwald, while the same was happening to farm-land in Norway and the ice was increasing on the seas about Iceland? Certainly, Andrew Fletcher of Saltoun in Midlothian knew very well the disastrous run of years of failed harvests, famine and death that had over-taken the upland parishes of Scotland when he presented his Second Discourse to the Scottish Parliament in 1698 and criticized the well-to-do and comfortable population of the eastern lowlands for their lack of concern.

In 1492 the pope of the day had written of his concern for the people in Greenland because of the extensive freezing of the seas. As is well known, the old Viking population in Greenland, cut off from Europe, ultimately died out or disappeared. In 1784, in the time of a renewed increase of the sea ice around Iceland and after a massive volcanic eruption in that country, the Danish government debated whether Iceland should be evacuated and the population resettled in Europe – an amazing proposal in relation to the resources available at that date. In the event, it was not attempted and ultimately proved unnecessary despite the immediate distress and loss of life.

CLIMATE VIEWED AS CONSTANT

The view, so widely held until recently, of climate as constant was perhaps no more than a premature conclusion from the first long records of weather observations made with standard meteorological instruments in the world's leading cities. Many of these records had covered a hundred years by about the end of the nineteenth century, and it so happened that between 1875 and 1895 the temperatures prevailing in Europe and eastern North America had reverted to values quite similar to those of just a century earlier. In between there had been some colder decades with important glacier advances – a major climax of the glaciers in the Alps about 1820–50 – followed by a warmer time, which was in fact the beginning of a general recession of the glaciers all over the world until around 1960 or even later.

The conclusion that climate is essentially constant, which at first seemed to be the verdict of scientific observation, though in fact the hundred-year record was not enough to establish it, was at odds with the acquired wisdom and experience of previous generations, It had actually been concern about 'the sudden variations in the behaviour of the seasons' to which the climate seemed 'more and more subject', and about possible effects on agricultural production and human health, that had led to the setting up of some of the first nation-wide networks of meteorological observations from 1775 onwards.

The assumption of constancy was, however, a convenient one for those practical operations using climatic statistics for planning. It implied that the average values and general ranges of temperature, rainfall, sunshine, etc., indicated by the meteorological observations of 'enough' years (say twenty or thirty years) should serve as a sound guide to the future. And no questions needed to be asked about *which years* the observations covered. It even meant that valuable statistical techniques could be developed to derive estimates of the ultimate extremes of temperature etc. and the average recurrence intervals (known as the 'return periods') of rare events, using the frequency distribution shown by all the observed values of this or that element of the climate during whatever sample period was available. Thus, construction engineers could be supplied with figures for the strongest gust of wind, the greatest flood or frost or the highest temperatures to be expected once in fifty, or even once in five hundred, years – all on the basis of a mere thirty years or so of observations. The only important criterion seemed to be that the instruments, and their exposure, and maintenance of the observation record, conformed to the prescribed standards of the scientific age. Climatology became essentially the book-keeping branch of meteorology – no more and no less.

A step in the direction of standardization was taken at the 1935 conference of the International Meteorological Organization (forerunner of the present World Meteorological Organization) when use of the observations of the years 1901–30 for all climatic purposes was recommended as the so-called 'climatic normal period'. Choice of the word 'normal' turned out to be unfortunate, but it has persisted in climatological practice. It spreads the impression that nature recognizes such a norm and that conditions should continually return to the regime of the chosen period. We now know that 1901–30 was a highly abnormal period, though it was surpassed by the following thirty years 1931–60, which were in due course substituted as the 'new normal period'. Globally, these were probably the warmest, and in many regions the moistest, periods of such length for centuries past!

The indications of climatic probabilities for future planning, particularly the occurrence of extremes and rare events, arrived at on this basis, have sometimes proved seriously misleading since about 1950. It is certainly unwise to specify return periods of supposedly rare events much longer than the length of the observation record which has been examined. This confronts us with the need to extend knowledge of the past record of

11

climate beyond the era for which instrument observations are available. The quest entails the use of various kinds of earlier documentary records and of 'proxy' data to which we will return in the next chapters.

The adoption 'for practical purposes' of a constant climate seems to have continued longer than it might have done for an understandable reason. It worked. But it has not worked so well in recent years. In fact, from about the beginning of this century up to 1940 a substantial climatic change was in progress, but it was in a direction which tended to make life easier and to reduce stresses for most activities and most people in most parts of the world. Average temperatures were rising, though without too many hot extremes, and they were rising most of all in the Arctic where the sea ice was receding. Europe enjoyed several decades of near-immunity from severe winters, and the variability of temperature from year to year was reduced. More rainfall was reaching the dry places in the interiors of the great continents (except in the Americas where the lee effect, or 'rain-shadow', of the Rocky Mountains and the Andes became more marked as the prevalence of westerly winds in middle latitudes increased). And the monsoons became more regular in India and west Africa. Planning on the climatic statistics of the preceding decades was in fact allowing wider safety margins for many activities than was apparent up to some time about 1950.

The almost four-and-a-half decades of near–immunity to very cold winters ended abruptly with Europe's notably severe war winters in 1940, 1941 and 1942 and another in 1947 which is still remembered for its great snowfalls and very low temperatures. Other cold winters followed in 1956 and most notably in 1963, which was a very long winter in many parts of the northern hemisphere and in England the coldest for over 200 years, since 1740. And later in the 1960s the Arctic sea ice returned to trouble the coasts of Iceland. There was another run of mild winters in both Europe and North America in the early to mid-1970s, but more European cold winters followed in 1979, 1982 and 1985, some of which affected North America also.

THE EFFECT ON RESEARCH AND THE DEVELOPMENT OF KNOWLEDGE

Not surprisingly, research into the longer-term behaviour of climate languished as long as things were satisfactory. The lead towards a more lively view of climatology as the science of the development of climate had, in fact, been given by the Swedish meteorologist Tor Bergeron in 1930,[1] but three decades were to pass before it was taken up.

Since about 1950 the climatic tendencies have changed. A global cooling, slight at first but very marked in the 1960s in the northern hemisphere, reversed the earlier upward trend of temperature. Obviously, a run of five or six mild winters in Europe after 1970, and three or four in eastern

North America about the same time, plus two very warm summers in the same regions in 1975 and 1976, caused judgement to hesitate and produced an impression that the spate of writing in the 1960s about climatic change had overstressed the subject. That was before the winters of record severity in parts of North America and Europe in the later 1970s. But planners concluded that the political uncertainties surrounding the supply of basic fuels had to be seen as a greater threat to the economy. There have, however, been very notable extreme seasons, famines and harvest shortfalls in various countries since 1960–70. And these have been not unconnected with the political difficulties of the immediately following years. This applies most obviously to the many years of drought in the Sahel and elsewhere in latitudes 10–20 °N and the 1973 revolution in Ethiopia which led on to international conflicts in the Horn of Africa. It has even been suggested that it was the world grain shortage following the droughts and harvest failures in 1972 that triggered the first great oil price rise in the following year, as the oil-rich desert nations sought to secure their ability to buy food. At all events, within a year the world price of wheat had doubled and that of oil had multiplied by four.

This is just one instance of how we find climate and its variability involved in the major problems of the present-day world. But it is, of course, the great growth of the population, plus the demand for an ever-rising standard of living everywhere, which are straining resources, particularly as regards food production and water supply – in some years already demanding more than nature (aided by human technology) can give. These experiences have created a demand for climate forecasting at a time when scientific knowledge is inadequate to meet it satisfactorily.

Until the 1960s improvements in agricultural technology, particularly the spread of harvesting machinery, aided by a run of some decades of benign climate, were reducing crop losses. At the same time, increasing use of fertilizer and pesticides and the development of new seed varieties were greatly increasing yields. Since then, however, demand has in most years outstripped production, so that by 1975 world grain reserves counted in days supply had fallen to less than a quarter of what they were in 1961. In at least five years between 1960 and 1979 droughts affecting the harvests in the Soviet Union and sometimes in China as well, and failures of the monsoon in India, drove those countries to make massive purchases of grain from the west, essentially from the North American surplus. In 1972 harvest shortfalls in all these areas together coincided also with the prolonged drought and starvation in the Sahel zone of Africa.

The burgeoning of the world's population and the expectations of higher living standards clearly increase our vulnerability to climate fluctuations. Vulnerability may also be increasing as a result of the rational-ization of agriculture and world trade, whereby huge areas concentrate on just one or two crops which supposedly grow best there. This in essence depends on a forecast constancy of climate; and when even in an individual year the weather conditions go beyond the expected range, the consequences may be drastic. The one-crop economy was at the root of many of the greatest famines of the past. In recent years, moreover, climates all over the world have shown once more an increased range of variability.

Adding to these problems, forward calculations of world population growth and energy demand have led to widely publicized forecasts of a drastic rise of the global temperature, leading to displacement of the agricultural belts. This is seen as an inescapable effect of the extra carbon dioxide introduced into the atmosphere by our burning of fossil fuels (coal and oil, etc.), as well as strange substances, including nitrogen oxides from our artificial fertilizers, and the waste heat from these and other processes (e.g. nuclear energy). Estimates published of the warming to be expected by the year 2100 range from 2 to 11°C, the more extreme ones implying that the level of the world's oceans should begin to rise rapidly as melting of the land-based icesheets in Greenland and Antarctica got under way, This is an opinion, seemingly founded on firm scientific knowledge, which has to be taken seriously, even though we may notice some grounds for doubt and scepticism.

It was against this background that Dr Henry Kissinger, who was at the time United States Secretary of State, in a speech at the United Nations General Assembly on 15 April 1974, mentioned the threat of climatic changes and pressed the appropriate international scientific organizations 'urgently to investigate this problem'. The World Meteorological Organization has for some years been organizing a Global Atmospheric Research Program (GARP) with the climate problem as one of its objectives. The United States took the lead in adopting by Act of Congress in 1978 a National Climate Research Program and urging designation of the twenty-year period 1980–2000 as International Climate Decades, to secure broad international co-operation in the collection and analysis of all available climatic data and study of the problem. In other countries there has been so far too much concentration on theoretical modelling, based on the observations made in just those recent years for which global coverage is available. It is also necessary to have whatever observational data can be gathered to cover a much longer period of time, long enough to survey a statistically useful number of repeats of all those natural processes of climatic change and fluctuation which may be important to our future planning periods.

Sir Crispin Tickell, a former Fellow of the Harvard Center for International Affairs and British representative at the United Nations, has put very clearly the awesome implications of the growth of the human population of the Earth and of the climatic changes which we may engender. In his Presidential Address to the Royal Geographical Society in 1991 he stated:[2]

The surface of the planet . . . is changing fast . . . the impact of industrial society has caused human population to multiply out of control . . . it has to come down sooner or later . . . the population was 2 billion when I was born (1930) and is now 5.3 billion and will probably be about 8 billion by 2025.

And he went on:

the rough carrying capacity of the Earth for people enjoying Western dietary standards is about 2.5 billion. . . . In short, we are on a roller-coaster to disaster if we do not grasp what is happening. . . . We have to work for a new broad equilibrium involving changes in our energy policy.

He returned to the theme in his Presidential Address in 1993:[3]

The prime engine of the dizzy-making rise in the human population and change generally is the industrial revolution. We have the misfortune to be perhaps the first generation in which . . . the global price to be paid is becoming manifest.

There are many aspects. . . . The most significant change is not the spread of brick, stone, concrete and urban sprawl, but . . . the destruction of forests world-wide and declining fertility of soils. . . . Disposal of chemical wastes is a world-wide problem. . . . No part of the world is now exempt from the wastes produced by industrial activity.

Examples of population declines in the past, some of them catastrophic and clearly responsible for ending whole chapters of history in the regions affected, are mentioned in this book. In some cases climatic events clearly played a part, commonly in connection with diseases and wars triggered by the stresses on human life and the economies on which the people depended. Probably the best known cases are the great plague which spread all over the Roman world and beyond in the emperor Justinian's reign, in AD 543–7, and that other great pestilence, known to us as the Black Death, which reached Europe from the Far East in 1348–9. In both these cases, disturbances of the accustomed weather regime seem to have been involved in the origins of the outbreak. Both occurred in times of intermittent great storminess and wet interspersed with some years of drought and great heat. In the late Middle Ages, the disastrously wet summers and failed harvests in Europe between AD 1310 and 1320, worst in 1315, had been followed by a similar run of years with great rains and river floods in China in the 1330s, notably in 1332. These conditions in the great river valleys and broad plains of China are believed to have destroyed the habitats of the rodents and therefore their ways of life, and set them roaming and scavenging in new areas. It may be significant that the earliest origin of the pestilence is set in inner China in 1333.

Abandoned irrigation works and cultivation systems in Asia Minor and the Arab lands from earlier times, as well as the mute archaeological evidence of great buildings and cities 'swallowed up' by later forest growth in central America and southeast Asia, point to other cases of vanished populations and drastic events and changes of the landscape in which climate presumably played a part. These surely underline that Tickell's warnings should be taken as realistic.

A population disaster that is now reasonably fully documented, in which climate played a key part, as the trigger which finally unleashed the calamity that had been prepared by several factors[4] working towards the same result, is provided by the terrible Irish potato famine in the 1840s. The warm damp weather which fostered the potato blight in 1845 and some of the very next summers affected much of western and northern Europe and caused potato blight in many countries. It was, moreover, a new disease for which no one was prepared by experience. It came fortuitously in a ship-load of potatoes from Latin America to Belgium and was wafted to Ireland by easterly breezes in July and August 1845. Other breezes spread it to other countries, as far as Scotland, Norway and Poland, though in some that summer was too cold for the disease to flourish. In Belgium and Holland, however, over three-quarters of the potato crop was destroyed in that year. What made its attack devastating in Ireland was the social situation there. The Irish population had been growing fast, from about 6½ million in 1820 to nearly 8½ million by 1845, nearly double what it has been since, until the late twentieth century. It was a largely rural population, living on tiny farm plots, mostly under 6 hectares and some as small as 1 hectare, following generations of dividing the inheritance. On such plots there was only one crop that could fill their bellies, and that was the potato, particularly the cheap Lumper variety, which unhappily proved to be especially vulnerable to the disease. The prevailing poverty was such that many families could not even afford salt to make the monotonous diet more palatable. Enormous numbers died and mass emigration on crowded ships began.

Moreover, Ireland's position on the edge of the Atlantic, where the southerly and southwesterly winds are warm and especially humid, meant that the disease recurred, to devastate the crop in several successive years, whereas in 1846 a much drier summer saved most countries farther east.

Reconstruction of the past record of climate is one of the most broadly interdisciplinary projects of research. Just as weather and climate touch almost every aspect of our lives and environment, so evidence of their past record turns up in a vast variety of places. Most branches of learning, from studies of the writings and inscriptions of Classical and pre-Classical antiquity to work with the isotopes of elements identified by nuclear physics, have something to contribute. Much of this work is cheap by comparison with the operation of the great meteorological computer laboratories used by the theoreticians to explore the performance of the atmosphere and oceans as simulated by their models. In reality, both types of research are needed and there must be continual collaboration and interplay between them. Commonly, however, the research funds made available have been of the order of twenty to fifty times as much to the theoretical work as to construction and analysis of the actual past record of climate. As in all science, observation of the phenomena to be explained is needed before theoretical understanding of them can be established.

This is now being increasingly recognized. In one connection, the theoretical laboratories should obviously have most to contribute: in exploring how the behaviour of the climate may be altered in future by man's possible inputs of pollution and heat which never occurred in nature. Even so, the theoretical results will only be trustworthy in so far as (a) the assumed quantities are realistic and (b) the ability of the models to simulate the real world has been tested by application to various climatic regimes which are known to have occurred.

We clearly need to know and understand more about climate.

It will not be amiss to point out some traps for the unwary in approaching this subject. We are all – professional scientists, clients seeking advice, and laymen alike – steeped in the practical experience and ways of thought of the age we live in. As a result, people considering the problem of climate and its current development commonly start from the following presumptions:

1 The basic observation data – in order to establish the facts beyond doubt – must be from a well-established network of observers with high-quality instruments, their calibration satisfactorily maintained and their exposure conforming to recognized modern standards.
2 The best answers to questions demanding prediction must surely come from the laboratories with the finest computers and most advanced mathematical models.
3 One can surely leave out of account the 'long, slow processes of climatic change'.
4 Any changes of the prevailing climatic regime observed today, or on time-scales of significance to forward planning, must surely be attributable to man's activities.

The first two of these suppositions are of course well-learnt lessons of the scientific age, having proved their value in many other connections. But we shall see that in coming to grips with the climate problem, all four items are prejudgements which need further examination. Let us consider them one by one.

17

1 From the time of invention of the basic meteorological instruments – barometer, thermometer and rain-gauge – in the seventeenth century,[5] and the gradual establishment of a network of observation points equipped with them, until around 1950, the climate was mostly changing in one direction, towards greater warmth. Some climatic processes and evolutions are therefore of long duration. The opposite change, which had introduced the colder conditions and swollen glaciers at the time of the beginning of the instrument record, obviously provides another case. If we wish to understand that change, and other regimes that have occurred in post-glacial times, we must find ways of reconstructing situations that existed before meteorological instruments were known. This may appear all the more necessary as the observed climatic trend over much of the world from about 1950 to the late 1970s at least has been a cooler tendency.

2 Computer models simulating climatic development may also be deficient for a variety of reasons, however skilled the mathematics used. In order to obtain the complete global observation coverage needed, the performance of the atmosphere and oceans which the models simulate is bound to be that of a rather short span of years since 1950 and is sometimes that of just a few recent years of specially arranged international observation effort, for example under the GARP. Items crucial to the development of climatic changes over periods covering a number of decades or longer, being not really known from beforehand, may well have been omitted. The models' ability to explain climates of earlier times needs to be tested. Moreover, many of the complex interactions within the atmosphere and between atmosphere and ocean, which must be simulated, involve so many unknowns (exchange coefficients and so on) that altering any of them must alter others and can lead to a range of virtually arbitrary results. This is particularly true of models which do not incorporate the changes in the ocean induced by events in the atmosphere and the reaction of the former upon the latter, or the changes of cloudiness induced and their reaction upon the atmosphere and oceans' heat budget.

In this present state of the science, actuarial estimates of frequency and probability of various occurrences, relying closely on the past observation record, are probably more generally acceptable as a basis for planning than the specific forecasts of the theoretician. But what past observation period shall we use? The frequency of a long spell of frost in England such as to freeze the rivers was twice in over forty years from about 1900 to the 1930s; but in the nineteenth century it had been two to four times a decade and since 1940 it has again been once or twice a decade. (The run of mild decades at the beginning of the century allowed the development in Britain of water supply and drainage pipe layouts in, or rather on, the outside walls of the nation's houses, which ignored the risk of frost. The electrification of British railways, begun in the same decades, adopted the third rail system, which was to become notorious for failures in frost.) Parallel changes affected the frequency of snowfalls sufficient to block many roads and halt work on the land. And we shall come across similar changes in this and other parts of the world concerning droughts and flooding. All these experiences affected planning and design in the periods concerned. Thus, seemingly objective statistical work may produce a variety of verdicts which are actually arbitrary in that they depend on the choice of observation period. So here we again encounter the need for greater scientific understanding. We must seek to select a datum period for the statistics we use that is really similar in the physical development of climate to the present and future period which we are planning for. In this the observational worker needs the help of the theoretician to understand the evolutions of atmospheric circulation and climate which he observes and to be sure that he identifies like sequences correctly. Observation and theory must advance hand in hand.

3 The processes of climatic change may be long – as we have already seen, some of them certainly are – but they include some step-like, *abrupt* changes. Thus, the level of Lake Victoria, which is fed by the equatorial rains over its catchment basin in east Africa, rose by more than 1 m within three months in 1961, beyond the entire range of the previous sixty years since the gauges were installed; and by the late 1970s it still had not reverted to its previous levels (fig. 1). Other great lakes in eastern equatorial Africa rose at the same time. It is now known that the lakes were higher in the 1870s and that their decline by the end of the century had also been rapid. (The low stand of these equatorial lakes in the earlier part of this century roughly coincided with the period of most persistent development of the prevailing westerly winds of middle latitudes, and seems therefore to have been bound up with a global change in the wind circulation and transport of water vapour.) An important question therefore concerns the rapidity of some climatic changes.

4 The objection to the notion that any changes of climate which may be observed nowadays, or in the nearer future, must be attributable to man is that it is unproven and, outside urban and industrial areas, it is probably

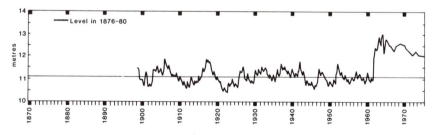

Fig. 1 Variations of the level of Lake Victoria in eastern equatorial Africa. Level in 1876–80 reported by the first European surveyors of the area. Gauge readings at Jinja since 1899 (monthly averages to 1961, thereafter yearly).

untrue. We shall have more to say on the questions involved in a later chapter. Nevertheless, in view of the increasing variety and scale of human 'insults' to the environment, there is clearly no room for complacency and every need for precautionary calculations and watchfulness.

Before we can go further in presenting what is now known of the past record of climate, or understanding its behaviour today, it is necessary to consider in outline how climate develops and the patterns of day-to-day weather are produced. Happily, when one investigates the working of the large-scale wind and ocean circulations, some encouragingly simple aspects are found. And, at least, it is clear that we can consider the climate system as a single global entity.

Part I

THE DEVELOPMENT OF CLIMATE

3

HOW CLIMATE WORKS

HOW CLIMATE IS GENERATED: BASIC MATTERS

Weather and climate are produced by the effects of heating and cooling of the surface of the Earth and the circulation of the atmosphere and oceans. This chapter is largely concerned with these circulations of wind and ocean currents as the 'mechanism' of climate. The items that together determine the climate of any place may be listed as:

1   The radiation balance – the balance between incoming energy originating in the sun and outgoing energy from the Earth. The gain and loss depend primarily on the latitude, but also on the aspect of the place (sloping towards north or south) and on the transparency of the atmosphere. Sunny south-facing slopes in northern lands enjoy the radiation climate of a lower latitude, and north-facing slopes the contrary, but as regards length of day both are governed by their actual latitude. Cloudiness, mist, haze and water vapour content as well as pollutants affect the atmosphere's transparency and are selective as between the sun's radiation and the long-wave radiation going out from the Earth.
2   The heat and moisture brought and carried away by the winds.
3   The heat and moisture stored in, transported by, and supplied from the sea and other water-bodies. In this, the ocean currents and their variations are important.
4   Characteristics of the locality and its surroundings, particularly the amount of water present in the soil and on the surface, the vegetation, the friction exerted on the winds by forests or buildings, friction and channelling of the winds by hills, mountains and sea coasts, all of which may also set up local wind circulations because of the local differences of heating and cooling. Of great importance, and sometimes most important of all, is the colour and reflective power of the surface: on this depends how much of the incoming heat is absorbed, while the dryness of the surface – a notable characteristic of paved urban areas and artificially drained soils – determines how much its temperature will go up for each unit of radiant energy absorbed.

23

These items are mostly subject both to short-term fluctuation and long-term changes. Some of the variations are regular, responding to the round of the day or year. Others are less regular or seemingly not at all. In 1 and 4 at least we encounter things that are affected by man.

THE HEAT SUPPLY

Taking the broadest overall view, about 2.4 times as much radiant energy from the sun is available over a year at the equator as at the poles. The ratio varies during the year: near the summer solstice 1.4 times as much solar radiation falls on the pole during its twenty-four hour day as is available at the equator, so that it is only because of the high proportion reflected by the persistent snow and ice, and the clouds, that even in summer less radiation is absorbed at the pole than at the equator. At this time the belt of maximum absorbed radiation crosses California and the central United States, then passes along the Mediterranean shore of the Sahara desert and over Iran. In the southern summer the corresponding belt crosses central-northern Chile and Argentina, near the Cape of Good Hope and over the Australian desert. At the winter solstice, the region of polar darkness with no direct solar radiation at all extends to about latitude 66½°. As the energy available at the equator varies less than 10 per cent during the year, the big changes over the higher latitudes mean that the gradient, or difference of heating, between low latitudes and places near the polar circle or beyond is greatest in mid-winter. At that season the effective gradient is intensified by the spreading snow and ice, which means that by late January or early February in the northern hemisphere little radiation is absorbed north of about latitude 45 °N. When the outgoing long-wave radiation from the Earth is taken into account, there is a net loss of radiation in mid-winter everywhere from near latitude 20° to the pole: this situation is much the same for the northern and southern hemispheres.

All the figures in the foregoing paragraph undergo some (minor) variation over periods of thousands and tens of thousands of years in the course of slow, cyclical variations in the Earth's orbit and the tilt of its polar axis. There is increasing evidence that, together with resulting changes in the extent in summer and autumn of the latest melted and earliest formed snow and icecover, which reflect away and waste much of the incoming radiation at those seasons, these variations have to do with the incidence of ice ages and warmer interglacial times such as we live in.

It is the large-scale circulation of the winds that is mainly responsible for redistributing the heat – and with it the moisture they pick up from the surface – about the Earth, and particularly to the higher latitudes. The much slower ocean circulation, and especially the surface ocean currents, account for the rest of the heat transport. The pattern is largely determined by the cumulative effect of the prevailing winds which drive the surface

24

water. Thus, if we wish to understand changes of climate and how the shifts in the patterns of rainfall and storm frequency, as well as of temperature, occur, we must first understand something about the wind circulation, its scale and strength and the variety of patterns in which it operates.

In the longer-term changes ocean currents must become increasingly important because of the much greater quantity of heat stored in the great mass of the ocean than in the atmosphere and the high specific heat of water: yet the pattern of the ocean currents must even in the longer run be a response to the drag of the winds. Variations of the temperature of

Fig. 2 Global cloud survey by satellite on 22 April 1978. Frontal cloud bands over the North and South Atlantic Oceans mark the feed of moisture and warm air into the cyclonic activity near south Greenland and Antarctica respectively, while numerous small flecks show the cumulus clouds in cold air moving towards lower latitudes. The equatorial cloud belt is rather weakly developed except over the Congo basin. (By courtesy of MacDonald, Dettwiler and Associates Ltd, Richmond, BC, Canada – Environmental Satellite Data Systems.)

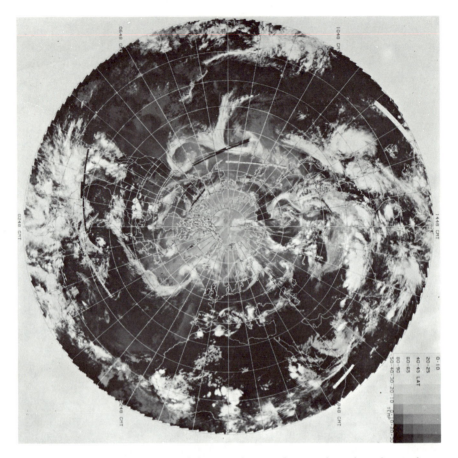

Fig. 3a Satellite cloud surveys of the complete northern and southern hemispheres on 27–28 June 1976. Frontal cloud belts, marking the feed of warm, moist air into the cyclonic activity over the higher latitudes, are seen in most sectors of both hemispheres. The steering of this activity over the central North Atlantic on this occasion is northwards: Europe was in the midst of a great drought and heat wave: temperatures over England reached 32–34 °C (90–94 °F) on both days.
(By courtesy of National Climatic Center, Washington, DC.)

land surfaces are of much less importance in connection with heat storage. The specific heat of rock and the matter of which the soil is constituted, apart from any moisture trapped in them, is much less and their conductivity of heat is much slower than for water bodies; so the temperature of dry ground surfaces changes quickly in response to every change of weather and sky conditions and little heat is stored.

Fig. 3b

THE WORLD'S WIND CIRCULATION

That the large-scale wind circulation can be viewed as a single huge convection system whereby the atmosphere is busily conveying heat and moisture – along mostly slanting paths – towards the poles can be appreciated
from the satellite imagery pictures which illustrate this chapter. The colder
air from high latitudes is carried equatorwards – with either clearer skies or
lowerlevel, and often broken, cumulus (heap-type) clouds – in the longitudes
between the main frontal cloud-bands which are the most prominent feature of the pictures (figs. 2, 3). We shall make clear the meaning of these
concepts in the description of the general wind circulation which follows.

It is the unequal heating of different zones of the Earth that sets the air
in motion. How this occurs and brings about the global wind circulation

27

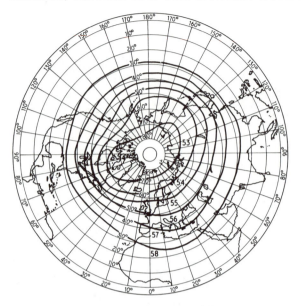

Fig. 4a Height of the level where atmospheric pressure is 500 millibars, about half its value at sea level. Average heights in hundreds of metres. The contours and gradients of the 500 millibar surface here seen have about the same significance as isobars and pressure gradients at the 5–6 km (17,000–20,000 ft) level here sampled. Under equilibrium conditions, the wind blows along the contours counter-clockwise around the north polar low-pressure region and clockwise around the south polar low-pressure region. The maps therefore show a prevailing pattern of upper westerly winds over both hemispheres, strongest over the middle latitudes, the circumpolar vortex. (Crown copyright. Reproduced by courtesy of the Controller of Her Majesty's Stationery Office, London.)

can be easily understood in outline. Air, like most other substances, expands when it is heated (that is unless it is in a confined space and extra pressure is exerted on it to prevent the expansion). Hence, the lower atmosphere over the warm regions of the Earth expands and the upper atmosphere is lifted. In the language of mechanics, the sun's energy does *work* upon the atmosphere by lifting its centre of gravity in the regions where there is most heat intake. Thereby potential energy is put in. By contrast, over the cold regions the lower atmosphere shrinks and the upper atmosphere descends somewhat. This means that at any level in the atmosphere above the more complicated effects near the Earth's surface, there is a pressure gradient with the greater pressure – because of the greater quantity – of the overlying atmosphere over the warmer parts of the Earth and lower pressure over the colder regions. This produces the very simple pressure distribution patterns which we see over the northern and southern hemispheres in fig. 4. The pressure gradient means that there is a force impelling

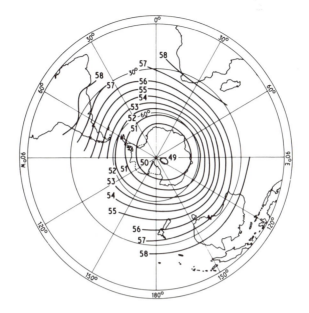

Fig. 4b

the air into motion. The potential energy is converted into kinetic energy, the energy of motion. However, because of the rotation of the Earth underneath the moving air, the motion turns out to be not down the pressure gradient but nearly along the lines of equal pressure. And so pressure maps such as here illustrated can be used as virtually presenting a picture of the prevailing wind flow at the levels to which they refer.

The circulation pattern which the maps reveal is a very simple one, a single great circumpolar flow of winds circuiting from west to east around the Earth over each hemisphere, mainly over the middle latitudes. It is called the *circumpolar vortex;* and at any given time the pattern of flow is in some detail similar through a great range of heights, from about 2 km (5000–7000 ft) to 15–20 km (over 50,000 ft, or 10–13 miles) above the Earth. Despite the decreasing density of the air with increase of height, this layer is so deep that it involves most of the mass of the atmosphere. And so the circumpolar vortex is in fact the main flow of the atmosphere, carrying most of the momentum. The flow is never strictly circular around the pole, but exhibits more or less prominent wave-like meanders, the so-called ridges and troughs in the pattern.

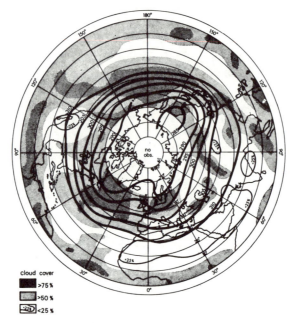

Fig. 5a Satellite survey of the average cloud cover over the northern hemisphere and mean configuration of the circumpolar vortex (in this case shown by the average height, in tens of metres, of the 700 mb pressure level) sampled at a height about 3 km (10,000 ft) during the spring months, March–May 1962. With the persistent cold trough in the circumpolar vortex over northern Europe, this was the coldest spring of the century so far in the British Isles and neighbouring countries. Notice particularly, however, the association of major regions of cloudiness with each of the cold troughs on the map and the tendency for the cloud to spread 'downstream' with the upper westerly winds. (This was historically the first season for which satellite cloud surveys were ever available, and as yet the satellites were incapable of covering the polar region.)

WEATHER SYSTEMS

It is at points of imbalance in the flow of the winds in the circumpolar vortex, as the air moves into regions where the pressure gradient is stronger (or weaker) and is accelerated (or decelerated, as the case may be), or into regions where its path becomes more (or less) curved, that the wind fails to conform so well to the pattern of the lines of equal pressure. In these regions therefore movements take place which *change* the atmospheric pressure over the lowest layers of the atmosphere, piling up rather more air over one region and removing some from somewhere else. In this way, the high and low pressure systems, *anticyclones* and *cyclones* (or *depressions*) respectively, the familiar features of day-to-day surface weather maps, are formed and intensified (or weakened and gradually suppressed). Thus, the

30

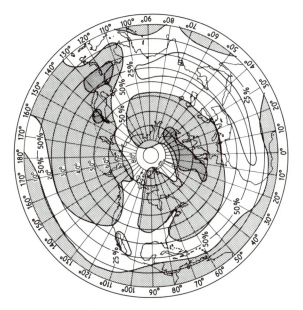

Fig. 5b Long-term average distribution of cloudiness in spring (March–May), corresponding to an average pattern of the circumpolar vortex much like that in fig. 4a.

places where these systems, which bring us our weather, are formed and decay are controlled by the pattern of the circumpolar vortex, and during their lives they are steered along paths controlled by the massive flow of the upper winds.

Among the places where surface low-pressure systems and the clouds and rain (or snow) associated with them form, the eastern sides of troughs in the upper wind flow are most important. The associations between the positions of these surface weather systems, and the cloudiness and precipitation that accompanies them, with the prevailing pattern of the circumpolar vortex during one sample season may be seen in fig. 5. Mostly high surface pressure and anticyclones are maintained along the warm side of the main upper wind flow. Low pressure at the surface and cyclonic systems prevail at most points near the cold side of the main flow in the circumpolar vortex. There are, however, certain areas in the flow pattern of the upper winds where the dispositions of surface low and high pressure tendency are reversed. Examples occur where the air is being accelerated near the entrance to a strong jet stream. There, cyclonic development occurs near the warm side of the jet, while an anticyclone may form on the poleward side.

Long-term average barometric pressure maps for sea level, like the ones in fig. 6, show a high-pressure belt in subtropical latitudes along the warm

31

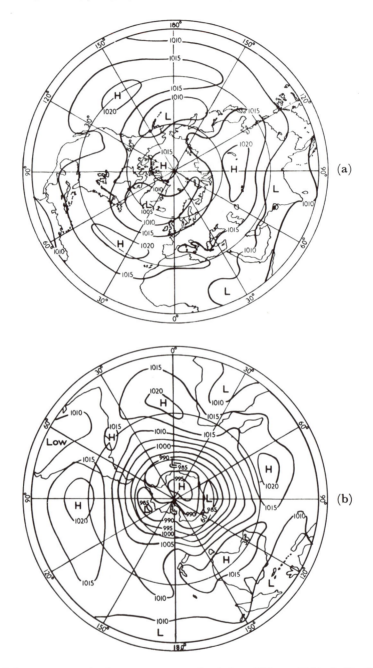

Fig. 6 Average atmospheric pressure at sea level, in millibars, 1951–4. Northern and southern hemispheres. (Crown copyright. Reproduced by permission of the Controller of Her Majesty's Stationery Office.)

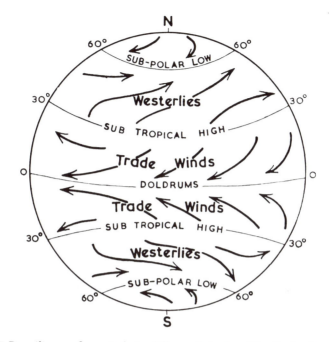

Fig. 7 Prevailing surface winds in different latitudes. (Idealized scheme.)

fringe of the strongest upper-level pressure gradient and strongest winds of the circumpolar vortex. Low pressure on the long-term average surface maps is seen in a belt near the poleward side of the strongest upper wind zone. The winds blow clockwise around (and at the surface somewhat outward from) the high-pressure systems in the northern hemisphere, and counter-clockwise around the low-pressure systems. The sense is reversed in the southern hemisphere. These average atmospheric pressure maps imply prevailing surface winds in different latitudes as shown in the simplified sketch scheme in fig. 7.

The size of even the biggest surface weather systems – the greatest cyclonic depressions up to 2000–3000 km (approximately 1200–2000 miles) in diameter, anticyclones up to 4000 km along their longest axis – means that surface winds of very different origin, from regions well to the north and south, are brought towards each other in some parts of the surface pattern. This may be seen in the lower left-hand part of the schematic map in fig. 8. The convergence of the warm and cold air creates a *front,* at which the warm air is forced to rise and clouds are formed in the process. The broken lines in the figure show the relationship of the depression, or cyclone, to the upper winds (or, strictly, to the pressure pattern of the circumpolar vortex). The distribution of the associated clouds and weather is indicated in fig. 9, which shows three stages of the cyclone's

33

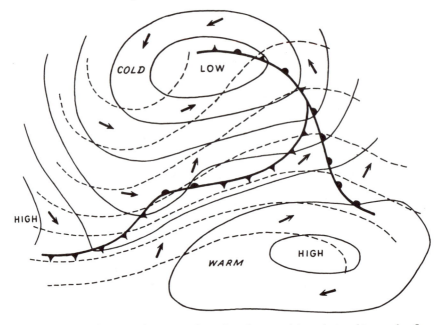

Fig. 8 A typical eastward moving frontal cyclone and its relationship to the flow in part of the circumpolar vortex (broken lines). Arrows indicate the surface winds at each part of the pattern. (Pictured in the northern hemisphere orientation: invert for southern hemisphere case.)

development. Vigorous cyclonic and anticyclonic systems developing in the lower atmosphere push warm air towards higher latitudes and cold air towards the tropical zone. This distorts the temperature distribution for the time being, and inevitably changes the pattern of flow of the upper winds to conform. The situation pictured in fig. 8 has already introduced a trough and a ridge into the pattern of the circumpolar vortex. These are mobile features, moving ahead with the depression unless and until the disturbance of the upper flow becomes so great that either the cold or the warm air moves nearly all round the system. When this happens, it produces a closed, or nearly closed, cyclonic circulation up to great heights. There is then no longer any steering current to move the system on, and it becomes stationary; it may weaken and die out gradually, as the masses of air involved become adapted in temperature to the latitudes they have arrived in, or it may be maintained by further cyclonic systems developing and being steered into its area, each bringing a renewal of the warm and cold air supply.

These surface weather systems, the developing depressions and fronts that we have described above and the anticyclones, can be looked upon as eddies – transient and often quite mobile eddies – which complicate during

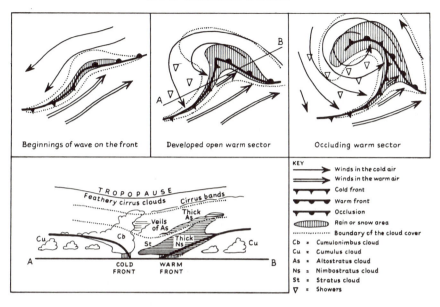

Fig. 9 Three stages in the life of a typical frontal cyclone and a vertical cross-section along the line AB through the cloud development, showing structure in the warm air above the frontal surfaces. (Northern hemisphere orientation.)

their lifetime the larger-scale simplicity of the circumpolar vortex. Both the large-scale mean flow and the eddies play a part in the poleward transport of heat that is the business of the global wind circulation. Because there are always 'waves', or meanders, in the path of the 'upper westerlies' around the circumpolar vortex, there are places where the air moves towards higher latitudes or colder regions. There it loses heat to the environment and, by radiation, to space. And where the air, having become colder, moves towards lower latitudes, it gains heat, thereby cooling the environment. The surface wind circulations that develop, and travel, each make a contribution to the the heat transport over the regions they traverse.

TRANSPORT OF MOISTURE AND POLLUTION BY THE WINDS

Not only heat is transported by these wind circulations. Moisture (and every form of pollution) that is picked up by the winds is carried along, until condensation produces droplets and ice crystals or snowflakes and the fallspeeds of these (and of the pollution particles) bring them down to earth. (In some cases, the pollution can be said to be washed out of the air by the drizzle, rain and snow.)

An interesting observation by two French investigators that the pollution by minute quantities of trace metals brought down on the Antarctic ice-cap was greater between about 1925 and 1940 than before or since indicates a point of some importance. For in that period the mean condition of the circumpolar vortex of upper westerly winds is known to have been particularly strong and regularly developed over both hemispheres. It seems that the winds that conveyed the pollution to the ice-cap in high latitudes must have been developed in the eddies, which presumably therefore grew to a larger size and were playing a bigger part in the poleward transport of heat, and moisture also, than in other epochs when the mean circulation pattern actually had more frequent and more wide-ranging meanders towards high and low latitudes. The same may be indicated by the fact that the deposit of snow on the ice-caps near the south pole and in north Greenland was also at a maximum in that period, when the mean upper wind circulation over middle latitudes was characterized by a notably frequent development of a smooth but particularly strong westerly 'zonal' pattern of the circumpolar vortex, with only small amplitude waves or meanders.

VARIATIONS OF THE WIND CIRCULATION

The circumpolar vortex is subject to variation between a smooth westerly 'zonal' form with little meandering and so-called 'meridional' patterns that are distinguished by large-amplitude waves or meanders (fig. 10). Sometimes the meanders get so big that the mainstream of the flow wanders from the lower middle latitudes to quite near the pole and back again, or it may get contorted until one or more closed loops are formed: such loops create a cut-off cyclone in middle latitudes (or even in the subtropics) or an anticyclone cut off in the higher latitudes, where normally pressure is low. Such situations are generally slow-moving or stationary, and the main features of the surface wind and pressure distributions become slow-moving or stationary too. In some parts of the map the high pressure in high latitudes reverses the normal surface winds, leading to easterly winds in middle latitudes. And because the situation is stationary, or nearly so, a long spell of these conditions may result. When the usually prevailing westerly winds are absent at places in middle latitudes, these are called 'blocking situations'. The westerlies, and the usual mobility of the travelling depressions and the rain belts that accompany their fronts, appear to be blocked.

Blocking situations bring abnormal weather and temperatures to many places; droughts may occur, with floods elsewhere, because the clouds and rain are steered away from the regions they usually frequent. Extremes of warmth or cold, wetness or dryness may be brought to different places if the unusual wind flow lasts long enough to warm or cool the seas and

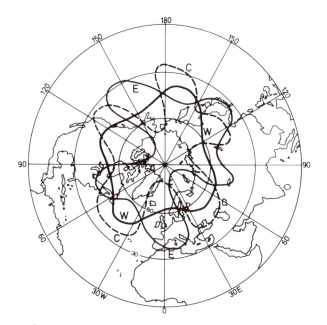

Fig. 10 Types of variation of the circumpolar vortex, illustrated by a sample flow line in the region of strongest winds around the hemisphere. (After a Russian classification by Girs and Wangenheim.) The flow patterns with the big meanders tend to produce slow-moving or stationary ('blocked' or 'meridional') situations, with more or less stationary extensions of the polar and subtropical regimes in different sectors in middle latitudes.

thoroughly wet or dry out the land surfaces and vegetation. Thus, in the persistent European drought in June–July 1976 the surface of parts of the continent became so parched that the winds brought temperatures between 32 and 35 °C to England on many successive days. A similar situation farther east brought temperatures up to 33 °C in Finland near the Arctic circle in July 1972. Equally, persistent northeasterly winds in February 1979 produced ice ('pancake ice' freezing on the surface of the sea) on the eastern part of the North Sea near the coast of Denmark, where fishing vessels foundered probably because of ice formation (from the cold spray) on their rigging.

Distortion of the main thermal gradient of the hemisphere, and of the circumpolar vortex accordingly, in these patterns with great meanders of the upper wind flow between low and high latitudes, distorts the steering of the surface weather systems. The distribution of prevalently cloudy skies, moisture transport and bad weather (rain and snow) is altered accordingly. Examples of this will be seen in the next chapter.

WORLD-WIDE RELATIONSHIPS OF WEATHER VARIATIONS

In addition to changes in the amount of meandering, and connected with them, changes occur in the strength of the main wind flow each year in the round of the seasons, from one spell of weather to another and from one climatic epoch to another. And when the upper westerly winds are most strongly developed around the globe, with only modest waves in the pattern of the circumpolar flow, the wave-length or spacing of these waves increases. The spacing also increases if and when the main stream is displaced towards higher latitudes without change of strength. Changes in the spacing of the troughs and ridges in the pattern of the circumpolar vortex mean changes in the positions around the hemisphere that are affected by extensions of the cold polar and warm subtropical regimes respectively. They also mean changes in the positions at which the cyclonic disturbances develop and the frequency, and speed, with which their rain belts and often stormy surface winds are steered along various paths. In this way great differences may arise between dry and wet, warm and cold weather prevailing in a given season or a given climatic epoch in different parts of the same latitude zone around the Earth. The situation depends on the wave-length and resulting longitude positions 'favoured' by the troughs and ridges in the flow of the upper winds.

It is the intimate relationship between the circumpolar vortex and the steering of the surface weather systems, together with the fact that its pattern is basically simple, and an entity subject to variations which are also of simply recognized types, that makes it possible to reconstruct global weather patterns of the past from fragmentary and scattered information just here and there around each hemisphere. From fossil evidence of the gross temperature distribution prevailing at the surface of the oceans and over land in any past epoch, we can reconstruct in outline the prevailing features of the circumpolar vortex and hence of the large-scale wind and ocean circulations at the surface.

Interrelationships between the northern and southern hemispheres also need to be studied, both as regards the large-scale wind circulation and other aspects of the climatic regime. It is a noticeable feature of figs. 4a and b (pp. 26–7) and 6a and b (p. 32) that in the present epoch the mean wind circulation over the southern hemisphere, with its glaciated continent in the high latitudes, is stronger than that over the northern hemisphere. On the other hand, the northern hemisphere circulation not infrequently develops much more 'meridional' patterns than are seen over the southern hemisphere: these are liable at times to push the interhemispheric convergence zone (the meteorological equator) far south across the equator over a narrow range of longitudes into Brazil, southern Africa or Australia, evidently assisted by its being drawn into the convection system developed over the heated continents. We thus find some evidence of impacts either way of the circulation over one hemisphere on that over the other. In the case of long-term changes of the climatic regime, some curious features come to light which have not received the attention they deserve and are not yet widely known or understood. Thus, although the whole Earth experienced the last ice age and now enjoys the present interglacial period, the timing of the changes shows some important differences between north and south. And within the last thousand years, the development of what has been reasonably called the Little Ice Age seems to have affected the whole Earth, as has the twentieth-century recovery from it; but when the ice on the Arctic seas extended farthest south, particularly in the Atlantic sector, all the climatic zones seem to have shifted south, including the storm activity of the Southern Ocean and the Antarctic fringe. This apparently broke up much of the Antarctic sea ice, enabling Captain Cook in the 1770s and Weddell in 1823 to sail farther south than ships have usually been able to reach in this century. The southward extension of open water would presumably result in some mildening of the regime not only over the ocean but some way into the interior of Antarctica, and this just when the world in general north of about 40 °S was experiencing a notably cold regime. Amongst the evidence which builds up this picture, at that time the winter rains failed to reach so far north over Chile. And radiocarbon dating of abandoned penguin rookeries on the Antarctic coast near 77½ °S, in the southernmost part of of the Ross Sea, suggests that there were periods of milder climate there about AD 1250–1450 and 1670–1840. These periods include the sharpest phases of development of the Little Ice Age climate in the northern hemisphere.

CONVECTION AND TEMPERATURE CHANGE WITH HEIGHT

Atmospheric convection systems on localized scales, in which the upward air motion is rendered visible by towering cumulus and cumulonimbus clouds and by rising smoke, and sometimes by swirling dust caught up by the wind, will be familiar to most readers. They arise wherever there is a specially strong lapse of temperature with height, as when land surfaces become hot in summer or Arctic cold air is blown swiftly over much warmer seas in winter. In their extreme forms violent up and down air motions (occasionally sufficient to damage aircraft flying through them) and violent weather – thunder and lightning, heavy rain, hail and wind squalls – result.

The basic condition for vertical convection to occur is that the rising air should be warmer, and therefore less dense, than its immediate environment. This will be so in the case of rising moist air with cloud forming, as long as the negative gradient of temperature with height exceeds -0.65 to -0.7 °C/100 m. As it happens, this is about the overall average variation of temperature with height: the average gradient or 'lapse rate', observed between stations at the foot and at the summit of Ben Nevis (1343 m), near the Atlantic coast of Scotland, is about -0.64 °C/100 m. Average gradients between the Alpine summits and the Swiss and Austrian lowlands are rather less than this, about -0.54 °C/100 m, owing to the cold conditions that frequently develop in the winter half of the year in the valleys. Sometimes in winter there is an inversion of temperature so strong that temperatures on the Alpine summits are higher than on the continental lowlands. And, owing to the fogs and low cloud in the valleys and over the European plain, there is more winter sunshine on the tops. But, at the other extreme, gradients of temperature with height at rates exceeding -1 °C/100 m (-5.5 °C/1000 ft) are liable to occur in fast-moving outbreaks of Arctic air in winter over warm seas and in air that is heated over the hottest land surfaces in summer.

TORNADOES

The severest air motions occur in tornadoes, in which the inflow of air at the surface required to supply a very rapidly rising column of air at the centre is organized by a spiralling inward cyclonic rotation of the surface wind. At a horizontal distance of only a few metres from the centre at the ground, wind speeds may be as much as 50 m/s (100 knots), or more, in the most swiftly moving ring of air, perhaps only a metre wide, within the vortex. With zero horizontal motion at the centre and speeds commonly only one-fifth of the maximum one or two hundred metres away, such a cross-section implies tremendous twisting forces: the torque commonly twists off the trunks of full-grown trees and bends strong metal objects. A reduction of atmospheric pressure takes place, which at the centre of the tornado vortex may be sufficient to raise a column of water 2 or 3 m and cause buildings and windows to burst outwards. Objects weighing many tons – such as a loaded railway truck – are sometimes lifted. Condensation of moisture facilitated by the low pressure in the rotating core of the vortex produces growth downwards from the main cloud base of a twisting funnel-shaped cloud that gives the tornado its most familiar and menacing appearance (fig. 11). The conditions in which tornadoes form are in moist air where there is a strong gradient of temperature with height, and the release of latent heat of condensation in rising air ensures its buoyancy: for without condensation the temperature of rising air drops more rapidly (about 1 °C/100 m). A further requirement seems to be a wind shear to start the rotation; this is commonly supplied by the proximity of a cold front with a very different air stream approaching, though it may be that differences of friction on the wind due to topography (for example, a line of hills) can also introduce the shear, and hence the rotation, when the situation is sufficiently unstable. The region of greatest frequency of tornadoes, and

Fig. 11 A tornado developing and decaying. Picture sequence showing the development down to the ground of a twisting 'funnel' cloud, at Edmonson, Texas, on 27 May 1978. (Originally published in *Weather*, August 1979 issue. Reproduced here by kind permission of the Editor of *Weather* and Professor R. E. Peterson of Texas Tech University and Mr Carl Holland of Plainsview, Texas, who took the photographs.)

the most violent, is over the great plains of North America east of the Rocky Mountains, where warm moist air from the Gulf of Mexico meets the fronts of cool air from the Pacifilc or Arctic air advancing from Canada. The worst situations seem to occur in spring, and perhaps especially after cold winters, when the air mass contrasts are greatest.

TROPICAL STORMS, TYPHOONS

Another class of convection system, which our description of the atmospheric motions so far has not included, is the tropical disturbances that sometimes grow into tropical cyclones, typhoons and hurricanes, These, together with tornadoes, are generally considered the most destructive wind systems on Earth. The ultimate prize for violence and destructiveness may however belong to systems of mixed type, in which the maximum energy release occurs in tornadoes within a larger storm or in a tropical hurricane which has become engaged with a trough in the circumpolar vortex and steered – the tracks are said to 'recurve' (see figs. 12 and 13) – into middle or higher latitudes, where its energy is intensified by drawing in polar air. Tropical storms first form over the warmest oceans in the world, surface water temperatures of 27 °C or above seeming to be a minimum condition. Deformations of the upper wind flow, which carry the intertropical convergence farther north and south of the equator than usual, seem also to be involved, and induce the initial rotation. Ultimately, general surface winds with speeds up to 50 m/s (about 100 knots) or more may be developed;

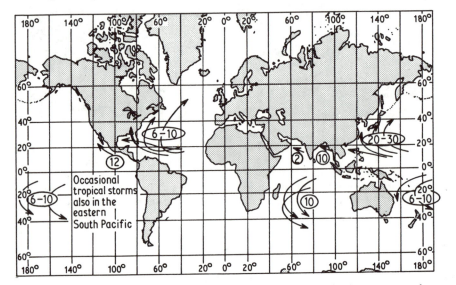

Fig. 12 World map of tropical cyclone incidence, with examples of recurving paths. Data from the most recent available 30–100 years of observation.

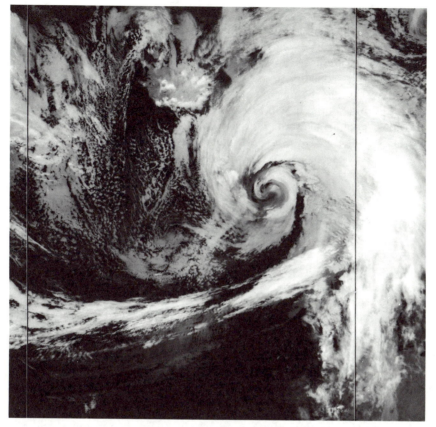

Fig. 13 Satellite view of tropical hurricane 'Flossie', after it had drawn in the main front of polar air over the North Atlantic, on 16 September 1978. The depression, at this stage with a central atmospheric pressure of 954 mb, is centred near 60 °N 4 °E. Iceland, the south and east coasts of Britain and the continental coastline from France to Denmark are seen in the picture. (Picture by courtesy of the Department of Electrical Engineering and Electronics, University of Dundee.)

but these storms gradually lose energy once they pass any great distance over land. Tropical cyclones seem to be most frequent in years, and at times such as late summer, when the circumpolar west wind belt is displaced towards higher latitudes: the general circulation over the lower latitudes is then correspondingly weakened so that the heat accumulates there. Tropical cyclones provide a release of some of this energy, which is thereby transported towards middle or higher latitudes and upwards from the surface towards the level of the cloud tops – i.e. towards the places where the excess energy may be lost by radiation to space. By contrast, when the circumpolar vortex is intensely developed and displaced somewhat nearer the equator than its usual position, it seems that the frontal cyclones and

43

anticyclones – the eddies associated with the upper westerlies – share in the conveyance of heat from the lower latitudes, and the distinctively tropical cyclones do not develop so often. It is also noticeable that there have been fewer tropical cyclones in those years when the circumpolar vortex continually developed great meanders and 'blocking' was frequent. Presumably at such times the mean wind circulation itself, with great 'meridional' northerly and southerly windstreams in different sectors around the globe, provides enough poleward transport of heat to reduce the risk of tropical storms – to 'defuse' the situation in the tropics and subtropics.

SEASONAL CHANGES

The regular yearly seasonal changes in the radiation situation and doubtless also in the development of the atmospheric circulation are greater than those which distinguish different climatic epochs, with the exception of the changes between the coldest and warmest phases of ice age and interglacial climates. Understanding of the seasonal changes may therefore teach us something about the major climatic changes. Nevertheless there are differences.

In the course of the seasons the zenith sun at noon moves (at present) from latitude 23½ °S to 23½ °N and back again. And the length of day, between sunrise and sunset, varies hardly at all at the equator, but changes from about 10¾ to 13½ hours at the tropic, from about 8 to 16½ hours near latitude 50°, and from zero to 24 hours at the polar circles and beyond. In the course of each year these changes are accompanied by changes in the atmospheric circulation, which can be most simply viewed as a northward and southward shift of the main wind zones. These are accompanied by changes in the strength of development of the winds and changes in the positions, spacing and size of the waves or meanders of the upper westerlies – and all the changes in the development and steering of our surface weather systems that go with that. But none of these changes are as regular as the shift of the zenith sun. Moreover, the regular seasonal movement of the wind zones north and south only amounts to some 8–10 degrees of latitude. Some of the shorter term irregular movements north or south, and back again, are much greater than this. For this reason, the progress of the seasonal warming and cooling, and storminess, is rarely steady or continuously in one direction without any setbacks. There are other changes seasonally, in the extent of snow and ice, of flooded marsh and desert, and in the colour and luxuriance of the vegetation, which affect the radiation budget and play a part in the differences of development of the wind and weather patterns from one year to another.

Land surfaces, particularly when dry, heat up much more quickly than water bodies. It takes more heat to change the temperature of water by one degree than in the case of almost any other substance. Moreover, there are possibilities of convection in the water, and the winds ensure that the

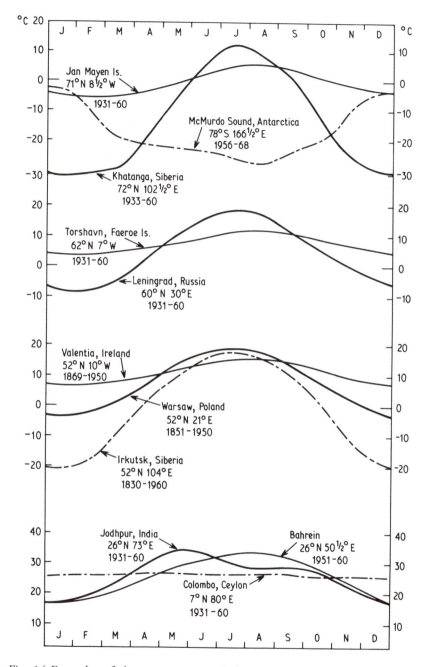

Fig. 14 Examples of the average seasonal change of temperature at island and continental situations in different latitudes. Month by month values, slightly smoothed.

uppermost layers of the ocean are well stirred, at least in the higher latitudes and in the winter. Thus, the seasonal range of temperature – and its daily range – is much greater over land than over water or near coasts. The highest temperatures in summer are reached on average within three or four weeks after the longest day over land, and the lowest temperatures tend to be just this much after the winter solstice, although in both cases there is a good deal of variability from year to year. The ocean, by contrast, does not reach its warmest until August in the northern hemisphere and is coldest on average in February, with similar delays in the southern hemisphere. These tendencies are reflected in the sample average temperature curves for island and continental situations in different latitudes shown in fig. 14. One Antarctic station is included to illustrate the flat-bottomed, or 'coreless', winter typical of high latitudes, especially in the far south.

The seas remain relatively warm in autumn and cold in spring. And it must be expected that most of the ocean in general remains relatively warm during climatic changes towards colder times and lags behind the warming of the lands with climatic changes in the opposite direction. There are clearly exceptions, however, where the advance of a cold ocean current or of a warm one plays an immediate part in the climatic shift and where there is a significant spreading or decline of the extent of sea ice. In these regions and at these times the ocean may be very far from acting as a thermostat or stabilizer of the climatic regime.

WORLD RAINFALL: DISTRIBUTION, SEASONAL CHANGES, MONSOONS

The distribution of rainfall by latitude over the globe at the present epoch is shown in fig. 15. The greatest total is yielded by the equatorial rains,

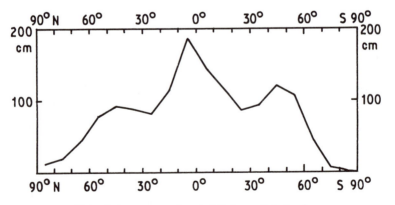

Fig. 15 Average yearly rainfall for each latitude.
(Observations about 1900 to 1950.)

produced in the massive uplift of air and towering clouds in the zone of colivergence between the wind systems of the two hemispheres – the so-called *intertropical convergence*. The moisture content of the air is greatest in this zone because of the enormous uptake by evaporation from the warmest oceans in the world (and also into the warm air from other water bodies and surface moisture). The secondary maxima occur in the rains and snows produced by the cyclonic activity over middle latitudes in each hemisphere. The average yearly figures are greatest over and near the oceans, especially on mountain slopes exposed to winds from the ocean.

These precipitation belts move north and south with the wind circulation zones that produce them. Averaged over the year, the overall mean latitude of the intertropical convergence zone, the meteorological equator, in the present epoch is 6 °N. This is related to the fact that the atmosphere over the glaciated continent of Antarctica is on average 11–12 °C colder than over the Arctic – at the Earth's surface the contrast is 20–30 °C – and the stronger thermal gradient produces a much stronger circumpolar wind system over the southern hemisphere. Hence also the prevailing temperatures at most latitudes are somewhat lower in the southern hemisphere than in the northern, and all the climatic zones are displaced somewhat towards the north. This is particularly so in the Indian Ocean sector, where the Antarctic continent itself reaches farthest towards the north. No doubt the summer heating of the great continent of Asia, and the high mountain wall of the Himalayas and Tibet largely barring the way to winds from the north, play a part in causing the equatorial rain system to move as far as 30 °N over the Indian subcontinent in summer. A belt of westerly or southwesterly winds develops in the lower atmosphere between the intertropical convergence and the equator. And so we have the main features of the Indian southwest monsoon. In winter the Siberian anticyclone and the massive build-up of cold air over Siberia drive winds from the northeast over the mountains and down over India, reversing the situation as the season changes. And in colder climatic epochs brief penetrations of northerly winds may frequently interrupt the summer monsoon. In the ice ages the meteorological equator doubtless kept well south of its present position, and was generally nearer to the geographical equator, restricting the monsoon.

Studies of the Indian monsoon, from the time when Sir Gilbert Walker was director of the meteorological service in India early in this century to the work of Dr C. Ramaswamy and others in recent decades, have consistently shown that each year's monsoon development is affected by that of the northern hemisphere belt of westerly winds farther north. The seasonal withdrawal of the westerlies towards higher latitudes, in its turn, is affected by the amount of snow put down in the winter and lingering on in the spring over the Himalayas and the Tibetan plateau. The development of the monsoon in any given year is also related to variations in the world-wide configuration of atmospheric pressure and the wind circulation, particularly over the lower latitudes of both hemispheres, There is a sort of slow see-saw oscillation, whereby pressure is lower than usual over Indonesia and the Indian Ocean in some years and higher than usual over Easter Island and the southeast Pacific; in other years the reverse is the case. The swings of this so-called 'Southern Oscillation' go on continually, one cycle being completed mostly in 2 to 2½ years but occasionally taking as long as 5 to 7 years. Sea temperatures in the equatorial Pacific are affected and to some extent those in the equatorial zone of the other oceans also. Some useful progress has been made using these discoveries in forecasting the yield of the monsoon in India each year before the season begins. It has been found too that, when the northern hemisphere circumpolar vortex develops a sharp trough near the longitude of India, the southerly wind component in the region of the eastern part of the trough tends to bring the monsoon system quickly north in that region. The northerly winds in the western part of the trough, and the surface northerlies near the axis of the trough, are equally capable of delaying the monsoon or producing breaks in the monsoon after it has already been established. Thus the monsoon situation over India, and presumably in like manner over east Asia, may be critically affected by the exact longitude in which a trough in the upper westerlies develops.

In the last two decades it has been shown that the behaviour of the monsoon over west Africa is also related to that of the westerlies in middle latitudes over that sector of the northern hemisphere: in periods when blocking anticyclones or northerly winds over western and northern Europe (especially in winter and spring) divert a branch of the upper westerlies and much of the cyclonic activity south into the Mediterranean, the monsoon commonly fails to penetrate so far north as usual, or is late, over west Africa and elsewhere south of the Sahara. In such years the zone across Africa from Senegal and the Sahel to Ethiopia is liable to be stricken by drought. The African monsoon, like that over southern Asia, represents the seasonal northward displacement of the convergence between the surface wind systems of both hemispheres and the accompanying equatorial rains. Over Africa the seasonal limit of northward penetration of the rains in modern times is seldom north of latitude 21 °N and often fails to reach 20 °N.

To complete our brief survey of seasonal changes in the present epoch, fig. 16 presents a few sample rainfall 'curves' for places each of which is representative of some well-defined regime. The example from western Iceland, with most precipitation in the winter and least in summer, illustrates a regime that is typical for islands and other places near the ocean in middle and higher latitudes. This seasonal distribution is determined by the greater cyclonic activity in winter, when the overall temperature gradients are strongest. The Iceland curve shows a further feature of some

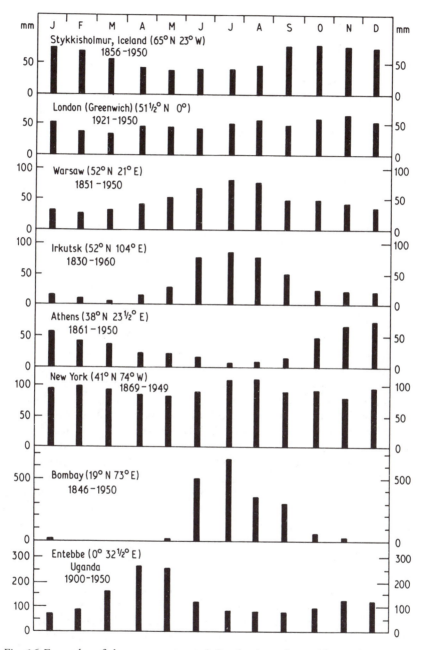

Fig. 16 Examples of the average seasonal distribution of monthly rainfall (or rainfall equivalent derived from the total down-put of rain and snow) at places in different latitudes.

interest: the abruptness of the increase of rainfall from August to September. This is due to a rather sudden change in the steering of most of the North Atlantic cyclones, away from the eastward paths which they commonly pursue in summer near latitude 60–63 °N to a path more northeastward into the Arctic – a switch of steering that is believed to be caused by the sharpening trough in the circumpolar vortex over Quebec and Labrador where the cooling season sets in early. Places in the eastern and southern parts of the British Isles show both winter and summer maxima of rainfall, neither of them very sharply defined: so any month may turn out to be the wettest month in some individual year. There is a modest minimum of rainfall in September in southeastern parts of Britain – more strongly marked in the middle weeks of the month – which is the counterpart of the increase registered around the same time in Iceland. This also applies to the abrupt decrease of rainfall in central and eastern Europe (e.g. at Warsaw) from August to September. Continental places in the middle and higher latitudes have their greatest rainfall in summer: this is partly due to thunderstorms at that season and partly because in winter the very cold air cannot contain enough moisture for such a heavy downput whether as rain or snow. The concentration on summer rain becomes more marked as one proceeds from Europe towards the heart of the great land-mass in Siberia with its extremely cold winters, This concentration of the rainfall in summer continues very marked all the way to Peking. New York, in a similar latitude on the eastern seaboard of North America, however, has not only a summer maximum but also a winter maximum associated with the frequent cyclonic activity off that coast. The Athens curve illustrates the Mediterranean regime that is characteristic for subtropical latitudes, between about 30 and 40 °N and S, near the fringe of the deserts in both hemispheres: the rains come in the colder months and particularly whenever incursions of very cold air arrive over the warm sea. The summer months, when the anticyclone belt moves towards higher latitudes, are commonly rainless. The curve for Bombay illustrates the Indian southwest monsoon, associated with the equatorial rain belt's seasonal movement north to latitudes 25–30 °N in this sector. Finally, Entebbe on the equator in east Africa (Uganda) is an example of the inner tropical regime with two rainy seasons each year as the convergence zone between the wind systems of the northern and southern hemispheres passes north and south over the place.

One principle of great importance in the distribution of rainfall is demonstrated by comparing the totals at places on the windward and leeward sides of great mountain ranges. Thus, the average annual total (1860s to 1940s) at Hokitika on the west coast of South Island, New Zealand was 2907 mm, and at Christchurch on the east coast the figure was 639 mm. Both places are near 43 °S, in the zone of prevailing west winds. And in Scotland in March 1938, when the winds were westerly on every day of the month, Kinlochquoich on the western side of the country had 1270 mm rainfall in the month, while Braemar in the shelter of the mountains on the eastern side had 5 mm. In occasional months when easterly winds prevail the rainfall distribution is reversed.

4

HOW CLIMATE
COMES TO FLUCTUATE
AND CHANGE

VARIATIONS OF THE PREVAILING WINDS AND THEIR EFFECTS

We have seen in the last chapter examples of the types of variation to which the world's wind circulation is liable. The circumpolar vortex over either hemisphere is a constant feature, as are the belts of prevailing high and low surface pressure that go with it. An oversimplified view of climatic variation might regard it as all a matter of 'expansion' of the circumpolar vortex in cold epochs, when the area of the polar regime expands, carrying the belt of westerlies to lower latitudes than before, and 'contraction' of the vortex in warm epochs when the polar cap contracts. There is some truth in this in so far as such expansions and contractions are indeed observed to take place in the course of the yearly round of the seasons and longer-term variations. But we have to consider other elements of the situation that change.

The strength and the wave patterns of the circumpolar vortex also vary, with changes in the wave-length (or spacing around the hemisphere) and in the amplitude of the waves. The positions of the troughs and ridges which constitute the waves or meanders vary in consequence, and the positions, orientation and intensity of development of the belts of prevailing cloudiness and disturbed weather, as well as the extent of the polar and tropical regimes, change accordingly. In particular, the development and extent of the very cold and rather calm surface layer of air over the polar ice and snow is affected. Doubtless the most fundamental question underlying the variations of the Earth's climate is the total energy taken in, plus what is released at any time from the heat stored in the oceans, to heat the surface and the atmosphere and to drive the winds. Great importance must attach to any variations in the heat available over the lower latitudes, where the Earth is so broad and where the greatest absorption takes place. The temperature variations are however very much amplified in the highest latitudes, where the extent of ice and snow varies in response to variations of the heat transport and the characteristics of ice and snow surfaces, if

52

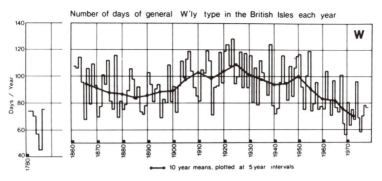

Number of days of general W'ly type in the British Isles each year

Fig. 17 Number of days each year with general westerly winds blowing over the British Isles, 1781–5 and 1861–1979.

present, produce very low temperatures. As a result, the seasonal and longer-term variations of the climatic regime are greatest in – and most clearly signalled by – the temperatures prevailing over the polar regions. The changes in the overall temperature gradient between low and high latitudes produce other signals in the form of variations of the strength and pattern of the wind circulation over middle latitudes.

The variations cover an enormous range of time-scales. We may begin to appreciate this from the long record of the frequency of westerly wind days over the British Isles in fig. 17. There were often big variations from year to year, sometimes with a sort of biennial rhythm. But the longer-term variations shown are also impressive. From the 1860s to the 1960s the overall average frequency was about 95 days/year, but for several decades in the early part of the twentieth century it was over 100 days, in the 1920s, 109; lately it has fallen to about 70 days/year, and in the 1780s it seems to have been only 60–65 days/year (in one year, 1785, only 45 days). There were compensating variations in the frequency of 'blocked' or stationary weather situations with quite different winds in this part of the world. We can detect still longer-lasting changes of the frequencies in the more distant past.

The climate of some places is particularly sensitive to changes in the prevailing winds. On windward coasts and at places on the windward side of hills and mountains much more cloud, more frequent rain or snowfall, and greater totals of precipitation prevail than on the sheltered, leeward side. Similar contrasts affect one and the same place or area when the winds change and these same places find themselves on the lee side. Good examples of this occur in Scotland, where the windswept Atlantic coast and slopes of the Highlands are characterized by extensive grass and heather moors, and peat-bogs, with average yearly rainfalls commonly around 2000 mm/year (and locally up to 4000 mm and more in the mountains). On the other side of the country, an area in the northeast around Nairn and

53

Elgin at latitude 57½ °N, sheltered by the mountains from the prevailing southwesterly winds, has an average rainfall of about 600 mm/year and sometimes experiences temperatures as high as 14 or 15 °C (nearly 60 °F) in mid-winter. There, the usually genial climate provides rich farming, a by-product of which may be seen in the magnificence of ancient buildings such as the ruined cathedral in Elgin; but when the wind blows from the north, snow may fall and temporarily cover even the low ground as late as early June. A similar, but more extreme case is provided by Trondheim (63 °N) in Norway and the extensive farmlands in the usually sheltered districts around the inner part of Trondheim fjord. In certain periods of history, when the southwest winds were less reliable and the north wind blew more frequently, the cultivable area in that district contracted and former settlements were abandoned and reconquered by the forest.

Other situations which are perhaps even more vulnerable to changes in the pattern of wind direction frequencies occur all around the edge of the Arctic and its cold or ice-covered seas. Thus, Archangel on the coast of the White Sea with an average (1851–1950) July temperature of 15.8 °C, close to the value for central England, has known a July (in 1938) with a mean temperature as high as 21.3 °C (compare Marseilles 22.5 °C long-term average) and another (in 1926) that had as cold a mean as 11.8 °C.

Similar vulnerability in respect of the rainfall needed for cultivation occurs near the great deserts in the subtropical and tropical fringes of the arid zone.

A still longer record of variations in the frequency of westerly winds in England, the longest such record for anywhere in the world, is indicated by the graph in fig. 18. There are signs in this curve of a repeating pattern, which may be related to other evidence of a cyclic process of about two hundred years length, and if so may be of some use in forecasting the

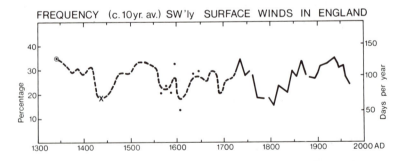

Fig. 18 Frequency of southwesterly surface winds in England, 1340–1978. From daily observations in the London area from 1669 to 1978 (ten-year averages). The earlier part of the curve is sketched from indirect indications, including various weather diaries e.g. in eastern England (Lincolnshire) 1340–44 and Denmark 1582–97.

climate over the next hundred years – unless its course be altered by the first major impact of human activities.

How the distribution of cloudiness and disturbed weather over the northern hemisphere shifts with the variations of the circumpolar vortex has been illustrated by fig. 5 in the last chapter. Fig. 5b (p. 31) showed an average situation for the present epoch with the cloud cover largely concentrated in two latitude belts: the equatorial rain system and the broad zone of cloudiness over middle latitudes, associated with the continual cyclonic activity around the fringe of the Arctic but broken by the lee effects of the Rocky Mountains and the mountains of Asia. Here we illustrate in figs. 19–22 the effects on the surface pressure and wind pattern, and hence on the prevailing temperatures, in winter and summer, of blocking or distortion of the upper wind flow. First, in figs. 19a, b, and c, we see an example of a winter month with little blocking but rather an invigorated, smooth flow of the upper westerly winds around the northern hemisphere. This brings more mild oceanic air than usual right across both great continents in middle latitudes. Only in the quiet region of the inner Arctic, away from the vigorous part of the wind circulation, and in areas penetrated by cold surface air from the Arctic, have temperatures much below normal developed. Some broad inland areas in the tropics are also colder than normal, owing to a surface wind pattern which brings them air ultimately drawn from the interiors of the northern continents. By contrast, fig. 20 shows a winter with much more distorted wind flow, such as is called blocking of the westerlies, over the Pacific, American, Atlantic and European sectors. Much bigger surface temperature anomalies, mostly negative, result. The warmth over northern Alaska is attributable to mild Pacific air from the south, driven over the mountains, and that over northeast Canada to more than usual drift in of air from the Atlantic Ocean and Davis Strait.

Fig. 21 illustrates the distorted pattern with the westerlies diverted well to the north over the northeast Atlantic and the Arctic coast of Europe, which in 1976 gave parts of Europe their warmest summer since the instrument records began. Yet in the region of the cold trough in the upper westerlies over Russia, where the surface winds were mainly northerly, temperatures averaged 3–4 °C below normal. That memorable summer seems, in fact, to have been a particularly cold one when considered over the northern hemisphere as a whole. Fig. 22 shows another summer month, in 1965, in which a distorted circumpolar vortex with stationary cold troughs in three sectors gave a very cold summer over most of Europe and eastern Canada. This was the year which halted the long recession of many of the glaciers in the Alps and, with surface northerly winds over the Norwegian-Greenland Sea, saw the return of the Arctic sea ice to Iceland.

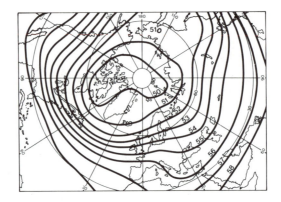

Fig. 19a Average pattern of the circumpolar vortex (heights in hundreds of metres of the 500 millibar pressure level) in January 1975. (The upper westerly winds on average follow the course of these contour lines.)

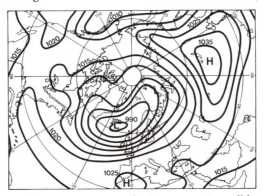

Fig. 19b Corresponding average atmospheric pressure in millibars at sea level in January 1975. (The surface winds again blow anticlockwise around the low pressure areas, but with some indraught across the lines towards the lower pressure side.)

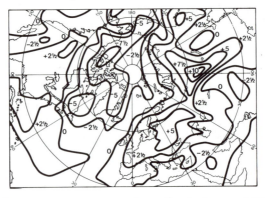

Fig. 19c How the mean temperatures at the surface in January 1975 departed (°C) from the average of the period 1931–60.

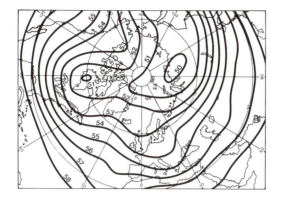

Fig. 20a Average height of the 500 millibar pressure level in January 1979.

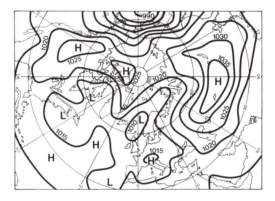

Fig. 20b Average pressure at sea level in January 1979.

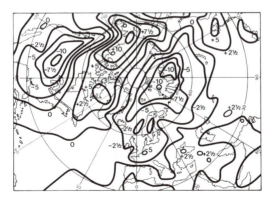

Fig. 20c Departure (°C) of the mean surface temperatures from the 1931–60 average in January 1979.

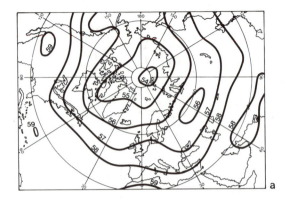

Fig. 21a Average height of the 500 millibar pressure level in July 1976.

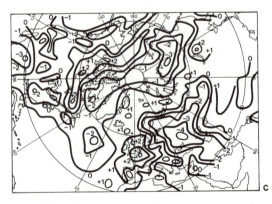

Fig. 21b Average pressure at sea level in July 1976.

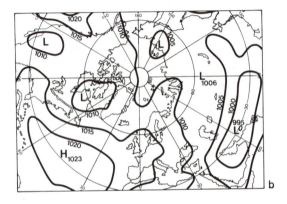

Fig. 21c Departures (°C) of the mean surface temperatures from the 1931–60 average in July 1976.

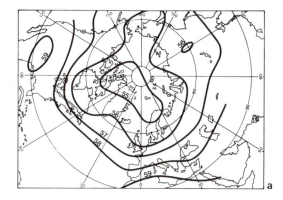

Fig. 22a Average height of the 500 millibar pressure level in July 1965.

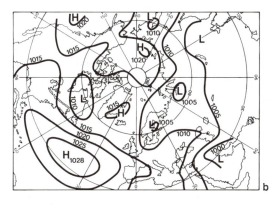

Fig. 22b Average pressure at sea level in July 1965.

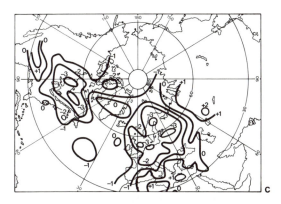

Fig. 22c Departures (°C) of the mean surface temperatures from the 1931–60 average in July 1965.

CHANGES IN THE OCEAN

The ocean circulation also plays a part in the transport of heat towards the poles. Within the tropics this is the main share of the total heat transport according to modern measurements. At latitude 20 °N it seems to account for 74 per cent of the total; at 50 °N it is about 30 per cent, and over the whole realm between the equator and 70 °N the oceans, contribution averages 40 per cent. But the figures must vary when different epochs are compared. The changes in the ocean surface, with which we are here mostly concerned, are greatest near the boundaries of different ocean currents and where there are shifts in the boundaries of ice on the polar seas and of cold upwelling water in the tropical or subtropical oceans. Fig. 23 illustrates how the boundary between water of Gulf Stream origin and the polar water at the surface of the North Atlantic Ocean has varied. The biggest temperature changes are found near the furthest advances of the cold water or ice replacing a previously warmer surface (or vice versa). Thus a large area between the Bay of Biscay and mid-Atlantic was 10–12 °C colder at the climax of the last ice age around twenty thousand years ago than in our own times. An area between Iceland and the Faeroe Islands (61 °N) seems to have been 5 °C colder than the modern average between about

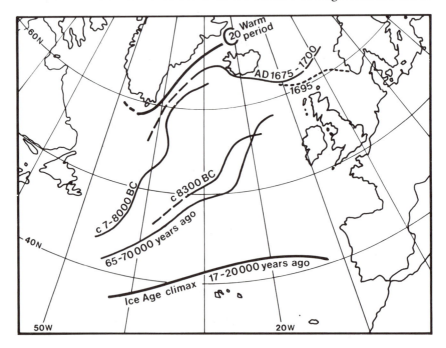

Fig. 23 Positions of the boundary in the ocean surface between water of Gulf Stream origin and the polar ocean current from near northeast Greenland in the twentieth-century warmest years and at various times past.

AD 1675 and 1705. Where shifts of the current boundaries such as this take place, the ocean fails to exercise the moderating influence on climatic variability that we otherwise expect of it. The southward limit of the polar water east of Iceland indicated for 1675–1705 in fig. 23 has sometimes been approached over short periods of up to a few weeks in recent years. For example, in April 1968 and again in 1969 this water-mass advanced to near the Faeroe Islands, and more briefly a number of times in the first half of 1979, but no comparable advances seem to have occurred for 40–50 years previously. Fluctuations of the boundaries and variations of the strength of the various ocean surface currents, the latter clearly related to anomalies in the large-scale wind circulation, are commonly observed; in consequence the ocean surface temperature is liable to show anomalies of up to 2–4 °C on either side of its usual value in such regions for some weeks or even for a few months. Another example occurred in the western North Atlantic where the northern limit of the Gulf Stream water was south of its usual position during much of 1968.

More is known now than when this book was first published about the exchanges that go on between the Arctic seas and the climate over wider areas and longer times (which we return to on p. 271). The great increase of ice on the East Greenland Sea in the mid-1960s, and the low salinity water that accompanied it, migrated from there slowly to affect the western Atlantic in the years that followed and, after a long clockwise circuit over the western ocean, was carried back into the Iceland region after fourteen more years. In the meantime there had been a rather warmer, more saline phase near Iceland. A study by Dickson and others in 1988[1] followed the progress of the anomaly and the cycle of changes that went with it over nearly two decades. And this inspired others, notably Mysak and his co-workers[2] at the Centre for Climate and Global Change Research and the Department of Meteorology at McGill University, Montreal, to trace the anomaly back through its earlier history before it first reached the Greenland Sea. Its origin is found to have been in a greatly boosted run-off into the Arctic Ocean north of Canada from the North American rivers, mainly the Mackenzie River system in the western part of northern Canada, in the early 1960s. This was associated with enhanced downput of rain and snow and sharper development of the upper (cold) trough in the winters over the Canadian Arctic and the islands. There was also a noteworthy tendency in those years to develop two other troughs, one north of Greenland towards the northwest Siberian coast and the other over the far eastern part of northern Siberia. These features amounted to a radical change in the configuration of the large-scale wind circulation – a 'climate jump', as defined by Knox et al.,[3] that would be associated with increased cyclonic activity and cloudiness in those parts of the high Arctic. The resulting flush of river water into the Arctic Ocean in the polar basin was accompanied by an increased supply of Pacific Ocean water, which is also of low salinity, through the Bering Strait. This is how what Dickson called the 'great salinity anomaly' came about. It extended the relatively fresh water on the surface of the Arctic Ocean, which arrived in the East Greenland Sea some four or five years later, increasing the ice there and bringing a dramatic fall in the Arctic temperatures.[4]

Searches for earlier examples of such a sequence by inspecting the very long, year-by-year, records of ice at the coasts of Iceland seem to suggest a tendency for somewhat irregular recurrences at about twenty- to thirty-year intervals, though the amplitude of the swings also varied from case to case. Statistical examination of the long series of Iceland ice records, using Koch's index, showed a peak frequency at about twenty-seven years, which fell just short of statistical significance, but there were significant peaks at frequencies of about 5 and 88 to 100 years.

The Canadian investigators consider that this evidence points to an Arctic (roughly) interdecadal cycle as of some importance, mentioning as examples the relatively warm (little-ice) years around 1900, the 1920s, the late 1930s, 1950s and 1970s, which were all followed by ice increases. The cycle is not symmetrical: the increases of North American run-off and of ice extent on the Arctic seas take place generally much more sharply than the subsequent 'returns to normal'. They also speculate that climatic jumps like the particularly sharp one about 1962 could either enhance or damp out altogether the global warming widely predicted as part of the 'carbon dioxide effect'.

MORE BASIC MATTERS

Volcanic dust in the atmosphere

As to the underlying causes of the variations of the wind and ocean circulations which we observe, these may be largely a matter of the global temperature level and the strength and positions of the main thermal gradients. This interpretation is basic to our understanding of the seasonal changes of the wind circulation each year from summer to winter and back again, and is supported at least in outline by mathematical modelling of the atmospheric circulation. It is demonstrated in the sequence of developments after great explosive volcanic eruptions, which leave a veil of sub-microscopic particles in the stratosphere for some years after the event. The dust veils screen off some of the sun's radiation, but allow the outgoing longer-wave Earth radiation to pass out to space. One must distinguish between volcanoes erupting in low latitudes and in the higher latitudes of either hemisphere. The winds in the stratosphere ensure that the dust soon (i.e. over a few weeks) encircles the Earth in about the latitude of the volcano. And, because of a slow net poleward drift of the stratospheric air, a veil of significant density is likely within some months to

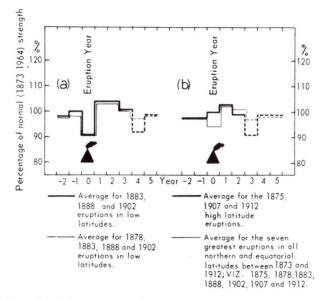

Fig. 24 Strength of the mean surface wind circulation over the North Atlantic Ocean in the years before and after great volcanic eruptions. (a) Averages in cases of eruptions in low attitudes. (b) Averages in cases of eruptions in high latitudes and (thin line) in all latitudes. (Circulation strength measured by the difference between the highest and lowest monthly mean atmospheric pressure values at the surface anywhere over the North Atlantic, averaged for all the months of the year.)

spread over the whole Earth from great eruptions that take place within about 20° north or south of the equator. From eruptions in the middle and higher latitudes usually only the hemisphere concerned is significantly affected. The dust seems to clear first in low latitudes, where the base of the stratosphere is highest. And after a year or more the veil is increasingly concentrated over the higher latitudes. The effects persist longest over the polar regions, where the reduced temperatures allow more sea ice to form.

The effect of volcanic dust veils on the world's wind circulation is illustrated in fig. 24. With global spread of the matter from great eruptions near the equator the wind circulation is at first weakened, but then changes over into an enhanced state after about two years, as a dust veil by then covering effectively only the higher latitudes enhances the difference of heating between low and high latitudes. The whole cycle takes typically three to four years after the eruption, but occasionally up to seven years, before conditions return to what they were. With high-latitude eruptions generally only the enhanced wind circulation phase occurs. There is also a tendency for the North Atlantic zone of main cyclonic activity to be shifted somewhat south in the summers after great eruptions, this accounting for many, perhaps most, of the coldest wettest summers of the last three hundred years in western Europe and eastern North America. The same probably applies to the North Pacific, where the four greatest famine-producing years of bad harvest since 1599 in the northern half of Japan were all years when great volcanic dust veils were present over the northern hemisphere.

Between 1980 and 1992 there were at least three great volcanic explosions: Mount St Helens in May 1980 in Washington State in north-western USA, El Chichon in Mexico in March–April 1982, and Mount Pinatubo in the Philippines in June 1991, all of them probably comparable with the greatest eruptions in the last two centuries. The last thirty to forty years have provided us with much greater facilities for observing the course of an eruption and the development of the cloud of 'dust' and vapours, the rocks, ash and aerosols ejected into the atmosphere, as well as new techniques for following them.

Among the new techniques exploited for the first time after the middle of this century was the direct sampling of eruption particles and gases captured in the atmosphere by high-flying aircraft. The particles could then be examined for particle sizes and their chemical nature. Also the chemical contribution and quantities of volcanic products from ancient, as well as modern, eruptions deposited on, and in, the buried layers of ice in Greenland and the other ice-sheets in high latitudes, north and south, began to be studied. This has produced extremely valuable chronologies. The intensely concentrated beams of light available in lasers since about 1960 have given us lidar, which penetrates the aerosol layers in the atmosphere and makes it possible to measure their height and optical thickness from the ground.[5] Other, related techniques have been developed more recently, using instruments mounted on satellites to observe, at up to fifteen successive sunrises and sunsets experienced by the satellite each day, the extinction times at different wave lengths of light affected by different gases in the atmosphere (thereby analysing the presence and quantities of such substances injected into the atmosphere).

In spite of all these advances in our technical capacity, it is surprising – and maybe salutary – that we have had to take note of a case in the winter of 1981–2 when an important 'dust' cloud appeared, and spread over all the lower latitudes of the Earth, without any reports of a great volcanic eruption that could account for it. Two possible sources have been suggested: eruptions around Christmas and mid-January 1982 of the Nyamuragira volcano near the equator in central Africa (Zaire), which certainly produced a sulphur dioxide cloud, and the explosive activity of Pagan in the Mariana Islands in the tropical Pacific from mid-January 1982 onwards. Cloudy weather had hindered detection of an eruption plume on the regular satellite photographs from either of those eruptions. If a major eruption can so escape detection in these days of continuous world watch, co-ordinated by the Smithsonian Institution, Washington DC, using all the techniques now available, we have to realize that our knowledge of past eruptions cannot be complete.

The three other great eruptions (named above) in recent years were instructively very different from each other. The differences must have ensured that any climatic effects would also be very different. The eruption of Mount St Helens, in the northern part of the United States Rocky Mountains in 1980, sent an ash volume which rose immediately 22 to 25 km, but the early expectations of a very great dust and aerosol veil in the stratosphere had to be drastically moderated when it turned out that the main blast was directed almost horizontally from the upper part of the volcano, killing the leader of one of the scientific teams on another height where the party had been stationed to observe and photograph the expected eruption. There was no great input of matter into the stratosphere.

The eruption of El Chichon in Yucatan, southern Mexico, two years later, was a very great one, but there were unusual features in the spread of the stratospheric veil over the Earth. These were probably associated with the seasonal development of the atmospheric circulation at the levels concerned. They affected the timing of the arrival of the eruption products over different regions of the Earth. Initially the column rose quickly to about 17 km and after about ten days up to 26 km, completing the first circuit of the globe within twenty-two days. Later in the summer, products of the eruption were found at up to 35 km or higher and enhanced sunset colours were seen from the Arctic to the Antarctic. But, in the main, the lateral spread of the materials remained curiously limited to latitudes between about 5 and 40 °N for about nine months, until the following December. The El Chichon volcanic veil proved to be a very long-lasting one which had some effect for about eight years until 1990.

The explosive eruption of Mount Pinatubo, near 15 °N 120½ °E on Luzon in the Philippine Islands began on 2 April 1991 and culminated in a great paroxysm on 14–16 June of that year. It sent a column of ejection products immediately to near 40 km height. Some of the tephra fell on Singapore, over 2,000 km away to the southwest. The distribution of wind-borne products was complicated by a typhoon. Over the first three months after the main eruption, atmospheric aerosols had spread over nearly twice the area covered by products of the El Chichon eruption at the same stage and by the end of 1991 their optical thickness measured over middle latitudes of the northern hemisphere was 60 to 80 per cent greater than after El Chichon. By the first months of 1992 aerosol concentrations in the atmosphere over the tropics were decreasing, and in latitudes 40 to 60 °S had peaked about November to December 1991, but over the zone 40 to 60 °N concentrations did not peak until April to May 1992.

One year after the eruption lidar measurements over the Mauna Loa mountain-top observatory in Hawaii indicated two to three times as much aerosol still present as at the same stage after El Chichon. By the end of 1992 aerosol optical thickness over the tropical oceans had declined to 15 per cent of its September 1991 peak. But over the higher latitudes the decline was much less, with actually very little change over 40 to 60 °S throughout 1992. The real decline of the Pinatubo loading of the stratosphere over the tropics set in at the same time of year as in the El Chichon case. Presumably the onset of its decline was connected with the regular seasonal change-over of the winds in the stratosphere. By June 1993 the tropical stratosphere was essentially back to the state it had been in before the 1991 eruption. The top of the aerosol layer had by then descended to 18–19 km over Germany, but many months later, in January 1994, it was still shown at well above 20 km by measurements over Hawaii and Cuba.

Some return of eruption products north across the equator during the 1992–3 winter was suspected and may have been repeated a year later. A fresh injection of materials from Mount Spurr, Alaska, which erupted in August 1992, continuing the very active volcanic period, complicated the picture.

There is no doubt that we have witnessed in the 1980s and 1990s two or three of the greatest explosions of volcanic matter into the stratosphere to have occurred in the last two centuries and some will claim that the incipient turn-down of global temperatures must be due to this.

Astronomical cycles affecting the heat supply

Another clearly identified cause of climatic change, applying to much longer-term events, lies in the changes of incoming solar radiation available at different latitudes and seasons which accompany regular cyclical variations in the Earth's orbital arrangements (fig. 25). There are changes in the ellipticity of the orbit and in the seasonal position of the Earth on its way round the ellipse; the tilt of the polar axis relative to the plane of the orbit also varies, changing the latitudes of the tropics and of the polar circles by a few degrees. These cyclical variations, over periods of about 100,000, 20,000 and 40,000 years respectively, affect our distance from the sun, and the strength (and, more slightly, the angle of incidence) of the solar beam, at different seasons. They are entirely calculable, and seem to explain in outline the many times repeated alternations between ice ages and warm interglacial climates, like those of today and of historical times, as well as the intermediate phases known as interstadials. During about 90 per cent of the last million years the climate has been more or less glacial, with extensive ice on land and on the polar seas, and much colder than today. Interstadials marked the culmination of lesser warmings with the

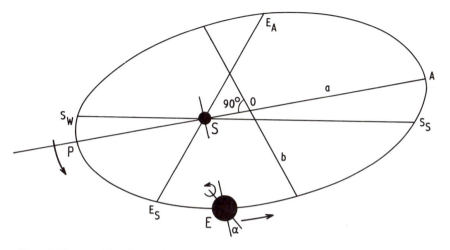

Fig. 25 The Earth's orbit and items which undergo long-term cyclic variations. In the sketch the orbit is seen from a slanting angle outside its plane: so the angle between the major and minor axes of the ellipse, a and b, does not appear as a right angle because of the perspective, The sun S is at one focus of the ellipse, and the ratio of its distance from the centre of the ellipse OS to SP measures the eccentricity e = OS/OP. Short lines are drawn through the sun S and the Earth E vertical to the plane of the orbit (the 'ecliptic'): The angle α between this vertical and the Earth's polar axis is called the obliquity. The present seasonal positions of the Earth in its orbit at the winter solstice, S_W, the spring equinox E_S, the summer solstice S_S, and the autumn equinox E_A are marked. The two other points marked are P where the Earth is at perihelion (i.e. nearest the sun) and A where the Earth is at aphelion (i.e. farthest from the sun). The seasonal points (shown joined by lines which are always at right angles to each other) slowly precess around the orbit so that the season at which the Earth is nearest to or farthest from the sun slowly changes, taking about 20,000 years to complete the circuit. The winter solstice S_W is at present slowly getting farther from P.

climate remaining colder than today and an intermediate type of vegetation. From analysis of cores taken from the sediments on the ocean bed covering the last one to two million years, the Quaternary geological period, and from the vegetation history derived from pollen analysis of undisturbed deposits on land (in one case continuous over the last 140,000 years), the nature and timing of the major climatic changes seems to fit very well the variations of radiation available, calculated on the basis of these orbital parameters.

The strength of the sun's beam and the total radiation available at each latitude on the Earth in each month and season of the year, disregarding the effects of any cloud cover that might be generated, have been calculated on the basis of these variations one million years back into the past and far into the future. The first such complete calculations were made by the Yugoslav scientist Milankovitch about 1930, and this approach is often referred to as the Milankovitch theory of ice ages, although this theoretical solution of the ice age problem was put forward by James Croll as long ago as 1864. More refined calculations have been done in recent years, notably by Vernekar in the United States and most recently (1979) by André Berger of Louvain-la-Neuve. It is now clear that the temperature changes resulting from the calculated variations of the radiation budget would not be sufficient to swing the Earth's climate into an ice age and back again without the amplifying effect of reflection away of the sun's radiation from an increased area of snow and ice. The time of strongest summer heating in the latitudes of the former North American and north European ice-sheets, about ten thousand years ago, coincided with the most rapid melting of the snow and ice. The highest general temperature level on the Earth was attained some five to six thousand years later, a lag imposed by the time needed to melt the accumulated ice and to warm the oceans to a considerable depth.

There are no other such regular cycles apparent in the long past record of climate as those we have mentioned: the round of the day, the year and the long cyclical evolutions of the Earth's orbital arrangements described above (and produced by the gravitational forces in our planetary system), together with the cycle of events which is set off at intervals by a great volcanic eruption and the resulting dust veil. Although the latter appear like random events at irregular intervals, it is now established that eruptions tend to occur, at maxima of the combined tidal force of sun and moon. And there are suggestions of a periodicity of about two hundred years length in the frequency of great volcanic eruptions, which may also turn out to be linked to gravitational forces.

Other cycles

Besides all these reasonably well-understood variations, however, spectral analyses of climatic records present indications of many other periodicities or 'quasi-periodicities', which are less regular in their operation but some of which may nevertheless be of some importance. Once they are better understood some of them may acquire a limited usefulness in forecasting. Fig. 17 (p. 53) has already introduced us to a (quasi-) biennial rhythm which is present in many meteorological data series. Thus, between 1880 and 1961 the summers of the odd-numbered years had a significantly better (meaning warmer) record in northern, central and western Europe than those of the even-numbered years. In London the overall difference of mean temperature was 0.5 °C, probably corresponding to about two weeks difference in the length of the growing season. (In various earlier periods of history the even-numbered years enjoyed the better record.) Other periodicities which seem to be of some interest include lengths about 5½ years (half the sunspot cycle?), 10–12 years, 22–23 years, about 50 years, 100 years, 180–250 years (perhaps really very close to 200 years), 1000 and 2000 years. In none of these cases has the physical origin been satisfactorily established, though it is commonly thought that several of them, and perhaps all, originate in (presumably slight) fluctuations of the energy output of the sun. Some may be associated with variations of the tidal force acting directly on the oceans and the atmosphere.

Cycles in the sun's activity

There are indeed indications that the sun's activity may vary on certain timescales in the range 100–10,000 years, as recently pointed out by R. K. Tavakol[6] of the Climatic Research Unit, Norwich. One of the sources of evidence is the errors discovered in radiocarbon dating of objects of certain known ages. The first assumption in this modern technique of dating archaeological and naturally occurring objects (such as peat or lake-bed deposits) laid down in the last fifty thousand years was that the proportion of radioactive carbon in the atmosphere's carbon dioxide should be constant, because this element is formed from the nitrogen atoms in the upper atmosphere by the steady bombardment by cosmic rays from the galaxy. Errors in the radiocarbon dates at various past periods are, however, revealed for instance by radiocarbon dating of tree rings. The errors are partly attributable to regular long-term variations in the Earth's magnetic field but, apparently, also demand variations on other period lengths in the sun's output of particles ('corpuscles') associated with powerful extensions of the solar magnetic field; both these solar influences deflect away from the Earth some of the continual bombardment by cosmic ray particles from elsewhere in the galaxy. This reduces at times the production of radioactive carbon atoms in the atmosphere. J. A. Eddy's work at the High Altitude Observatory of the US National Center for Atmospheric Research at Boulder, Colorado has stressed the coincidence of the two great minima of sunspot activity, the Spörer Minimum from AD 1400 to 1510 and the Maunder Minimum from 1645 to 1715, when almost no sunspots were observed and few (if any) polar lights (auroras) were reported, with the two greatest periods of climatic stress on the Earth in the so-called Little Ice Age. The long record derived of radiocarbon fluctuations and the sketchier record of phenomena directly related to the sunspot cycle point to a variability of the sun on the suggested time-scales. The evidence is not conclusive, however, and more knowledge is needed of how various types of solar fluctuation can affect the motion patterns of the Earth's atmosphere and hence our climate.

The heating pattern and reconstruction of past climates

In the last chapter we indicated the intimate association between the thermal pattern, generated in any season or in any climatic epoch by the distribution and intensity of heating of the Earth, and the shape and intensity of the circumpolar vortex. This relationship can be used to derive the prevailing features of the large-scale wind circulation, including storm tracks and regions of cloudiness, prevailing surface winds and calmer anticyclonic areas, in past epochs from fossil evidence of the prevailing temperature levels. Such reconstructions show us the type of general regime which would be in equilibrium with the known temperature distribution. And in many cases the pattern can be at least partly verified by seeing how the available fossil data, or historical manuscript evidence, registering the prevailing winds and weather in various areas, fit the map. The climatic patterns so derived, however, tell us nothing of the events which produced the change of climate from one regime to the next. Evidence is accumulating from studies of the great climatic shift which took place in Europe within the last millennium, from the warmth of the high Middle Ages to the cold of the late sixteenth and seventeenth centuries and a recurrence of the latter around 1800–50, that the great glacier advances and times of advance of the polar sea ice were concentrated in just three to six well-separated periods of about ten to twenty years. One or two of these, but probably not all, were associated with world-wide 'bursts' of volcanic activity. It seems certain that great anomalies of the wind circulation, involving high frequency of blocking situations, occurred in these decades. There is evidence, both from the yearly growth ring sequences in trees, and in actual temperature measurements in the late seventeenth and eighteenth centuries, of greatly increased year-to-year variability in the decades of most rapid ice advance, which seems to confirm the diagnosis of frequent blocking of the 'normal' middle latitudes westerlies.

THE RAPIDITY OF SOME CLIMATIC CHANGES

Several climatologists, notably Bryson, Flohn and Manley, have drawn attention to the apparently great rapidity of a number of major climatic changes in the past. For example, three major coolings covering perhaps half, and in two cases more than half, the range between present (or interglacial) prevailing temperatures and the ice age climax temperatures seem to have taken place in and around the North Atlantic Ocean, in Europe and the Mediterranean, in the later stages of the last interglacial. These coolings, dated at roughly 115,000, 90,000 and 70,000 years ago, took place within about one thousand years and possibly within a century or so.[7] The first two seem to have lasted only two to five thousand years and were followed by rapid recovery to temperatures only a little below those previously prevailing in the ten thousand years of the warmest part of the interglacial. The third occurrence introduced fifty thousand years of colder climates, including the main phases of the last ice age. The rapidity of these coolings and warmings indicates that no great accumulations of ice were formed in the two abortive ice age onset events. The last recurrence phase of the ice age around 10,800 years ago was similarly abrupt: at that time the prevailing summer temperatures in England, which had become as warm as (or a little warmer than) today's, dropped 4–5 °C, probably within about fifty years, and small glaciers reappeared in the Lake District in northwest England. The colder climate lasted about six hundred years. In this case, however, the magnitude of the reduction of temperatures in so short a time is made more readily understandable by taking account of the nearness to England of a then still-extant ice-sheet in Scandinavia and a much smaller one in Scotland.

Well-dated evidence of shifts of the vegetation boundaries in post-glacial times, particularly in North America where there were no mountain barriers in the way of northward advance, make it clear that some of the warming episodes were also very rapid once the ocean had warmed up. The rapidity is especially clear when one makes the allowance that must be made for the time required for the vegetation to respond. There must be a lower limit of about a hundred years before standing pine forest can be replaced by forest dominated by oak, though the process may be speeded by whole-sale disease and death of the old forest, followed by fire. Even so, some of these changes seem to have taken place so rapidly that the time elapsed was less than the error margin of the radiocarbon dates of organic matter from before and after the change.

One gets the impression that in many of the major climatic changes we have referred to the change of behaviour of the wind circulation is more or less instantaneous. The temperatures prevailing at the surface of the Arctic ice adjust themselves within a few years at most to either a calmer regime than before or to one with stronger winds and ocean currents which import more heat from other latitudes. Such changes – in both directions – have been observed within the present century. Shifts of the wind zones bring also displacements of the main belts of cloudiness. Through these and the changed pattern of wind and ocean heat transport the prevailing temperatures in other latitudes will also become adjusted to the new regime.

It therefore remains to explain the sudden changes of the wind regime which produce periods of perhaps ten to fifty or seventy years of great prevalence of blocking patterns, with wide-ranging meanders of the flow pattern of the circumpolar vortex aloft and stationary anticyclones and cyclones at the surface in middle latitudes. These in their turn produce persistent, or repeated, northerly and southerly surface winds, calms and sometimes easterly winds, dominating different sectors in middle latitudes.

The result is a reduction of the frequency of mobile westerly situations, with their continual interchange of warmer and colder air-masses in those latitudes. The stationary features of a blocking situation may be so placed as to maintain an abnormal extension of snow cover over land or to 'waste' the snow in the ocean, where it has no effect in changing the surface characteristics and their response to solar radiation. Better understanding is needed of the controls which determine any long-term preference for particular positions and especially whether the blocking anticyclones are mostly over Greenland or Scandinavia. The position of such features in summer may either accelerate or hinder the melting of the previous winter's snow over broad regions in latitudes north of 45–60 °N. And the maintenance of cloud cover, or of abnormal upwelling or evaporation from the ocean surface, in lower latitudes may have similar effects upon the overall heating of the Earth. If they come at a critical stage of a long slow upward or downward trend of the energy budget, such periods of prevalent blocking patterns in the wind circulation may make the switch to a warmer or colder global climate very rapid.

There is a school of thought in meteorology, articulated in recent years by Lorenz, which maintains that such changes of the wind circulation need no external cause. It is suggested that various alternative regimes of wind and climate are possible without any change in the external influences, so that at any time the pattern may 'click over' from one regime to another. On this view, even the ice ages may have no external cause. This is a philosophy of pessimism, so far as further understanding and prospects of forecasting are concerned. It is an unnecessary pessimism in cases where external causes can be demonstrated,[8] as seems clear at least in the cases of the effects of the Earth's orbital changes and the sequels to great volcanic outbursts. And there are hints that some variations in the frequency of stationary features in the large-scale wind circulation may be related to a number of variables in the motion of the Earth and to tidal forces on the sun associated with the alignments of the other planets, which have not hitherto been taken into account. Pioneer work by Bryson in the United States and Maksimov in the Soviet Union invokes the slow migration of the Earth's polar axis over distances of at most a few hundred metres under the influence of sun and moon and the even smaller 'Chandler wobble' of the Earth's polar axis. The latter is still unexplained but agreed to entail adjustments of the angular momentum of Earth and atmosphere. Unpublished work by Mörth in England further invokes effects upon the angular momentum induced by dispositions of the other planets in the solar system, as well as variations of the Earth's magnetic field and its effects upon electrically charged droplets in the atmosphere. All these suggestions are still controversial, but attempts to prove or disprove them should be pursued because most of the variables in question are predictable or partly so.

The question about whether side-effects of man's activities may now, or soon, begin to modify the global climate – and how they have modified some local climates – is another story, to be dealt with in later sections of this book dealing with the world today and the outlook for the future. We shall also deal briefly with some of the things which have been, or may be, done in the way of deliberate action to modify climate either locally or on a global scale.

5

HOW WE CAN RECONSTRUCT THE PAST RECORD OF CLIMATE

METEOROLOGICAL INSTRUMENT RECORDS

Part of the past history of climate can, of course, be known from the records of meteorological instrument measurements from the time when they first came into use. Unfortunately it is only a small part of the story, though an interesting one, stretching back in just one or two places as much as three hundred years, but over much of the world covering no more than the last few decades. Moreover, a great deal of painstaking work is involved in making sure that the figures derived from readings of the old instruments are rendered truly comparable with those from modern instruments in standard exposures. This task has so far only been thoroughly carried out for a limited selection of observation points in various parts of the world.

The barometer and the thermometer were invented in Italy, by Torricelli and Galileo respectively, in the first half of the seventeenth century. The wind-vane and the rain-gauge are earlier, but the earliest surviving rain measurements are from the late 1600s. There were already some reliable instruments about by 1700 or earlier, but the problems of exposing the thermometer and rain-gauge so as to obtain representative measurements were not solved until much later. The barometer presented fewer problems and was already being used for important scientific measurements by Blaise Pascal in France and Robert Boyle in England within a few years of its invention. But thermometers were for many decades commonly exposed in an unheated north-facing room (or, at one famous observatory until recently, outside on the north wall of a building); and their calibration gave trouble because of the ageing of the glass. Fahrenheit's real claim to distinction was as the maker of thermometers with the best glass. Early rain-gauges in Europe were sometimes exposed on the roofs of houses, where the catch is reduced by splashing out of the gauge and by evaporation in the wind; but this position enabled Richard Towneley – who produced the first gauge in England in 1676 – to lead a pipe down through the house so that he could measure the rain in the comfort of his bedroom!

74

Other problems in using early instrument records lie in the miscellany of different units and scales used. (Sir Isaac Newton already in the seventeenth century had a thermometer scale that defined 0° as the freezing point of water, but took as its other fixed point the normal temperature of the human body, labelled as 12°.) A good deal is known of the early thermometers and their scales, for instance from the comparison and calibration tests reported by van Swinden in 1792 and Libri in 1830. Van Swinden listed no fewer than seventy-seven different scales. By the time that anyone seeking to use early instrument records to create long time series of temperature or rainfall measurements, and barometric pressure and wind maps, has worked through such material and coped with inches, feet and miles which had different lengths in every part of Europe (and sometimes even within one country), he is readily convinced of the virtue of the uniformity of the metric system!

Our knowledge of the climatic history of the last two or three hundred years owes a great debt to those who – like Birkeland in Norway, Labrijn in Holland, Manley in England, and Landsberg and his co-workers in the eastern United States – by close study of the instrument records with their occasional changes of site and exposure, entailing endless comparisons of overlapping records at places not too far apart, have produced apparently reliable long series of values. The longest of these is the series of mean temperatures for each month of every year from 1659 to the 1970s at a typical lowland site in central England, produced by Professor Gordon Manley. Comparisons with other places in Europe, and with the reported weather and wind patterns, suggests that the values in this series may be reliable to within about 0.2 °C from 1720 onwards and to within 1°C in the earliest years. Rainfall series present more difficulties because of the real differences of measured catch over quite short distances. These are due to the effects of even the gentlest topography – as well as buildings and trees – and the differences in any one year in the random distribution of showers and thunderstorms. The early collection by G. J. Symons of monthly rainfall measurements in England and Wales averaged into a single long series from 1727 can therefore be criticized because of the size and diversity of the terrain and doubts about the representativeness of the measurement sites. Lately, however, J. M. Craddock has produced a very thoroughly researched series of monthly values for each year from 1726 representative of the much smaller and more homogeneous area of the East Midlands of England. Labrijn's series for Holland is of about the same length, and there is a composite series for London from 1697.

This work is tedious and time-consuming, so that it has not been easy to find willing and suitable people or to get the funds needed to do it, but it is rewarding. Manley emphasized, in a letter to *Weather* written shortly before his death in 1980, his concern that, because the magnitude of the changes of climate shown by monthly means must be small and 'commonly in decimals of a degree if taken over decades', to produce a reliable record of temperature (or rainfall) using the observations made in early years every detail that can be discovered about the instruments and their exposure, the observer's technique and observation hours, must be painstakingly examined. Several European countries now have temperature and rainfall series which offer a reliable basis for agriculturists, water engineers and others to study in connection with growing season length, drought risks, energy demand and so on. And these series also provide a firm basis on which the scientist can proceed in his diagnosis of the nature of the climatic evolution observed. It is a privilege to be able at last to turn to fruition the patient work day in, day out, over many years, of the ancient observers who hoped in their day to understand something of the vagaries of climate and the workings of the atmospheric system.

The observers themselves included some of the wide-ranging pioneers of science; though these seldom managed to concentrate on this one field for as many years as the learned clergymen, medical doctors, a few among the landed gentry, and others indulging a persistent interest in this aspect of our changing environment despite the inevitably slow progress towards understanding. Some of these pioneers were, however, shrewd enough to gain some valuable insights into what was going on. Thus, Thomas Barker of Lyndon, Rutland (England), whose daily weather observations were maintained from 1733 to 1798, writing in 1775, noted a general increase of rainfall since the 1740s and, in particular, a remarkable change in the character of the weather that accompanied east winds: in the 1740s such weather had been generally cold and dry, stopping the vegetation in spring, 'but for the last ten years the East winds have often been very wet; many of the greatest summer floods came by rains out of that quarter'. In a similar shaft of light in 1846, Sabine recommended research into the cause of 'the remarkably mild winters which occasionally occur in England', noting that in the case experienced in 1845–6, as in 1821–2 and in November 1776, 'the warm water of the Gulf Stream spread itself beyond its [then] usual bounds . . . to the coast of Europe'.

Our ability to go further in the interpretation of climatic development owes most to modern knowledge of the behaviour of the global wind circulation. This knowledge could only come with the daily upper air probes made possible since the 1940s by radio-sounding balloons and since around 1960 by the global cloud surveys provided by meteorological satellites.

Fig. 26 indicates the growth of the world network of meteorological instrument observations and the extent of the reasonably precise wind circulation analysis, which can be derived. It is only with the satellite era since 1960 that we can lay claim to complete observational coverage of the Earth including even the Southern Ocean and Antarctica. But luckily the area which can be covered as far back as the 1750s provides a sample of conditions in the main northern hemisphere belt of the prevailing westerlies

76

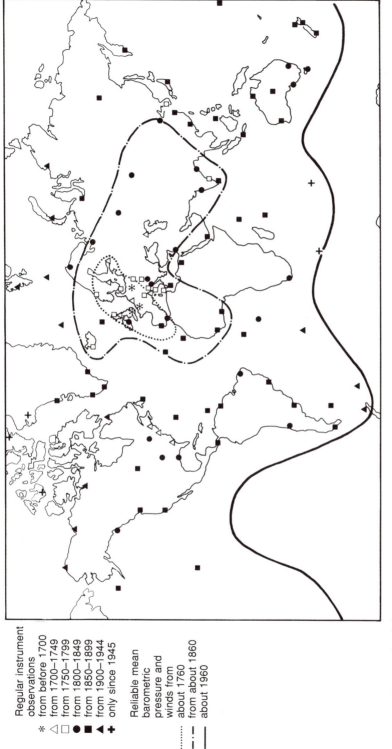

Regular instrument
observations
* from before 1700
△□ from 1700–1749
● from 1750–1799
■ from 1800–1849
▲ from 1850–1899
+ from 1900–1944
  only since 1945

Reliable mean
barometric
pressure and
winds from
.... about 1760
–·–· from about 1860
―― about 1960

Fig. 26 Growth of the world network of regular observations with meteorological instruments. The growth is illustrated by the dates of start of just the first one or two stations in each major region. The lines delimit the area over which monthly and yearly average barometric pressure and wind maps can be drawn with acceptable accuracy. A complete series of daily maps could be produced for the area of Europe bounded by the innermost line, and extended to include Iceland, from about 1780 (and perhaps earlier).

of middle latitudes. The sample weather map illustrated here (fig. 27) for a day in March 1785 indicates the sort of observation coverage that is available. A sketchy analysis of the wind circulation over the same area can be tentatively extended back in terms of monthly or seasonal summary maps to the late seventeenth century, perhaps to the 1660s, on the basis of daily weather and wind observations at a number of points (most of them provided by the logs of naval ships in various ports or patrolling certain stretches of coast) supported by observations made with just a few of the earliest barometers.

For a first sample of the results of all the effort on homogenizing the available meteorological observations, fig. 28a summarizes the temperature record from 1659 to modern times in central England already referred to. Over the 100 years since 1870 the successive five-year values of average temperature in England have been highly significantly correlated with the best estimates of the averages for the whole northern hemisphere and for the whole Earth. This conformity doubtless has to do with England's position in middle latitudes at a point where the prevailing winds have come from an extensive ocean. The correlations may not hold for the last ice age when the winds and ocean currents were different and the temperatures in this part of the world were depressed very much more than the global average, but they probably mean that over the last three centuries the central England temperatures provide a reasonable indication of the tendency of the *global* climatic regime. The smoothed curves in fig. 28a show that a good deal happens besides the familiar differences from year to year (which the ten-year averages eliminate). There are, for instance, fairly plain signs of a tendency to oscillate on a roughly twenty-year (actually twenty-three-year) period, There has also been an obvious 'improvement', or warming, of the climate from the late 1600s to the present century. This has affected all seasons of the year, and is enough to have lengthened the average growing season by about a month from the 1690s to the apparent climax which the averages for the whole year place around the 1940s.

The fact that almost the entire record since instrument measurements began is occupied by a long-term trend in one direction, towards warming, giving us therefore only a very limited view, means that we must seek evidence to extend the record to earlier times. Fortunately, there are many kinds of evidence available. These make it clear that the colder climate in the seventeenth century was a more or less world-wide event, and they also provide information about other periods in the past when the tendency of world climate was in the other direction, towards cooling. Indeed, there are signs in many parts of the world that the cooling phase in the late Middle Ages – and the changes of rainfall and shifts of storm tracks, etc. which accompanied it – had important effects on human history.

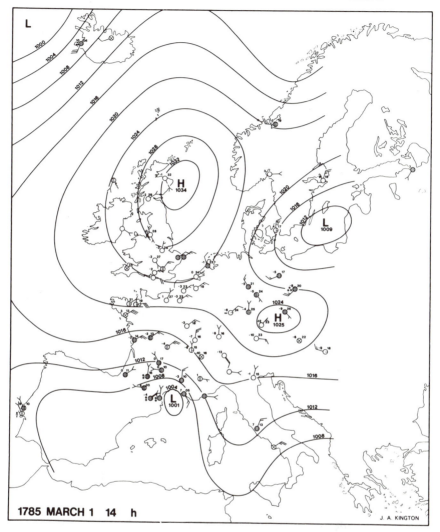

Fig. 27 Synoptic weather situation in the early afternoon of 1 March 1785, reconstructed from archived weather observation reports. The reports plotted on the map show the air temperature in °C, the barometric pressure in millibars, the wind direction and force, and the cloud cover at each place. Note the coverage of observation reports. The month thus beginning was to prove the coldest March in Britain and continental Europe in the entire 200 to 300 years of the observation records. Notice the warmth in Iceland, where the southerly wind is bringing a temperature of 10°, a higher temperature than those prevailing at the time in Rome and the south of France.

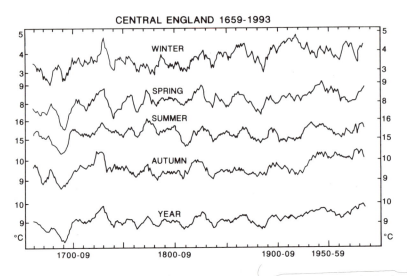

Fig. 28a Prevailing temperatures at typical lowland sites in central England, averaged for each season of the year and for the whole year, running ten-year averages from 1659–68 to 1984–93. (After Professor G. Manley, updated and reproduced by kind permission. I am indebted to my colleague Dr P. D. Jones of the Climate Research Unit, University of East Anglia, for supplying the up-to-date figures.)

The types of evidence from which past climate can be reconstructed may be summarized as follows:

1   The meteorological instrument record.
2   Earlier weather diaries and descriptive accounts of the weather, particularly the prevailing character of the seasons of individual years, reports of floods, droughts, great frosts and snows, etc., recorded by the people who experienced them.
3   Many kinds of physical and biological data, which provide 'fossil' evidence of the effects of past weather. Such material is commonly called 'proxy data' of past weather and climate.

OTHER RECORDS OF PAST CLIMATES

Diaries, annals, chronicles, etc.

In spite of the importance of reports of the weather by the people of the time – particularly any obviously seriously intentioned weather diaries, farm and estate records, mentions of weather difficulties in audited accounts and reports of building operations and repairs, etc. – little or no attempt seems

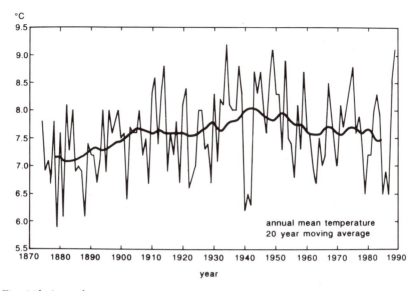

Fig. 28b Annual mean temperature in Denmark 1874–1989. (I am indebted to Dr K. Frydendahl of the Danish Meteorological Institute for permission to reproduce this figure.)

to have been made to submit them to meteorological analysis until quite recent years. Doubtless the main reason for this neglect has been the formidable size of the task. A massive amount of material could ultimately be collected for Europe and the Far East and a few other regions; but it is dispersed in many kinds of archives and literature and problems are presented by ancient styles of handwriting and earlier forms of the many languages used, establishing the trustworthiness of the reporter, interpreting the dates, and so on. (Fig. 29 shows part of a fourteenth-century scroll, an 'account roll', from the manor of Knightsbridge, London, referring to a great drought in the summer of 1342.) The dates were commonly given in state and other papers as the *x*th year of the reign of this or that prince, often princes of territories which have long since been merged or absorbed in the bigger countries of present-day Europe. The difficulties produced by numerous differences of practice over the date from which each new year started, and in the time at which various countries introduced the modern (Gregorian) calendar, have also led to errors in ascribing the year of certain events or misplacing their exact date by several days in the texts derived by historians and others from the original sources. Nevertheless, the wealth of material is such that every season of any kind of dramatic character in Europe since about AD 1100 is probably either known already or could finally be determined, especially now that the possibilities of verification from various kinds of proxy data have been greatly enlarged by modern

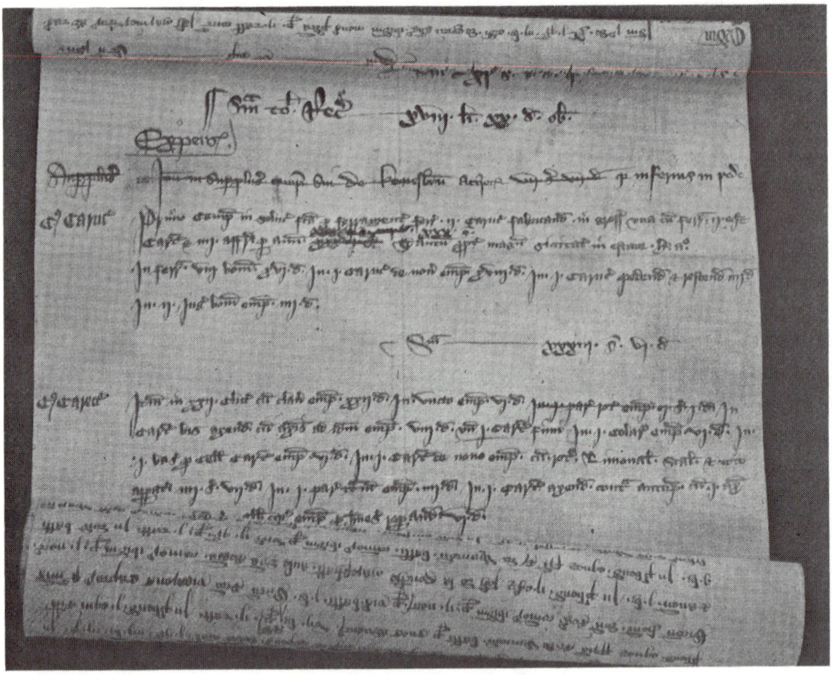

Fig. 29 An account roll from the medieval manor of Knightsbridge – now part of London – stating that iron for ploughs and horseshoes to an extra cost of thirty shillings was required in 1342 because of the 'great drought' in the summer of that year. (Picture reproduced by kind permission of Dr T. Williamson of the Science Museum, London and the *Geographical Magazine*.)

techniques. More documentary reports could certainly be unearthed if the effort could only be organized. Many great archives have hardly yet been tapped for this purpose. And where lengthy historical manuscripts have been printed and published, any weather information has commonly been omitted in the interest of economy, so that the original manuscripts should be sought out. Reports exist, in fact, of a considerable number of remarkable seasons and severe events in much earlier times – the inscribed stones of ancient Babylon and the Near East and many centuries of records from China and India also await attention – and these, too, are open to comparisons with relevant fossil evidence.

The full list of historical sources of weather reports and references to 'parameteorological' events, such as great floods, parched ground, shipwrecks and damage to coasts is a long one. It includes medieval monastic chronicles such as those of the Venerable Bede at Durham in Saxon England and Matthew Paris at St Albans (20 miles north of London), the audited accounts of great estates and their farms, legal and government papers –

82

e.g. reporting harvest difficulties, hunger and cattle raids, epidemics of disease and so on – harbour records, bridge repairs and many others. The use of audited accounts is valuable: in documents such as these, where the weather is only mentioned as an explanation or excuse for expenditure and losses, false reports might be entered unless there was a system of checking by persons who could not be misled while the real character of the seasons concerned was still common knowledge. The manors belonging to one famous English abbey were indeed found to have falsified their accounts in this way in the fourteenth century.

A few meteorologists and others of a historical turn of mind have published collections of these reports covering various parts of Europe, Iceland, eastern North America, Chile, China and Japan, some of which go back over many centuries. A few published compilations of this sort appeared from the sixteenth century onwards in England and central Europe. Famous later ones included Thomas Short's *A General Chronological History of the Air, Weather, Seasons, Meteors, in Sundry Places and Different Times etc.* (London, 1749), a huge collection by the French meteorologist F. Arago published by the Academy of Sciences in 1858, R. Hennig's *Katalog bemerkenswerter Witterungsereignisse von den ältesten Zeiten bis zum Jahre 1800* (Berlin, Royal Prussian Academy, 1904) and C. Easton's *Les hivers dans l'Europe occidentale* (Leyden, 1928). The latest in the series are the many-volumed work by C. Weickinn *Quellentexte zur Witterungsgeschichte Europas von der Zeitwende bis zum Jahre 1850* (Berlin, 1958–63) and the most thoroughly researched and verified compilation by M. K. E. Gottschalk of North Sea storm floods and river floods in the Netherlands from early times to the year 1700, in three volumes, published 1971–5 (Assen, van Gorcum).

All these compilations, except perhaps the last, contain mistakes, and in using them care is particularly needed to detect any multiple entries of the same flood or frost, etc. due to errors in the transcription of dates by the compilers or in earlier works that the compilers used. On the other hand, the British compiler C. E. Britton (*A Meteorological Chronology to AD 1450*, London, Meteorological Office, 1937), confronted with the numerous reports of severe weather in various parts of the British Isles in the winters of the 1430s, addressed himself to the problem of *which* winter they all meant, and he seems never to have considered seriously the possibility that there might have been such frequent severe spells that the dates reported might be right. Mapping the reports together with those from the continent appears in fact to substantiate spells of several weeks of widespread severe weather in seven, perhaps even eight, of the winters of that decade – an experience matched only in the 1690s so far as present knowledge goes. The only safe recourse is to check the reports quoted in the compilations back to the original source documents, but the quantity of data available for Europe is such that placing the reports for each given season

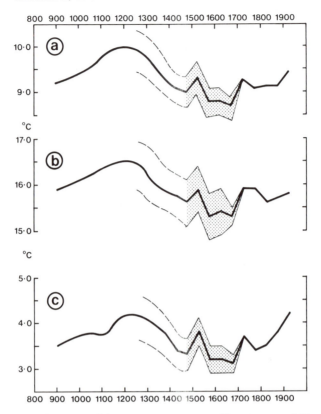

Fig. 30 Estimated course of the temperatures prevailing in central England since AD 800. Probable fifty-year averages: (a) for the whole year; (b) for the high summer months, July and August; (c) for the winter months, December, January and February. The shaded area indicates the range of apparent uncertainty of the derived values.

on maps shows up many of the faulty ones. Despite the need for critical scrutiny, the compilers have not only effectively revealed the possibility of ultimately producing maps of the prevailing weather, season by season, each year back to the Middle Ages in Europe (and, perhaps, for eastern North America from the early 1600s); they have also provided the basis on which reasonable estimates of the prevailing temperatures and rainfall in successive half-century periods can be achieved.

Numerical indices were designed, and calculated, expressing for each decade the relative numbers (best expressed as a ratio) of reports of mild or cold months in winter and wet or dry months in summer. The decade values of the winter and summer indices were then compared with the measured temperatures and rainfalls in the winters and summers in England in the case of the decades when instrument observations were available,

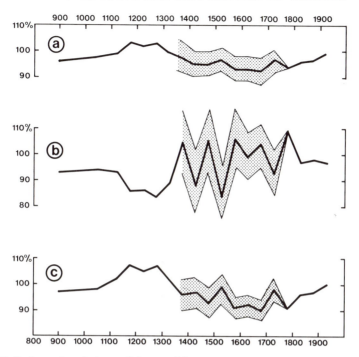

Fig. 31 Estimated variations of the rainfall over England and Wales since AD 800. Percentages of twentieth-century values: probable fifty-year averages: (a) for the whole year; (b) for the high summer months, July and August; (c) for the rest of the year, September to June inclusive. The shaded area indicates the range of apparent uncertainty of the derived values.

after 1700. From the statistical relationships revealed, temperature values and rainfall estimates could be derived for each decade back to the early Middle Ages. Because of uncertainty about the completeness and accuracy of the reporting, however, it is unwise to rely on the figures for the individual decades, though the more extreme decades are probably correctly identified. Taken in blocks of fifty years, the results are thought to be reliable as a first approximation to the temperature and rainfall history of England since AD 1100.

These derived histories are shown in figs. 30 and 31, on which estimates of the 100-year means for the eleventh century and the 200-year means for the period AD 800–1000 (derived in the same way but from scarcer reports) have been added. The diagrams in this form are designed to show the range of uncertainty of the individual values.[1] What stands out is the certainty of a warmer period that lasted several centuries in the high Middle Ages and of an equally long period of colder climate culminating in the seventeenth century. The rapidity of the decline in the fourteenth to

85

fifteenth centuries and again in the late sixteenth century is verified as a striking feature. Neither the decline between the years 1300 and 1600 nor the recovery from around 1700 to the present century were smooth, uninterrupted processes. Even in the fifty-year means a time of easier conditions around 1500–50 and a reversion to colder conditions around 1800 come clearly to light. The rainfall changes seem to indicate a long period when the totals were about 10 per cent below modern experience: this may be explained by colder seas and therefore less water vapour taken up into the atmosphere during the time of colder climate. During the medieval warmth up to 7 per cent higher average yearly total rainfalls are indicated, but usually lower rainfall in the summer-time: this probably means that the warm summers were commonly influenced by the anticyclone belt, as is true of the occasional warm summers we get in England nowadays. There is also an appearance of an interesting oscillation, whereby the summers of the second half of most centuries were wetter than those of the first half.

Other methods of converting descriptive reports to numerical values have led to long series of data for parts of the world outside Europe. Thus, 'content analysis' applied by Catchpole and Moodie at the University of Manitoba to the linguistic terms used in the journals kept by the Hudson's Bay Company's staff at their fur trading posts near river mouths in the southern and southwestern parts of the Bay from the early eighteenth century onwards, and a little later on other rivers in northern Alberta, Saskatchewan and Manitoba, has produced year-by-year series of the dates of formation and break-up of the ice on the rivers.[2] The trend of the dates, averaged over five years at a time, shows quite good correspondence with the trend of the temperatures observed in Europe, including the warmth of the 1730s and the cold climax around 1880; though, unlike the experience in Europe, the decade 1810–19 seems to have produced mostly mild winters. Rather similarly, H. Arakawa succeeded in extracting from Japanese documents a year-by-year series of the dates of freezing of the small Lake Suwa in central Japan from 1440. This Japanese series shows less close correspondence with Europe than do the series from central Canada, though the fifty-year means are correlated significantly with those of Europe (the central England temperatures).

Just how much detail of the individual years can be taken as trustworthy from the descriptive reports available in any part of the world must be established either by checking back to the sources – and, preferably, some knowledge of the character and motives of the observer – or by comparison with other independent reports for the same season on a map capable of meteorological analysis. A further alternative check may be provided by support from some kind of fossil data. A number of handwritten journals which can more or less be described as weather diaries and some strictly daily registers of the weather, often with wind directions reported, are available: the quality and completeness of some of them, such as that kept by

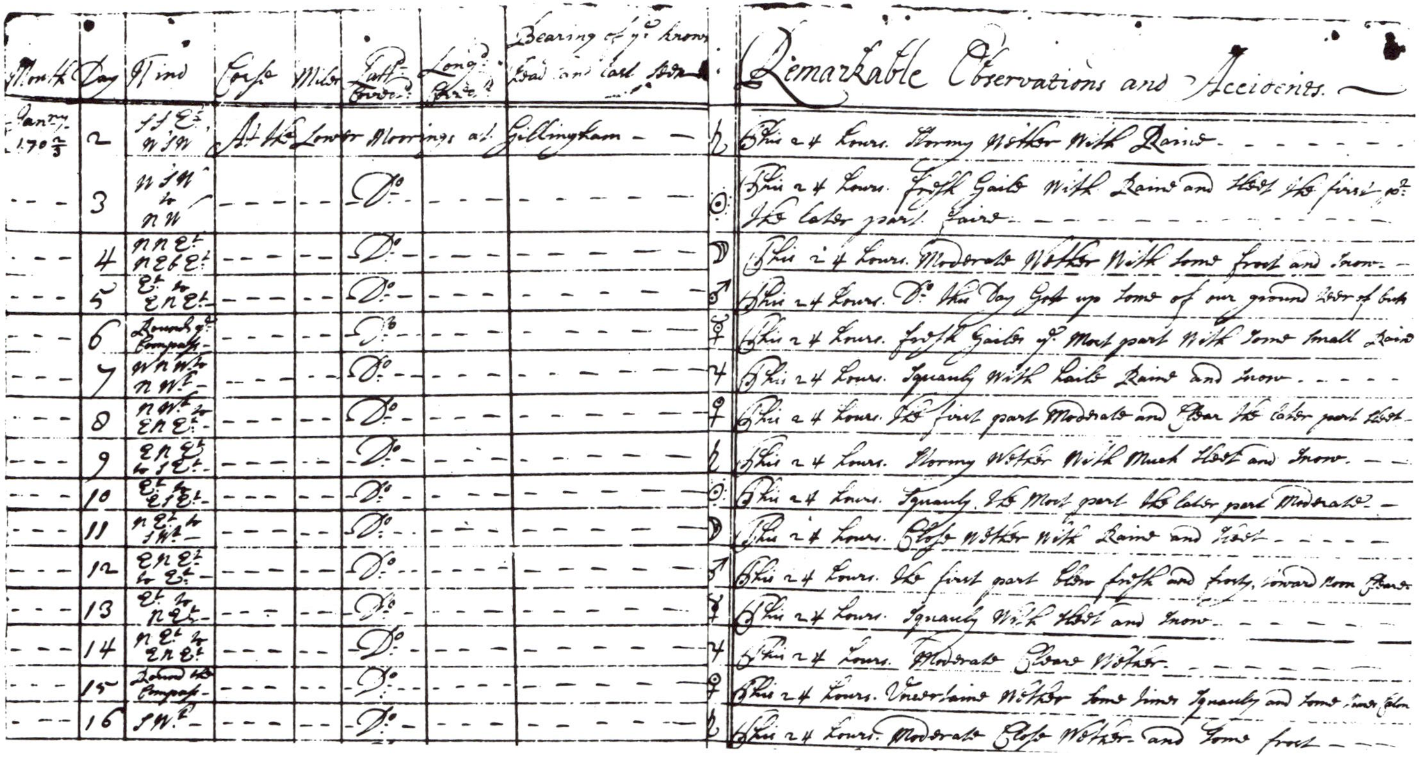

Fig. 32 Example of two pages from an early ship's log; that of the British naval ship *HMS Association* from 2 to 27 January 1702/3 (Old Style) – i.e. 13 January to 7 February 1703 on the modern (New Style) calendar – moored at Gillingham, Kent. (Reproduced by kind permission of the Public Record Office, London.)

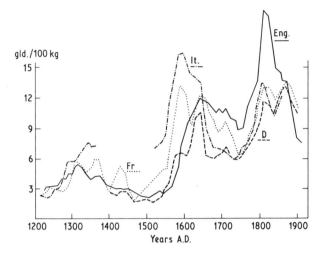

Fig. 33a Wheat prices prevailing (25-year averages) in western England (Exeter), France, Italy and the Netherlands (curve D on the diagram) from about AD 1200 to the nineteenth century, The prices are here expressed in Dutch guilders per 100 kg wheat, as given in *De Landbouw in Brabants Westhoek in het midden van de achtiende eeuw* (Wageningen, Netherlands, Veenman). (Graphical presentation by L. M. Libby, reproduced by kind permission.)

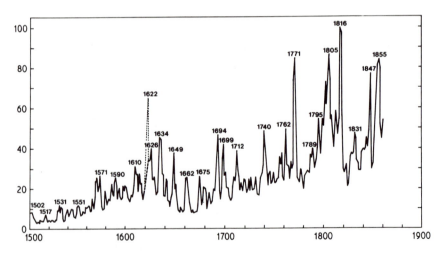

Fig. 33b Price of rye in Germany year by year from 1500 to the 1850s, preliminary estimates of the yearly averages, expressed as an index by H. Flohn. (Reproduced by kind permission.)

the Reverend Father Merle mostly at Driby in Lincolnshire (but also on his travels to and from Oxford) from 1337 to 1344, and that of the Danish astronomer, Tycho Brahe, from 1582 to 1597, are a testimony to the serious intention of the observer, so that errors, exaggeration or falsification seem most unlikely. From the mid-sixteenth century onwards there are daily reports of the weather from somewhere or other in western or central Europe with few years omitted. And from about 1670 there are in various archives many daily reports in the logs of ships in diverse ports and on the high seas (fig. 32). For climatic research this is a vast treasure trove waiting to be used – for instance, to improve our knowledge and understanding of the severest phase of our climate in recent centuries, in the 1690s.

Grain prices records

Another type of unbroken year-by-year series is provided by the prices of wheat and sometimes other crops in various countries. From early times until the agricultural advances in the eighteenth century – in some countries until about 1800 or after – the fluctuations in the price of grain, or of bread, responded essentially to the yield of the latest harvest. This means that they registered each year's weather, except when the harvest was affected by wars, civil commotion or other disasters such as loss of manpower through disease epidemics. W. H. Beveridge in the 1920s collected and analysed the wheat prices in England (at Exeter) from AD 1316 to 1820. He found evidence of certain periodicities or cyclical tendencies of which, for instance, the statistically significant one around 5.1 years length and most of the others suspected (those of 12, 20 and 55 years) are close to period lengths commonly identified since in long series of climatic data. Even in the smoothed version of the long histories of wheat prices seen in fig. 33a, the major price rises around AD 1300 and 1550–1650 can probably be largely attributed to climate, and that around 1800 probably owes something to climatic difficulties as well as to the Napoleonic wars. In the year-by-year record of the German rye prices (fig. 33b) the peaks nearly all correspond to years of particularly unfavourable weather, the troughs to runs of good harvests: this is true even during and after the Napoleonic period. (The remarkable year 1816, when the sun was dimmed by a thick veil of volcanic dust from the great eruption of Tamboro in the East Indies the previous year, became known in Europe and North America as 'the year without a summer'.) But in grain prices we are using what must really be classified as proxy data.

Varieties of 'fossil' records showing yearly layers

There is no space here for any approach to a full coverage of the many kinds of indirect data from which information about climate in the past may be derived. Because weather and climate have an impact on nearly every aspect of our environment, the lines of evidence of past events are innumerable. The reader who wishes for a more comprehensive survey must be referred to other works, such as the author's treatise *Climate: Present, Past and Future – Volume 2: Climatic History and the Future*, pp. 1–279 (London, Methuen, 1977) or M. Schwarzbach's *Das Klima der Vorzeit* (Stuttgart, Enke, 1961). A very brief over-view of the main categories of evidence of past climate, including instrumental measurements, historical manuscripts and fossil data, both physical and biological, is given here in table 1 at the end of this chapter.

Some of the pieces of evidence left by past climates that differed from today's could individually be interpreted in other ways, even attributed to primitive man's disturbance of the landscape. But many items, such as old moraines left by glaciers which have since shrunk or disappeared and old shorelines, are unequivocal. Even in these cases, however, there are complexities to be unravelled: was the former glacier expansion mainly due to lower temperatures or to increased snowfall on the heights? And has the lake or sea level in the area where the old shoreline is found been affected by warping of the Earth's crust, that is by land-sinking or the reverse? Problems arise with the interpretation of tree ring variations: at the site concerned how much was due to temperature and how much to moisture variations? In all cases there is a problem to be solved by mathematical or statistical techniques of how far we can arrive at clear numerical estimates of the past temperatures, rainfalls, etc. within definable limits of error from the types of evidence available. But, taking all the independent types of evidence together, there is no longer any doubt about the major features of the climatic history which they reveal.

Here we can only enlarge a little on a few things. Particular importance must be attached to those items which register, however indirectly, the weather of each individual year – tree rings, year-layers in ice-sheets and glaciers, also year-layers (varves) in lake bed deposits – and to techniques of dating evidence of whatever kind. Evidence of the year-by-year sequences just before, and during, the times of most rapid climatic change should be of great interest; but little has so far been done in this line of research. On the other hand, special importance must also be accorded to those types of evidence from which most has already been learnt of the long history of climate; these include most notably pollen analysis, marine-biological studies of the deposits on the ocean bed, and oxygen isotope studies.

The oldest year-by-year record that has come down to us is the flood levels of the River Nile in lower Egypt, a variable which depends mostly on the summer monsoon rains over Ethiopia. Yearly gauge readings at Cairo are available from the time of Mahomet, and some records inscribed on stone go back to the first dynasty of the pharaohs around 3100 BC. There are problems, particularly in connection with the prolonged silting of the river bed which effectively changes the zero level, but enough data exist to show that, there was a sharp drop about 2800 BC to persistently lower flood levels than before. The floods seem also to have been generally lower between the AD 600s and 1400, and again for a while around 1500, than in the seventeenth, eighteenth and nineteenth centuries before falling again to the present century. From AD 622 the yearly gauge readings also record the seasonal low level of the Nile, which registers the flow of the White Nile fed by the equatorial rains over east Africa. This seems to have been lowest in the AD 700s and around the 1600s and particularly high in and around the 1100s, 1450–1500 and about 1840–90.

Another splendid series is provided by the measurements of the thickness of the yearly mud layers in the bed of the small Lake Saki in the Crimea (near 45 °N 33½ °E) from about 2300 BC; these are considered to register more or less directly variations in the (mainly summer) rainfall in that region. A smoothed version of this record is seen in fig. 34. A somewhat irregular fluctuation of about 200 years length seems to be an element in the story, but a more striking feature is the evidence of wetness in the Crimea that accompanied the periods of warmer and more genial climate in west and northwest Europe (and apparently over most of the northern hemisphere) before 2000 BC and in the early Middle Ages. (The fluctuations of rainfall and run-off in the Crimea and much of southeast Europe are believed to be generally inverse to those over most of northern, western and central Europe.) A number of features of the individual years' record in the Crimea also look important. There are signs of a fifty-year cycle, and perhaps a 20–23 year one, both of which are known to be present in the incidence of blocking anticyclones in the northeast Atlantic–northern Europe sector; this is a development which certainly affects rainfall over Russia. But also extraordinary deviations in individual years in the Crimea seem to occur at times of major, long-lasting change of regime. Just before the first big drop of the curve there were two years (2177 and 2150 BC) that produced mud layers of respectively five and ten times the normal thickness which had been typical of the wet regime that was just ending.

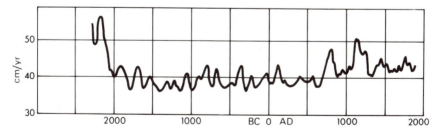

Fig. 34 Thickness of the yearly mud layers (varves) in the bed of Lake Saki in the Crimea, indicating rainfall variations in the area since 2300 BC. (After W. B. Schostakovitch, with a rainfall scale suggested by the late Dr C. E. P. Brooks.)

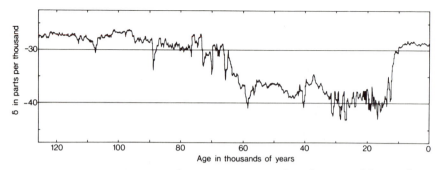

Fig. 35 Variations in the ratio of oxygen isotopes in the substance of the ice-sheet in the far northwest of Greenland (77 °N 56 °W) over the last 125,000 years, expressed as deviations from the mean ratio of oxygen-18 to oxygen-16. This curve can be considered as a fossil record of the prevailing temperature. The lower the temperature, the lower the curve. (After W. Dansgaard, reproduced by kind permission.)

The further decline after 2000 BC may have been signalled by two runs of eight to eleven consecutive years in the nineteenth century BC producing extraordinarily thin mud layers. Similarly extreme deviations in either direction – in AD 805 the only other occurrence of a thick layer approaching that of 2150 BC and a run of very thin layers in the AD 1280s – mark off the beginning and end of the medieval period of moist climate in the Crimea.

Even longer series of year-layers, or varves, in lake sediments are available for study from the beds of lakes and former lakes that have since disappeared, which formed around the margin of the former North American and Scandinavian ice-sheets during the post-glacial melting. Some of these series in Wisconsin and northern Sweden are nine thousand to ten thousand years long. The original Swedish series, worked out by De Geer many years ego, spans the whole fifteen thousand years or so of the ice retreat and postglacial time: it is, however, a composite structure, built up from shorter overlapping series derived from lakes which individually had a shorter history; some uncertainty and controversy has developed about its structure.

The year-layers in the great land-based ice-sheets still present in Antarctica and Greenland also provide material for study. The layers are made identifiable by seasonal changes in the density and texture of the snow, caused by temperature changes and wind. Most simply, the thickness of each layer – after allowing for compression by the overlying younger ice – indicates the amount of snow which accumulated each year. Allowance must be made for the slow movement of the ice, such that the older layers (except near the ice-sheet crest) have arrived by plastic flow from some distance from their present site, naturally from a somewhat higher level.

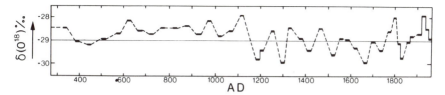

Fig. 36 Oxygen isotope variations at the same site from AD 300, arranged to be read as equivalent to a temperature curve. (After W. Dansgaard, reproduced by kind permission.)

Measurements of the proportion of the heavy isotope of oxygen, the form with an atomic weight of 18 instead of the usual 16, present in the H_2O of which the ice is composed, indicate the temperature of the snow at the time (however long ago) when it formed by condensation from the water vapour in the atmosphere. Oxygen-18 measurements therefore make it possible to recognize the seasonal changes of temperature, and so they can help identification of the year-layers in cores taken from the ice-sheet. This has been done for about the last 1500 years of the fossil record. They also identify long-lasting temperature changes in the course of changes of climate and make possible estimates of the magnitude and rate of change. The Greenland record in fig. 35 indicates the whole course of the last ice age and post-glacial time – mostly dated by other methods – as well as much of the last interglacial. A close analysis of the last thousand years of the record from this site in far northwest Greenland indicates a great deal about the changes of temperature and the downput of snow there, but it must not be assumed that the temperature sequence is identical with that anywhere in Europe. The medieval warmth clearly reached its climax, and also ended, earlier there than in Europe. The Little Ice Age affected north Greenland too, as fig. 36 indicates, but there were some differences of phasing. Indeed, the climatic history obtained by various techniques from central longitudes of Canada appears to parallel that in western and northern Europe more closely than that from Greenland, This is a discovery which was not altogether unexpected on meteorological grounds: it is doubtless related to the big waves in the westerlies.

Radiocarbon and its role in dating evidence

For times beyond the range of identifiable year-layers in ice-sheets and lake sediments, and beyond the longest yearly tree ring chronologies or human records in any part of the world, dates can be estimated by radiometric methods or, in the case of material in sediments, by assuming a broad constancy of sedimentation rate. For the periods with which human history and archaeology are concerned the most important of the radiometric methods is radiocarbon dating. This depends upon precise measurement

93

of the radioactivity produced by the minute amount of the unstable isotope of carbon present in the carbon constituent of organic matter. This radioactive isotope ^{14}C, distinguished by an atomic weight of 14 – normal carbon has atomic weight 12 – is produced in the atmosphere by the effect of cosmic ray bombardment on the nitrogen atoms with which they collide. Carbon 14 is assimilated into the structure of the living vegetation with the carbon dioxide breathed in from the atmosphere. About 1 per cent of the carbon in living wood is the unstable, radioactive isotope; and the atoms of it decay, producing on average about fifteen disintegrations per minute per gram of carbon present. After the death of the vegetation, which means cessation of the absorption of atmospheric carbon dioxide, its store of radioactivity is no longer renewed. The activity therefore decays. In scientific language, we say that the half-life of radiocarbon (^{14}C) is 5730 years: this means that the activity falls by a half every 5730 years. In practice therefore the amount of radioactivity dwindles and ultimately becomes very difficult to measure – and the errors produced by any contamination become greater – the older the material to be dated. The effective limit of radiocarbon dating is about fifty thousand years. Estimates of the margins of error arising from experimental difficulties are always quoted. There is an additional source of error, however, established by radiocarbon dating of objects of known age and attributed to the fact that the amount of radioactive carbon in the atmosphere has evidently not been precisely constant down the ages. The variations give rise to an error amounting to 500–1000 years in middle post-glacial times and to over 100 years in another period as recent as about AD 1400–1800. These errors can, however, in most cases be corrected by using a calibration curve relating apparent radiocarbon ages to true ages obtained from specimens of the wood of bristlecone pine dated by its rings.

Because of these difficulties dates in the times before Christ based on radiocarbon tests with the results unadjusted for the calibration errors referred to are commonly distinguished in the newer literature by small letters, as for instance a date of 930 ± 220 bc for the latest occurrence so far identified of a pine-tree stump significantly higher than the present upper treeline on the mountains of Scotland. In this notation, bc means before Christ and the ± 220 years is the estimate of the range of experimental error about the indicated date of 930. Capital letters BC are used for firm calendar dates not resting on radiocarbon or other deductive methods and also for radiocarbon dates which have been corrected by applying the bristlecone pine calibration. With the latter the estimated range of experimental error will be quoted. In specifying the age of radiocarbon-tested material, the letters bp (or BP where the calibration has been applied) mean years 'before present', actually years before AD 1950 which has been adopted as a fixed datum at the beginning of radiocarbon work.

What is learnt from studying past glacier variations by dating organic matter and tree-stumps, etc. buried in old moraines will be referred to at appropriate points in the text of later chapters. Here it is sufficient to note that there have been great variations of the mountain glaciers in all parts of the world during post-glacial time. The course of major fluctuations eight to ten thousand years ago during the melting of the former ice-sheets can be followed. Later moraines which may have been formed during the warmest post-glacial times have generally been obliterated by glacier advances since the time of minimum extent, which in some places was nearly six thousand years ago but more generally around 2000 BC.

Pollen analysis and vegetation history

By far the most of what was known of post-glacial climatic history until 1950 was contributed by pollen analysis. As long ago as 1876 the Norwegian botanist Axel Blytt first detected the broad post-glacial sequence of vegetation history, and this and the succession of climatic regimes which it points to were outlined in his *Essay on the Immigration of the Norwegian Flora* (Christiania). As the ice melted, at first tundra vegetation was established, spreading from regions which had always been south of the ice, then birch trees infiltrated the tundra, and later birch and pine forest was established. This in its turn was replaced by the mixed forest of broad-leafed trees in and beyond all the regions now occupied by this forest type, which is now known to have been dominated by the warmth-loving elm and lime (linden) trees for some long time before about 3500 BC in the zone of present-day oak forest. Each forest type presumably corresponded to the same sort of climate as that in which we find it today, except that in Europe generally there seem to have been delays of up to some thousands of years in the arrival of each tree species and forest type into territories that had earlier been too cold for it. Further decades, centuries or longer had to pass before each new forest type became dominant in all the areas which had become suitable. We can detect these delays by the quicker response of the insect populations to climatic change. The evidence of the beetle species present at each stage is particularly well preserved. No similar delays seem to have occurred over the broad plains of North America east of the Rocky Mountains, and it is clear that in Europe the advance of the successive forest types was hindered because the ice age refuges of the trees were south of the great mountain barriers. Other complications in interpreting the results of pollen analysis concern the later retreat of the vegetation types from the greatest extent which they attained towards the north and on the heights in mid post-glacial times. Soil deterioration set in and peat-bogs replaced the forest in many persistently wet places, a development which can be attributed to the cumulative effect of a long period of wet climatic regime preceding. In many places it does not imply a change of climate at the time when bog was replacing the forest. In some places, moreover, this change of the landscape may have been assisted by man's activities, making clearings in forest that was already under stress and grazing his animals on the cleared areas. In southwest England and west Wales the times of clearance and cultivation by Neolithic farmers, evidenced by charcoal layers and the pollen of weeds like plantain associated with cultivation, often seem to coincide with the earliest layers of peat formation. And from that time on, roughly covering the last five thousand years, in this part of Europe the problem of making deductions about climate from evidence of the prevailing vegetation character is increasingly complicated by man's management of the land.

Another limitation of pollen analysis as a tool for reconstructing the climatic record is that, apart from the pollens in varved sediments and a few deposits of peat or lake bed sediments which may have grown very fast, it is seldom possible to achieve a time resolution, or fix, closer than a hundred years. On the other hand, some very long records can be produced. Two from Europe, one from Alsace and one from Macedonia, have provided records which start more than 125,000 years ago, indicating the entire course of the last interglacial from its early stages, and continuing right through the last ice age and post-glacial time. The correspondence of the large-scale features of the sequence in Europe with the Greenland isotope curve, including the early shocks which heralded the end of the interglacial and beginning of the ice age, is highly satisfactory. Professor R. G. West and his co-workers at Cambridge have performed similar analyses of the vegetation history of eastern England through several previous interglacial periods, making it possible to identify common features of the climatic development and, in very broad terms, its timing.

A model of how some specific indications of prevailing temperatures may be derived from evidence of plant distributions was demonstrated in a classic work by J. Iversen of the Danish Geological Survey in 1944.[3] By plotting the long-term average temperatures of the warmest and coldest months of the year (scaled along the x and y axes of graph paper) at all sites where a given plant was present (or, plotting in a different colour where it was absent) in the present climate of Denmark, Iversen was able to show in the case of holly (*Ilex aquifolium*), ivy (*Hedera helix*) and mistletoe (*Viscum album*) that the limits were quite strictly defined by certain temperature values.

The post-glacial record and evidence from beetles

The smooth curves in fig. 37, which outline the history of the prevailing temperatures in England since the depths of the last ice age about twenty thousand years ago, are an example of what may be deduced about the climate by way of pollen analysis used to reconstruct the vegetation history.

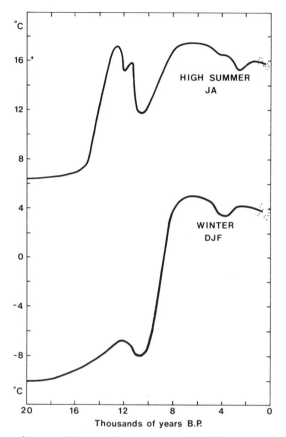

Fig. 37 Supposed course (500–1000-year averages) of the temperatures prevailing in central England over the last 20,000 years, as derived largely from pollen analysis. JA = July and August; DJF = December, January and February. Dots show the summer and winter temperatures derived for the last twelve centuries from fig. 30.

These curves are thought to be a reasonable approximation to the 500- or 1000-year mean temperatures of the high summer and winter months. Points added about the right-hand end of each curve indicate the amount of deviation apparently distinguishing the last twelve individual centuries. A similar variability may have occurred in earlier millennia, and indeed we shall notice evidence of it within the last three thousand years. Indications derived from the more rapidly changing beetle faunas have been used to fix the timing and magnitude of the sharpest climatic changes at the end of the last ice age. Many fascinating studies by G. R. Coope at Birmingham have demonstrated the value of the well-preserved remains of the abundant beetle faunas and their rapid responses to climatic change in this field of research. The main features of the curves in fig. 37 are indeed the rapidity of the changeover to post-glacial temperatures, particularly the early peak of the summer temperatures, and the drastic reversion to a glacial climate for a few centuries in the eleventh millennium before Christ. The warmest post-glacial times appear as a broad hump of the curves between eight thousand and four thousand years ago, followed by an undulating decline to our present climate. The variations which we can determine within the last millennium or more include some sharp fluctuations of shorter time-scale.

Archaeology

The archaeology of man and the larger mammals is on the whole less informative about past climates than the record of the insects and pollens and micro-organisms in the sea that are at the mercy of their immediate surroundings. The mobility and adaptability of man and the larger animals allow them to roam widely, experimenting with unfamiliar foods when necessity drives, and straying at least temporarily into environments which might in the long run prove hostile. There are exceptions where deductions can be made about past climate, as for instance where routes of travel were established across terrain that at other times was closed by snow and ice, desert or marsh. But one type of find deserves mention here, that of the men of about two thousand years ago whose whole bodies have been found preserved in the peat-bogs of Denmark (and comparable finds in the peat of the Hebrides and of bodies released by melting of the Greenland ice): in some of these cases the preservation has been so good that analysis of the stomach contents tells us about the food they lived on. In the Danish cases these results have been tied in with near-by pollen analysis studies to confirm the cultivation of Iron Age fields in the neighbourhood and the weeds that were eaten along with the primitive barley and linseed. We also know of a Danish alcoholic drink of two thousand years ago, made from barley, cranberry and bog myrtle: this from analysis of residues in pottery of the period.[4]

Tree rings

The potential of the yearly growth rings of trees to indicate a detailed record of climate was for long little exploited outside North America. Ring widths in trees occurring at the upper altitude limit and at the poleward limit are relatively straightforward to interpret climatically. (Good growth years are warm years, poor growth years are years with cold summers in these positions.) The same applies, though apparently to some less extent, to trees near the warm-arid margin of their occurrence, where it is failure of the rain that explains narrow growth rings. The situation is much more complicated in the middle of the habitat of the tree species, though useful possibilities of interpretation may exist where the tree is known to have grown on a dry site where moisture stress might often be important or on a permanently damp site where temperature fluctuations might be the only important variable. H. C. Fritts at the Laboratory of Tree Ring Research at the University of Arizona in Tucson has studied in some detail how the width of each ring is related to the weather of the preceding fifteen months. His colleague V. C. La Marche has built up a chronology of ring widths in the very long-lived bristlecone pine trees on the heights of the White Mountains on the California–Nevada border. This record extends back to 3431 BC and its dating has been demonstrated as sound by repetition. This long series at the upper tree line essentially registers summer temperature. It is of interest that from AD 800 to the present century its hundred-year averages are correlated, in a statistically significant degree, with the temperatures derived for central England. They tend therefore to provide independent support for the English temperature history presented in fig. 30.

The tree ring method, or 'dendroclimatology', is progressively coming into much more extended use in climatic research. Finds of well-preserved treetrunks buried in peat in Ireland, in eastern England and in river gravels in central Germany promise to produce an ultimately unbroken chronology extending back at least four thousand years and possibly much longer. The large size of the oaks which grew in the fenlands of eastern England in the warmest post-glacial times attracted the attention of H. Godwin already many years ago.

Estimates of the temperatures prevailing in those times may be derived from observation of the greater height of the upper tree limit on the mountains of Europe (and other parts of the world) and the northward displacement of the northern limits of various species. Curves representing the post-glacial history of the upper tree line in various parts of the world tend to parallel the last ten thousand years of the temperature curves shown in fig. 37.

In the closer detail which we can survey in the last thousand years. Dr J. Fletcher of the Research Laboratory for Archaeology and the History of Art, Oxford, has compared the tree ring sequences shown by English and German oaks, using for the former a chronology built up from English oak chests in Westminster Abbey. He finds that for some centuries in the Middle Ages prior to AD 1250 there was 70–75 per cent agreement between the sequences of broader and narrower years, growth rings in the English and German trees, but after 1400 the agreement fell to only 50–55 per cent. Presumably westerly winds sweeping across both countries were much commoner in the former period than in the latter. This is not altogether surprising, since the latter period was when the colder climate of the Little Ice Age was setting in and winds from the north and east are known to have increased in frequency.

Other new possibilities in tree ring work are offered by measuring isotope ratios and by X-ray techniques for measuring wood density and examining cell sizes. The ratios of the stable isotopes of oxygen, and similarly of the stable isotopes of carbon, hydrogen and nitrogen, in the substance of the wood may each lead to estimates of the temperatures prevailing when the wood was formed. Examination of the cell structure of the wood, as is being pioneered by F. Schweingruber and his colleagues at the Forest Research Institute at Birmensdorf near Zürich, may be even more promising. The large cells of the early season's growth are followed by a darker pack of dense, smaller cells in the late summer wood; but sometimes even variations within the season can be established. Furthermore, better correlations have been obtained across wide areas of Europe between the density measurements than in the case of ring widths.

Ocean bed deposits

There is one valuable source of information about past climates which we have not so far mentioned. Rather as pollen counts at different levels – effectively different ages – in cores taken from peat and lake bed deposits on land (and on former lands now submerged by the sea) can give us a very long record of vegetation and climate history, but with limited time resolution, so have counts of the representatives of various marine-biological (micro-) species in the deposits on the ocean bed yielded extremely long records from which sea-surface temperature histories can be derived. And such conclusions can be verified by oxygen isotope measurements applied to the calcium carbonate ($CaCO_3$) in the skeletal remains concerned. Most ocean bed deposits, which also include inorganic mineral dust blown by the winds, accumulate extremely slowly: 1–4 cm per thousand years seems to be typical, compared with 1–4 cm per century for peat in west European peatbogs. This means that the deposits sampled on undisturbed parts of the ocean bed may yield extremely long records. The longest so far, from the equatorial Pacific, analysed by N. J. Shackleton of Cambridge, goes back over more than two million years and indicates that ice ages have occurred approximately every hundred thousand years during the last million years or more. Again comparisons of the course of climatic development in each interglacial and its timing are possible. Such slow sedimentation means, however, that one cannot distinguish variations of shorter duration than some thousands of years. In any case, the minute burrowing animals that live within the sediment blur the record, mixing at any given time the uppermost layer of the ooze laid down over about the last two thousand years. This is a limitation on the time resolution of data from ocean sediments, which means that while they are of great importance to our knowledge of the long-term and gross-scale changes of climate between ice ages and interglacials, and to verify what we have derived from other evidence, they offer little of the shorter-term detail with which human history is largely concerned.

Oceanographic research in the realms which here interest us should therefore concern itself with places where the sediment is laid down ten or more times as rapidly as the world average – and remains undisturbed – and, at the other end of the scale, with periods that can be covered by fisheries and whaling records, etc.

Table 1 Data for reconstruction of the past record of climate

Type of data	Climatic elements concerned	Time resolution of the observations	Lags in response	Beginning of record	Areas covered
Standard meteorological instrument observations: barometer, thermometer, rain-gauge	Surface pressure, temperature, rainfall, wind flow, etc. (humidity observations start later)	Virtually instantaneous	Insignificant	1650s–70s in parts of Europe	Extended to much of Antarctica only since 1956: Southern Ocean still largely uncovered except by satellite observations
Upper air instrument measurements	Upper air temperature, humidity, pressure values, winds	Virtually instantaneous	Minutes	1930s in parts of Europe and N. America. Fragments much earlier, e.g. St Gotthard Pass 1781–92, and some mountain top stations in Europe 1883–85	Northern hemisphere from 1949; southern hemisphere from 1957
Ship-borne instruments	Sea temperatures. (Salinity and ocean current observations start later)	Virtually instantaneous	Minutes	1850s. Fragments from 1780 sufficient to deduce some 40- to 50-year means	Mainly Atlantic Ocean for the first 50–80 years

Table 1 continued

Type of data	Climatic elements concerned	Time resolution of the observations	Lags in response	Beginning of record	Areas covered
Descriptive weather registers, weather diaries	Wind, weather, rain and snow frequencies, etc.	Daily	–	Earliest examples include E. England (Lincolnshire) 1337–44, Zurich 1546–76; Danish Sound (Tycho Brahe) 1582–97	Parts of Europe. Scattered data from early expeditions in the eastern half of N. America and elsewhere
Ships' logs (mainly useful in port or patrolling short sections of coast)	Wind, weather, rain and snow frequencies, etc.	Once or several times a day	–	1670–1700. Isolated much earlier voyages	European waters and some long voyages, e.g. to the Indies and the Far East
Annals, chronicles, audited account books, state and local documents, farm and estate management reports, accounts of military campaigns, etc.	Weather, especially extremes and long spells of weather, droughts, floods, frost, snow, great heat, great cold, harvest results, etc.	Month or season, sometimes to the specific day	–	About AD 1100. Occasional reports much earlier, e.g. Italy from 400 BC, Britain from 55 BC, central Europe from about AD 500	European record of 'dramatic' spells of weather probably could be made complete from AD 1100

Table 1 continued

Type of data	Climatic elements concerned	Time resolution of the observations	Lags in response	Beginning of record	Areas covered
River flood levels	Rainfall and evaporation (also soil moisture)	Month or year, sometimes to the specific day	Varies from a few hours to half a year (Nile)	AD 622. Fragments much earlier, from 3100 BC	Earliest record River Nile
Lake levels	Rainfall and evaporation (also soil and subsoil moisture)	A few years	Up to 15 years?	About AD 1650 generally	Earliest reports from Siberian lakes; others (including much earlier evidence) from dating of ancient beaches
Tree rings	Temperature, rainfall	Ring width 1 year, cell structure 1–5 weeks	Ring width depends on up to 15 months previous weather, cell structure a few days?	4000–6000 BC in the southwestern United States	Continuous records from AD 200 to 500 in central Europe, AD 1180 in Lapland. Ultimately a 10,000 year record may be developed for parts of Europe, perhaps also places in the temperate zone of the southern hemisphere

Table 1 continued

Type of data	Climatic elements concerned	Time resolution of the observations	Lags in response	Beginning of record	Areas covered
Varves (year-layers in lake bed and a few river estuary and sea bed sediments)	Stream flow, rainfall	1 year (more difficulty than with tree rings in eliminating dating errors and uncertainties)	Days or weeks	About 8000 BC	Sweden and northern USA, also Japan
				2200 BC	Crimea (many more lakes with varve series may yet be found; little prospecting has so far been done)
Year-layers in ice-sheets	Snowfall	1 year	–	About AD 1000 AD 1760	N. Greenland, South Pole
Glaciers (advances and retreats, reported, old dated moraines etc.)	Temperature, duration of the melting season, sunshine and cloudiness, snowfall	Determined by the time resolution of the dating techniques used – e.g. radiocarbon or (better) recognizing tree ring sequences in trees killed by the glacier advance	About 10–20 years characteristic depending on size and shape (slope) of terrain	Last ice age maximum advance, usually 17,000–22,000 years ago	Most of the world's mountain regions, and lower levels in latitudes poleward of 40–45°

Table 1 continued

Type of data	Climatic elements concerned	Time resolution of the observations	Lags in response	Beginning of record	Areas covered
Stable isotope measurements, especially oxygen-18					
(a) on ice-sheets	Temperature, snowfall	At best a few days or weeks, aids recognition of year-layers	–	5000–10,000 years ago (with coarser time resolution unlimited)	Mainly Greenland and Antarctica
(b) on tree rings	Temperature, rainfall (interpretation problems still controversial)	At best perhaps 30 years because of smearing by the sap movement	–	AD 200 AD 1350	California Central Europe (Comparatively little work has so far been done. Many more and older records may be obtained.)
(c) on organic carbon in $CACO_3$ sediments, e.g. on the ocean bed	Temperatures. Amount of H_2O in glacier ice (i.e. removed from the oceans)	Ranges from about 100 to 2500 years depending on the rate of deposition at the site	–	Unlimited age	Samples available from every ocean and all latitudes.

Table 1 continued

Type of data	Climatic elements concerned	Time resolution of the observations	Lags in response	Beginning of record	Areas covered
Pollen analysis	Temperature and rainfall criteria for vegetation boundaries	About 100 years. (In a few places where a sediment was laid down rapidly, or with rapid bog growth, a resolution of 20–50 years has been possible)	Quick response to adverse conditions, up to 5000 years' lag in recolonization of northern Europe after ice age	From 125,000 years ago continuous record at one or two sites. Fragmentary records from much earlier times	All the world's land areas and some ocean bed deposits
Insect faunas	Temperature and rainfall criteria for species population boundaries	About 100 years	In some cases negligible. Probably never more than a few decades or at most centuries	At least 300,000 years ago and perhaps much more	Limited coverage so far: most work in England
Marine microfauna (Foraminifera, Radiolaria, etc.) and calcareous algae	Surface and deep water temperatures (according to species habitat)	Ranges from 100 to 2500 years depending on rate of deposition on the ocean bed at the site	–	500,000 to one million years ago	Samples from every ocean and all latitudes

Part II

CLIMATE
AND HISTORY

6

CLIMATE AT THE
DAWN OF HISTORY

The last three chapters have indicated in outline what is now known of the past record of climate and what more, given further research, we may hope to reconstruct. Further details of present knowledge will be mentioned at appropriate points in this and later chapters. Let us now look at the ever-changing scene of climate and environment as the stage on which the history of mankind has been played.

THE ICE AGE WORLD AND THE PEOPLING OF THE AMERICAS AND AUSTRALIA

The earliest gleam of man's own record of his story comes to us in the sketches and paintings on the walls of caves, the record left to us by the inhabitants of central France and northern Spain during the last ice age of the environment which they knew between forty thousand and fifteen thousand years ago (fig. 38). It is a world of bison and other wild cattle, mammoths, rhinoceros, horses and deer, hunted down with arrows and spears, in a treeless landscape. The areas concerned were, of course, always beyond the range of the great ice-sheets. But we see man living in caves in the rock walls of valleys that were still habitable in the Dordogne, the Pyrenees and Cantabria, adapted to and exploiting a landscape and a fauna that differed from today's.

Similar cave paintings, all executed in red, found in the Kapovaia cave in the southern Urals show seven mammoths and two rhinoceroses as well is a number of horses. A big encampment of mammoth hunters has been identified at Vyzovsk, at 65 °N on the Pechora river in the northeast of European Russia: 98 per cent of the bones dumped were mammoth, and the remains of a dwelling built of mammoth bones were found. Another find, near Vladimir, east of Moscow, indicates the clothing worn by people of the once common European Cro-Magnon type thirty thousand years ago. The articles had been decorated with ivory beads, which still traced the form of a pullover shirt with a round neck and trousers with boots, also a head covering.

Archaeological finds in North America point to human beings living in ice-free areas of Alaska, north of the ice-sheets,[1] during the last ice age and probably therefore roaming to and fro across the dry plain which then linked Alaska to Siberia. This dry land existed because world sea level was lowered about 100 m by the loss of the water constituting the expanded glaciers and ice-sheets. The Mongoloid traits of the American Indians suggest that their ancestors came from Asia, most likely exploiting the dry land connection where the Bering Strait now exists. The distribution of earliest dated archaeological traces of human occupation of the Americas suggests arrival during the ice age; and there is some probability that an ice-free corridor through Alberta, between the 'Cordilleran' ice-sheet over the Rocky Mountains and the huge 'Laurentide' ice-sheet centred where Hudson Bay is now, was used for the migration south.[2] This corridor existed for many thousands of years during relatively milder phases in the middle of the ice age and it re-appeared about twelve thousand years ago, towards the end of the glaciation, when the ice-sheets were dwindling. The dating of finds associated with human activity hints at migrations by this route in both these periods.

Radiocarbon dates also indicate the arrival of the first human population in Australia during the ice age, perhaps forty thousand years ago, in the time when the lowered sea level created great stretches of dry land almost linking Australia to Asia. But there, there do seem to have remained some open water straits which the people somehow managed to cross.

Thus, we see the early hunting and gathering communities of human beings living often, though not everywhere or in all cases, in sparsely distributed groups, restricted in their range by the barriers of ice and ocean and high mountains but also exploiting the opportunities that the ice age world offered.

An aspect of the ice age world that has not been much written about is the enormously greater extent of many lakes and inland seas in temperate and lower latitudes. They were there because of shifts of the main rainfall belts and the reduced evaporation resulting from lower temperatures than now and increased cloudiness. The Caspian Sea spread far to the northwest and north of its present shores into the central and eastern part of European Russia, and attained over twice its present size. And in early post-glacial times the Arctic Ocean waters invaded much of northwest Siberia from the north, where the land had been depressed by the ice load. Lake Chad, which is at present but a remnant in the southern fringe of the Sahara, became a great inland sea in ice age times as big as the present Caspian. And in North America west of the main watershed, the continental divide, there were numerous lakes, the biggest of which was another great inland sea, Lake Bonneville: this spread out from the present Great Salt Lake of Utah to attain an area of over 50,000 km² (as big as the present Aral Sea in central Asia) and a depth of over 300 m. Other lakes

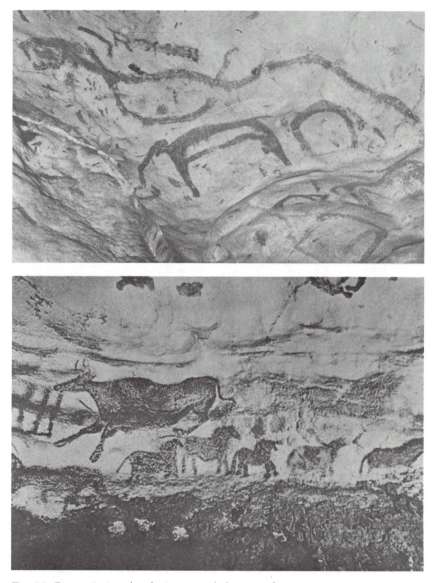

Fig. 38 Cave paintings by the ice-age inhabitants of France, 15,000 or more years ago, of the animals they knew and hunted. (a) Mammoth and other lines drawn on the roof of a cave at La Baume Latrone near Nîmes in southern France. (b) Ox and a row of small horses drawn in the cave complex at Lascaux in the Dordogne. France. Both pictures originally featured in a book entitled *Höhlen-malerei* published by the Verlag Brüder Rosenbaum in Vienna and are reproduced here by their kind permission.

in the same general region included Lake Lahontan with an area of about 25,000 km² in northwestern Nevada and Searles Lake and the Salton Lake in southeastern California. This watery landscape in the western mountain region of the present United States continued in existence into early post-glacial times, as long as the dwindling Laurentide ice-sheet still covered much of Canada. In Australia, too, there were lakes where there are none today.

THE END OF THE ICE AGE WORLD

The ending of the ice age brought great changes in the landscape, not just the melting of the mountains of ice and the gradual disappearance of many lakes but the rise of sea level as the melt water returned to the oceans, and the beginning of the prolonged rise, or rebound, of those land areas that had been weighed down by the masses of ice. The land around the north-ernmost end of the Baltic, the head of the Gulf of Bothnia, where the former north European ice-sheet was centred, is still rising about one metre per hundred years. The total rise of this part of Scandinavia since the ice disappeared is estimated to have been 270–300 m. And besides all this, there came the advance of the forest over vast tracts that had been tundra or grassy plains.

These were drastic changes for the people and animals then living, whose way of life was adjusted to the ice age world. In various regions – around the Mediterranean, about the North Sea, and the Great Australian Bight, to name but a few – and perhaps in most parts of the world, the early populations seem to have lived near the sea, probably because of the oppor-tunities of catching fish in the estuaries and evaporating sea water to get salt to preserve the food they caught on land and in the water. It seems likely therefore that the centres of gravity of the ice age populations were often in areas now submerged by the sea. It has been suggested that the end of the ice age, and the continued rise of sea level that followed, may have greatly reduced the total numbers of mankind – an event rare in history – and may have given rise to many of the legends of a great flood in ancient times.

The most distinctive feature of early post-glacial times was, of course, the globally increasing warmth. In most parts of the world the climate between 5000 BC or earlier and 3000 BC seems to have been generally warmer by 1–3 °C than it is today. In the northernmost parts of North America, where remnants of the former ice-sheet lingered longest, and also in Greenland, the warmest time was not reached until nearly 2000 BC. And, of course, it was the melting of the land-based ice-sheets which caused the level of the seas to rise.

114

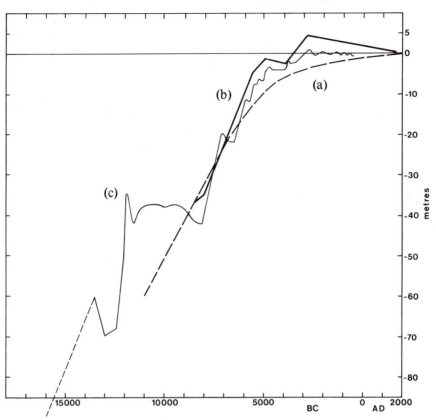

Fig. 39 The rise of world sea level as the former ice-sheets melted and through post-glacial times. Three reconstructions: (a) a very smooth mathematical 'best fit' curve, eschewing all detail; (b) a highly regarded reconstruction of the main stages, calculated from numerous dated shorelines in the Baltic after subtracting the steady rise of the region as the land recovered from the former ice-load; and (c) a carefully calculated more detailed curve, due to Mörner, supported by points of agreement between reconstructions in different regions. (The dates are corrected radiocarbon dates.)

THE RISING SEA LEVEL AND ITS EFFECTS

The post-glacial history of world sea level is plotted out in fig. 39. The rise began before 15,000 BC,[3] as soon as the ice-sheets began to recede. Various assessments have been made, and the diagram indicates the range within which the reasonable estimates lie. It also shows one of the more detailed curves that have been attempted. Of course, the details are less certain than the overall trend, but there is considerable agreement that the most rapid phases were between about 8000 and 5000 BC, also that the rise of general water level was effectively over by about 2000 BC, when it may have stood a metre or two higher than today. There were one or two drastic stages, as with the rapid melting of the Scandinavian ice-sheet after about 8200 BC, until there were only small remnants not much greater than today's ice-caps in Norway by 6000 BC, and the entry of the sea into Hudson Bay around 6000 BC followed by quick reduction of the great North American ice-sheet: by about 3000 BC the last remnants of the latter had gone, apart from the ice still present on Baffin Island and the Canadian Arctic islands. At times the rate of rise of the ocean was even overtaking the land rise in the Baltic region and in places like Scotland and Hudson Bay, where the former weight of ice had been centred. But in those regions the emergence of more land from the water has dominated in the last five thousand years.

Fig. 40 shows the geography of land and sea in the North Sea basin about 8000 BC. Over the following centuries the coast receded rapidly, forming complicated patterns of channels and islands. The Strait of Dover began to open perhaps as early as 7600 BC, and by 5000 BC the map of the coasts in the region resembled today's. The isolation of the British Isles from the continent by the rising sea level naturally cut off the way for immigration of those plants and animals which had not already made their way back after the ice age with the exception of any which could – themselves or their seed – be air or water-borne or be brought by birds. This accounts for the narrower range of species represented in the British Isles than on the European continent. The numbers of species are even more restricted in Ireland and on most of the smaller islands, which were cut off by water from Britain sooner.

During the same millennia the geography of the Baltic and the course of its outlets changed several times, and low-lying coasts in other parts of the world far from the former ice-sheets must also have receded fast before the advancing tide.

It may be imagined that even the most rapid post-glacial rise of mean sea level, averaging between one and five metres per century, would have drowned nobody. But this is a misunderstanding. The history of disasters near the low-lying coasts of the North Sea in the last thousand years teaches that recession of the coasts does not take place as a gradual process but in sudden advances of the sea at times of great storms which coincide with an exceptional tide heightened by the storm surge.

HUMAN MIGRATIONS

Evidence of the former human population of the Northsealand plain includes the spread of the Maglemos culture from Denmark to the Star Carr camp in Yorkshire in the eighth millennium BC. These people were hunters, living in a clearing in the early birch forest, who also used boats for fishing in a lake near their camp. Already some thousands of years

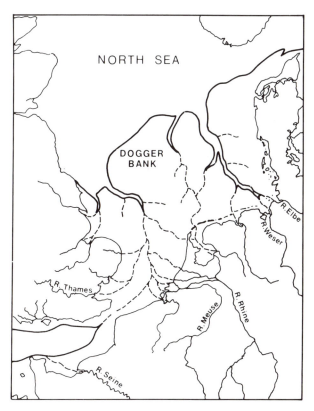

Fig. 40 The North Sea basin about 8000 BC: supposed geography of the then existing Northsealand. (This map was first presented by Clement Reid in 1902, and has never been superseded, though it is now recognized that the situation changed rapidly with the development of channels connecting the North Sea to the south during the following centuries.)

earlier, when the ice edge was still in Denmark, reindeer hunters roamed as far north as Hamburg, while others, living in skin tents, camped in the tundra near where Copenhagen now is. Research by the Arkeologisk Museum in Stavanger (funded by the Norwegian hydroelectric development undertaking) and by others working farther north in southern Norway has revealed that reindeer hunters were present on the 1000-m high plateaux within one thousand years of the disappearance of the ice sheet. Among about seventy radiocarbon dating tests, the results at some sites on Hardanger Vidda indicated the hunters' presence before 7000–6500 BC. By 5000 BC the deer were being systematically trapped where their favourite migration routes passed along narrow places between rocky slopes and the numerous water bodies not only on Hardanger Vidda but on the Ryfylke heights west of Setesdal in the south and on the Oppdal fells of eastern

117

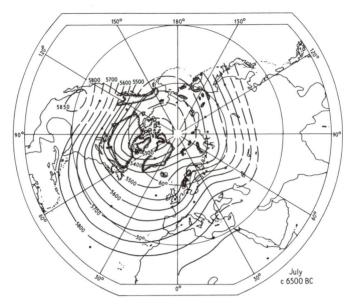

Fig. 41a Reconstructed pattern of the circumpolar vortex, derived from the strongly distorted thermal pattern prevailing, about 6500 BC: derived mean heights of the 500 millibar atmospheric pressure level in July. The map also shows the geography of land, sea and land-based ice-sheets around that time.

central Norway. In those early times there seems to have been more human activity on the plateaux and in the upper forest fringe than in the deep, shaded valleys, where the pine and birch forest may already have left few routes easily passable.

Reconstruction of the prevailing character of the atmospheric circulation can be made with some confidence for early post-glacial times, when there was still extensive ice covering northern North America while Europe and the North Atlantic were already ice-free to near their present extent. The strongly distorted character of this thermal distribution must have steered the Atlantic storm activity strongly to the northeast and north, towards both sides of Greenland and the Arctic Ocean, leaving Europe rather dry and anticyclonic except in the far north. Fig. 41 shows a proposed reconstruction of the average barometric pressure pattern and winds in July in the seventh millennium BC. For this date all reconstructions of the general wind circulation are broadly agreed. The regime depicted would be warm in summer, but still cold in the winters. The pattern seems to be verified by the established early arrival of warmth in Iceland and winds bearing pollens to western Greenland from much farther south in the forested zone of the present United States: in other words, prevalent southerly and south-westerly winds over the western and northern Atlantic.

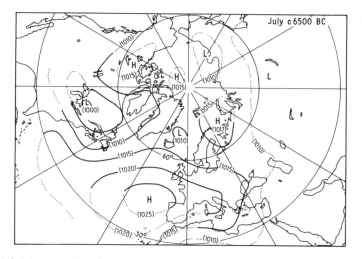

Fig. 41b The probable distribution of mean barometric pressure at sea level and implied surface winds prevailing in July around 6500 BC, corresponding to (a). In these maps note particularly the ice-sheets still existing in North America, the southerly winds prevailing over the western Atlantic towards Greenland and Iceland, and the anticyclonic (fair-weather) conditions prevailing over most of Europe.

Another part of the world with its own distinctive indications of a movement of population inland during the great post-glacial rise of sea level is the eastern Mediterranean. Around 10,000 BC the coastal plain of Syria and Palestine was still much wider than it is today, but it was already narrowing fast and, apart from a hesitation in the ninth millennium BC, continued to do so. M. R. Bloch of the Arid Zone Research Institute in the Negev, Israel,[4] has pointed out that the appearance, by about 9000 BC, of the world's first 'city' at Jericho occurs fairly near the beginning of the rapid rise of world sea level over the coastal plains and is coupled with evident exploitation of the rich salt deposits of the inland Dead Sea near by, which was then drying up and falling away from its ice age high stand: the salt seems to have been used for tanning leather, and for some kind of bread, as well as for preserving food.

THE BEGINNINGS OF AGRICULTURE AND THE HERDING OF ANIMALS

A little farther north in the Near East, archaeological evidence from the Shanidar cave, and pollen analysis from the lakes, in the Zagros Mountains on the borders of Turkey, Syria, Iraq and Iran indicate that about the same time man was beginning to domesticate animals and food grains. Until 10,000 BC the human inhabitants of the area had been living in caves and hunting wild game, largely sheep and goats. Then sheep became more important in their diet than other animals, and the high proportion of young sheep bones suggests that they were beginning to be herded. By 9000 BC the former glaciers in the Zagros Mountains[5] were confined to the heights and the vegetation zones were advancing up into the mountains. Wild barley had been present in the steppe at the mountains' foot, and wild einkorn wheat in the interior plateaux and among the mountains; but it is not certain that there had been any wild emmer wheat nearer than Palestine. Professor Herbert E. Wright of the University of Minnesota, who has long worked in the area, discussing these findings and the people's use of obsidian tools from over the mountains in eastern Turkey, sees evidence of seasonal migrations already in the tenth millennium BC and probably the fetching of some food grains and perhaps plants. At that stage a large number of grinding tools, found at the Shanidar cave, indicate the use of plant grains but no clear evidence of actual cultivation of them. In the next stage, however, on the Mesopotamian steppe near the mountains, wheat and barley occur where neither the climate nor the terrain were suited to the wild forms of those grains. Clearly cultivation had begun. Moreover, the population, accompanied by goats, sheep, pigs and dogs, was inhabiting houses at permanent sites, the first agricultural villages.

Wright, summing up these observations, wrote:

> All this implies that prehistoric man in the Zagros Highlands before 11,000 years ago was familiar only with einkorn among the wild grains, and that emmer and barley immigrated subsequently. . . . If we assume then that game animals before 11,000 years ago were more abundant in the mountains than in the plains . . . and that man the hunter also lived mostly in the mountains because of the availability of wild game for food and of caves for shelter, then with the change to a warmer climate and the immigration of wild grains we had . . . the combination of circumstances most favorable for domestication of animals as well as of emmer and barley, accompanied by a shift from caves in the mountains to open living sites in the foothills, where the ground was more favorable for cultivation.
>
> Although I have always felt that cultural evolution – gradual refinement of tools and techniques for controlling the environment – is a stronger force than climatic determinism in the development of early cultures, the chronological coincidence of important environmental and cultural change in this area during the initial phases of domestication . . . cannot be ignored.[6]

THE SHIFT OF THE VEGETATION ZONES AND THEIR FAUNAS, THE RANGES OF BIRDS AND OF FISH IN THE SEA

Of course, it was not only in the Zagros Mountains and the Mesopotamian plain that the vegetation zones were on the move from eleven thousand years ago onwards. In Europe the plains where bison and other animals had wandered were being gradually invaded, and then taken over, first by birch thickets, later by extensive birch and pine forests; later still hazel infiltrated, then oak, elm and lime took over the landscape, with alder in the wetter places. These woodland types were accompanied by the insect life and the birds and animals that thrive in each respective habitat. And, little by little, the birds and bigger animals were extending their ranges – as were the fish in the ocean – and establishing their seasonal migration routes, some of which must have had different starting points as well as different destinations from those that are so well established today. Men and animals had to adapt to a changing world. And it seems likely that many extinctions took place – for example, the horse and mammoth in North America and, perhaps, the mammoth in Eurasia too – because of man's greater skills and ability to cope with an unfamiliar world. It seems to the writer that the likeliest explanation of the sudden death of those mammoths that have been found well preserved in permanently frozen ground to modern times is that they were among the last survivors of their species, which strayed or fled from human hunters into swampy, near-frozen wastes in the tundra in some of the last of the warmest summers of post-glacial times four or five thousand years ago. Since then the permafrost, or permanently frozen subsoil, has advanced again and the animals have been preserved, though some have been released from time to time and have gone floating down the River Lena and other rivers in northeast Siberia in occasional exceptionally warm summers. One such case was vividly described by the captain of a vessel operating up the Lena in 1846, and in the last years of the previous century another mammoth was found rotting on the shore of the Arctic Ocean near the river mouth.

CLIMATE AND CULTURAL CHANGES IN PREHISTORIC TIMES

Discussion of the general question of the impact of climate in leading to cultural changes in human history, or in its most extreme form climatic determinism, has been carried an interesting stage further in recent years by a few authors, notably Wendland and Bryson writing in the journal *Quaternary Research* in 1974.[7] Since the climatic record presents an appearance of a series of more or less stable regimes separated by quite rapid transitions, radiocarbon dating tests have commonly been applied to obtaining the dates of related decisive changes observed in the environment (changes in the make-up of pollen assemblages, corresponding to changes in the composition of the vegetation; transgressions of the sea; maxima of glacier advances, etc.). These researchers therefore statistically examined the time distribution of the whole catalogue of radiocarbon dates of sharp environmental changes during post-glacial times, published in the journal *Radiocarbon* (which is the official organ for such reports). Over eight hundred such dating tests, performed on organic material from anywhere in the world, were examined. Their time distribution was then compared with the results of about 3700 radiocarbon dating tests used to identify times of cultural changes in all parts of the world. The analysis revealed five major post-glacial epochs of environmental change and five major epochs of cultural change: the dates of cultural change were in each case close to the dates of environmental change, following the environmental change in each case by apparent lags of the order of fifty to a hundred years. (It is necessary to say 'apparent lags' because of the margins of error of the dating tests, though these are reduced by taking averages of a large number of datings.) Geyh and Jäkel in Germany have performed a similar analysis concentrated on the present Sahara desert region, and Karlstrom and others have found from tree ring dating that the long history of cultural and population changes among the American Indians on the plateaux in Colorado during the last 2500 years seems repeatedly to have been triggered by the stresses of changes in the environment.

THE WARMEST POST-GLACIAL TIMES IN NORTH AFRICA: THE MOIST SAHARA AND ITS ENDING; CONTEMPORARY CHANGES IN EUROPE AND NORTH AMERICA

Proceeding now to areas farther south than we have so far discussed in this chapter, in the Sahara we once more find a record in rock drawings and paintings of a fauna and a human life-style which betoken an environment different from today's. The approximate dates of the pictures can be established from radiocarbon dating of related cultural material in the surrounding region, as far as the Nile valley. The earliest drawings of the fauna left by the hunters, who evidently operated in the Sahara, go back at least to 5000 or 6000 BC (corrected radiocarbon dates). Elephants, rhinoceroses, buffalo, hippopotami, crocodiles, antelopes, giraffes and fallow deer are all pictured. Examples of these species, which are today unknown in the region, are depicted even in the central Sahara. The paintings came later and continue up to the time of the early dynastic period in Egypt.

The most striking items in these Saharan rock pictures are those which belong to a watery environment, including (fig. 42) even some kind of

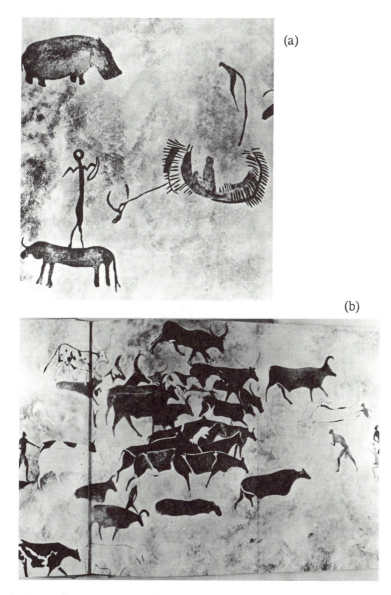

Fig. 42 Parts of cave paintings of around 3500 8C from Aounrhet in the Sahara illustrating (a) a hippopotamus hunt with canoes, clearly implying a much wetter climate than today's in the region, and (b) cattle herding. (From H. Lhote's *The Search for the Tassili Frescoes: The Story of the Prehistoric Rock-Paintings of the Sahara*, London, Hutchinson, 1959. Originally published as *À la découverte des fresques du Tassili*, Paris and Grenoble, B. Arthand.)

boat or canoe. This assemblage of artistic evidence clearly implies that there were enough moist places in the Sahara in those times to sustain life, so that animals and men could roam about, and cross, what is now the world's greatest desert. This conclusion is supported by evidence of the dated former levels of Lake Chad, which in the ice age had been an extensive inland sea and which, despite a severe fluctuation about the beginning of post-glacial times, had a water level 30–40 m higher than today until around 3000 BC. The old Mega-Chad lake has left its traces not only in old shore lines but a wide distribution of fish remains.

K. W. Butzer, surveying what is known from the more abundant material from Egypt in dynastic times, has concluded that elephants and giraffes were already becoming rare there in the centuries before 2900 BC and that between then and 2600 BC elephants, giraffes and rhinoceroses disappeared altogether from Egypt. The elephants which were still present in Algeria in Hannibal's time were evidently a remnant already isolated from the main stock in central Africa by the Sahara Desert. There were others near the Atlantic coast of Africa and in the forests at the foot of the Atlas Mountains. But the last of them seem to have died out in the third century AD.

The drying of the desert region from between about 3500 and 2800 BC onwards, which these considerations imply, and the concurrent decline in the recorded levels of the yearly Nile floods fed by the rains over Ethiopia, noted in the last chapter, seem to have been related to a climatic development of hemispheric, and probably global, extent. The timing coincides with changes in the composition of the forests in Europe and in a belt across North America from Minnesota to New England: specifically a marked decline of the elm and apparently also of the linden or lime, the two more warmth-demanding species in the zone of broad-leafed trees. There was also a marked advance of the glaciers in the Alps, which has become known as the Piora Oscillation, apparently the first noteworthy advance after some millennia of the warmest post-glacial times. It has been concluded by some that the rise of Egypt, and the organized cultivation of the Nile valley by use of the yearly flood for irrigation, may have been a necessary response to the great contraction of the habitable terrain in northern Africa at the time. It was also made possible, of course, by the knowledge of agricultural techniques which must have been spreading already for some time in the Near East.

7

IN THE TIMES OF THE EARLY CIVILIZATIONS

THE WARMEST POST-GLACIAL TIMES: DEVELOPMENT OF THE DESERTS OF ASIA AND THE RIVER VALLEY CIVILIZATIONS

As with Egypt and the Nile valley, so it may be also that the civilizations organized in the third millennium before Christ in the Tigris and Euphrates lowland, in the Indus valley and in China were at least in part a necessary development to feed a more concentrated population at a time when huge areas outside those valleys in Arabia, in Afghanistan and Rajastan, and in the Gobi and Sinkiang, were becoming more desert-like. If pastures and stocks of wild game for hunting were failing, the advantages of cultivation in more or less reliably irrigated valleys would be more obvious to those faced with abandoning an age-old way of life.[1] The Japanese meteorologist and geographer Hideo Suzuki has made the interesting suggestion that it was the refugee herdsmen and farmers from the increasingly desert regions round about who were fated to become the slaves who made possible the intensive agriculture and the great building works for which ancient Egypt and the other river valley civilizations are famous. But even as man learnt to produce controlled environments for agriculture, he still had to work with the conditions that nature provided.

The course of the prevailing temperatures through post-glacial time[2] is illustrated by the history of the upper limit of trees on the mountains in various parts of the world (fig. 43). This record has been established by examining those types of remains of former trees (stumps, seeds, etc.) unlikely to have been moved far from the sites where they grew. The limit is essentially controlled by the prevailing summer temperatures. The curves so derived run parallel with those resulting from locating the poleward limits of the various vegetation zones at different epochs by pollen analysis and from the indications of glacier variations. The tree line curves probably follow the actual temperature changes with a lag of no more than fifty to a hundred years, i.e. less than the uncertainty of radiocarbon dating. By contrast, the history of sea level (fig. 39, p. 115) must have an important

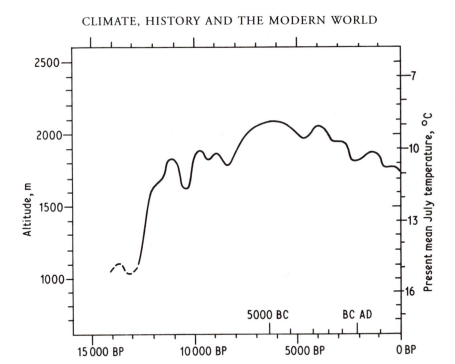

Fig. 43 Average height of the upper tree line on the mountains in temperate latitudes since the last ice age. (From work by V. Markgraf. Reproduced by kind permission.)

lag, since one should expect that (other things being equal) sea level would be highest about the end of the warmest climatic regime, when the melting of glaciers had run its full course.

It is interesting to notice that the first canal made to link the Mediterranean to the Red Sea at Suez, actually a fresh-water canal from the Nile delta to near where Suez now stands, was dug in the twentieth century BC under Pharaoh Sesostris I, about the time when world sea level was probably just at its highest. Perhaps it was this that made the project seem feasible, because the land barrier to be passed would be shortest then. Sea level must have continued nearly equally high for several centuries thereafter, though probably fluctuating more after about 1600 BC. A second Suez canal, or possibly a reconstitution of the first one, dates from Rameses II (1304–1237 BC), whose reign seems to have been in one of the last of the fairly prolonged spells of more or less the full warmth of the warmest postglacial times. Sea level should accordingly have again been close to its highest stand. It may well be that the escape of the early Israelites out of Egypt about 1230 BC, described in Exodus, is explained by a short-term fluctuation of the Red Sea waters (as with a storm surge or tidal wave) over the shallow sands at a time when the sea was normally higher than now.

126

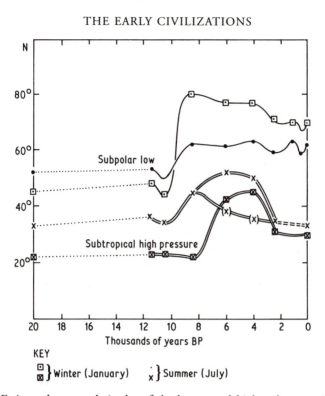

Fig. 44 Estimated average latitudes of the lowest and highest barometric pressure (at sea level) in the European sector of the northern hemisphere since the latter part of the last ice age.

   Reconstruction of the prevailing atmospheric circulation during each of the main stages of post-glacial climatic history produces a record of the probable latitude of the North Atlantic–north European storm zone (the concentration of paths of the low pressure centres) and of the Azores–European–north African anticyclone zone (fig. 44) that runs broadly parallel to the prevailing temperatures. These curves are of interest in relation to aspects of the history of the environment mentioned in the last chapter and others with which we are concerned here. The indicated more northern position of the subtropical anticyclone belt between about nine thousand and three thousand years ago than either before or since, culminating with prevailing positions around latitude 40 °N through the middle of that period, is consistent with summer monsoon rains penetrating much farther north than now over the Sahara region. It seems likely that occasional erratic cyclonic activity from the south could have reached right across what is now the desert to the Mediterranean in the summers of the warmest post-glacial times (and probably at any season of the year). With such a pattern, it is no wonder that there was enough vegetation and surface water in the Sahara for animals and men to wander and that Lake Chad was much bigger than now between about 7000 BC, or earlier, and 3000 BC. Estimates of the average annual rainfall in the hyperarid desert centre area between Kufra and Tibesti, which at present rarely sees any rain, range from 200 to 400 mm before 6000 BC to around 50–150 mm towards 3000 BC, and in the region of Lake Chad nearer the edge of the Sahara two to five times these amounts. Moreover with much more cloud than now it is also obvious that evaporation was reduced. From the Tibesti Mountains permanent rivers flowed.

The water level in Lake Rudolf, like other lakes in East Africa, was also some tens of metres higher than now, and it had an overflow into the Nile. Recent research has described the Saharan scene in the times before 3000 BC as a dry savanna with trees along the permanent, or nearly permanent, river courses. There is evidence that this moist regime had been interrupted by something less than a thousand years of marked desiccation, probably just before 5000 BC. This climatic disturbance interestingly seems to coincide in time with the entry of the sea into Hudson Bay, followed by rapid degeneration of the great North American ice-sheet: events likely to have distorted the hemisphere's thermal pattern and wind circulation for a time. A hint that this brief drier episode had some importance in the connections that concern us here may be seen in the findings of archaeology in Palestine, where the spread of human settlements into the driest regions seems to have had two high watermarks, around 6000 and 4000 to 3000 BC, before and after the drier period recognized in the Sahara.

A moister regime than now seems to have prevailed through the warmest post-glacial times in other lands in the same latitude zone. Indeed, farther east in Rajasthan (northwest India) and in China no interruption such as that noted in north Africa in the sixth millennium BC has been reported. It seems likely that, at some stage within this period of 'over-developed' monsoons, events took place which gave rise to other flood legends, distinctively rain-produced floods, including the biblical Noah's Flood. Archaeology reports flooding episodes between 4000 and perhaps as late as 2400 BC at Ur, Kish and Nineveh.

A more serious break in the climatic regime came between about 3500 and 3000 BC. There was at least one fluctuation in that interval of greater magnitude than had occurred for a very long time. The glaciers in the Alps advanced and the forest retreated somewhat from the heights: this was the so-called Piora Oscillation, named after the Val Piora where the first evidence was found and established by pollen analysis as indicating a cold episode. In the temperate forest zone all across Europe and in parts of North America the more warmth-demanding trees, the elms and the linden (or socalled lime) trees, declined and never regained their former, probably dominant, position in the forests. It is not certain whether, or to what extent, human interference or browsing cattle played a part in this, but the phenomenon seems to have been too widespread for this to be the main explanation. For a time in northern Europe the oak declined too, and the hazel withdrew for good from its northernmost limits.

The duration of this colder episode seems to have been quite short, at most four centuries, but traces of it or of parallel vegetation changes extend to Alaska and the upper forest limit in the Colombian Andes and on the mountains of Kenya. There was evidently some disturbance of the global regime. Moreover, it marked the end of the most stable warm climate of post-glacial times, a regime which had been associated with great prevalence of the westerly winds in middle and sub-Arctic latitudes. That regime is referred to, appropriately therefore, in the older European literature as the 'Atlantic' climatic period. With prevailingly mild winters and warm summers and with the storm belt far away to the north, in high latitudes, overall mean temperatures in Europe seem to have been up to 2 °C higher than in recent times. Indeed most of the world seems to have been almost that much warmer.

Among the evidence of these higher temperatures is a northward displacement of the limits of plant and forest species and an upward extension of their ranges beyond their present limits on the hills in many parts of the world. And the insect and animal species which go with each vegetation type had their ranges extended too. In Europe a pond tortoise (*Emys orbicularis*) whose present range in France and Germany seems to be limited by a requirement of mean July temperatures as high as 18.5–20 °C was present in Denmark and East Anglia in the warmest post-glacial times. And it is known that the Dalmatian species of pelican (*Pelecanus crispus*) migrated as far north as Denmark.[3]

The so-called 'sub-Boreal' climatic period, which covered the following two to three thousand years, until well into the last millennium before Christ, included some periods which were as warm as any since the last ice age. But it is clear from many kinds of evidence of vegetation history and peat growth, as well as lake deposits such as that of the Crimean lake seen in fig. 34 (p. 91), and some later glacier advances, that the variability of the climate was greater than before. The year-to-year and longer-term variations seem to have been particularly marked in rainfall, as may be recognized in the tree rings in central Europe and bog growth in many parts of the temperate zone of the northern hemisphere. The fluctuations of the river levels in Egypt, northern India and China seem to fit into this pattern; but in those latitudes, as in the Crimea, the net effect after 3000 BC seems to have been a step-wise progression towards generally lower river and lake levels than before.

It is noticeable that the beginning of the New Stone Age cultures, and the rapid spread of the first agriculture across middle and northern Europe, approximately coincided in their timing with the climatic upset which ended the steady 'Atlantic' regime. This is a remarkable coincidence. It is tempting to suppose that some disruption of established ways, which the climatic events caused, provided the challenge and the stimulus to undertake some deliberate cultivation and new tools. What exactly these events were, and what the stimulus was, we do not know: but it is likely that some familiar pastures failed at this time, and that some wild fruits and grains which the earlier economy had relied on gathering also became harder to find or less reliable in their cropping.

Even after 3000 BC conditions in the areas of high civilization evidently in the main continued for a long time somewhat moister than now. It is not certain how much of the difference can be attributed to a higher water table and, as yet, more numerous and extensive oases in the growing deserts as a dwindling legacy from the earlier climatic regime. The beginnings of the practice of irrigation in Mesopotamia go right back to before 5000 BC, and there is evidence of settlements in what is now the Saudi Arabian desert in the earlier part of the period. Down to around 2000 BC cultivation in Mesopotamia still extended some 50 km north of the present limit of feasibility of any such activity and a density of human population seems to have grown up that could no longer be sustained there today.

THE INDUS VALLEY AND ITS CIVILIZATION

The cities of Harappa and Mohenjodaro in the Indus valley in northwest India flourished between about 2500 and 1700 BC. Wheat, barley, melons, dates, and perhaps cotton, were grown in what is now the Thar desert of Rajasthan; and there were elephants, rhinoceroses and water buffalo there. The normal yearly rainfall in the region at that time has been estimated as between 400 and 800 mm. Some breaks in the area's cultural development in those times have been attributed to flooding episodes. One of the century-to-century, and sometimes longer, fluctuations which seem to have been particularly marked all through the 'sub-Boreal' climatic period, between about 3500–3000 BC and 800 BC or after, may have had something to do with the emergence of the Indus civilization. In this case it was probably a reversion to stronger northward development of the summer monsoon rainfall, and displacement of the temperate zone rains and winter snows farther north towards and into the Arctic, with consequent drought in central Asia. The rise of Harappa by about 2500 BC seems to have corresponded to an interruption in the record of human settlement in a wide region stretching from central Asia (Turkmenia) to eastern Iran and southwestern Afghanistan and a severance of trading links from the south with central Asia and across central Asia.

At its height the Indus civilization cultivated an area greater than the Nile valley and Mesopotamian civilizations combined. The final decline,

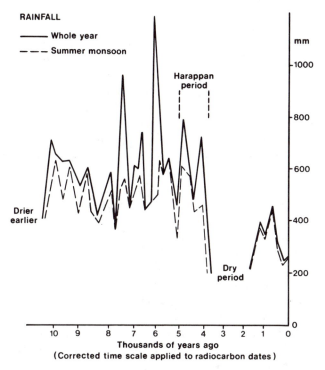

Fig. 45 Estimated variations of the rainfall in Rajasthan, northwest India over the last 10,000 years, from lake levels and botanical (pollen analysis) evidence. (Adapted from a diagram by R. A. Bryson, based on the work of G. Singh)

when it came, was at first gradual, though with dislocation of the developed urban life after about 1900 BC. Later the decline became rapid. It can reasonably be attributed to increasing drought. (A botanist's history of rainfall in the region is shown in fig. 45.) There seems to have been a gap of several centuries before the arrival of the Aryans, the next inhabitants of the region, and they seem not to have spread out their settlements away from the rivers as the Harappans had done. The Aryan settlement declined in its turn, and many of the rivers have since disappeared. It seems certain that, even when Alexander the Great marched his armies across southwest Asia to the Indus, between 330 and 323 BC, there were still more water sources than now but that they were a relic, left over from the earlier climates of the region. There was residual vegetation too – the army crossed the Indus in boats made from timber growing along the river banks. Ptolemy (Claudius Ptolemaeus), writing in the second century AD, mentioned five rivers also in Arabia and trade routes which had formerly been in use which were by his time impassable. There were Roman bridges across wadis which are now dry.

131

ANCIENT CHINA

In China the researches summarized by the late Dr Chu Ko-chen suggest that the departure of the prevailing temperature level from present conditions in the warmest post-glacial times was somewhat greater than in other parts of the world. The overall mean departure may not have been so very different, around 2 °C warmer than now, but the mid-winter temperatures seem to have been about 5 °C above today's. Snow fell very rarely in central China, and rice could be sown a whole month earlier than it is normally sown today. Bamboo groves were much more extensive than they are now in the central lowlands as far north as the Yellow River (Hoang-ho) basin. The northern limit of the natural distribution of bamboo, which today follows the mean position of the 0 °C January isotherm, was three degrees of latitude farther north around 3000 BC than it is now. Its retraction does not seem to have begun in earnest until about 1100 BC, when the average January temperature was probably still 3 °C higher than it is today. The many uses of this valuable plant, its shoots providing food and its full-grown stems providing building material, besides its ready use for making hats and writing materials, furniture and musical instruments, suggest that it may have played an important part in the early development of a high civilization in China. (At some stage the papyrus reed also disappeared from lower Egypt, though this must presumably be put down to the drying trend of the climate there.) The Far East is relatively isolated from the other centres of early civilization, and it seems likely that the beginnings of agriculture and the domestication of animals had an independent centre of original development in China. The first Chinese Neolithic agricultural civilization grew up on the fertile loess soils of northern China in the belt between Kansu province in the northwest and the lower part of the Yellow River basin. But the suddenness of the advance to a Bronze Age culture, cultivating wheat and millet with the aid of irrigation, seems to suggest contact across central Asia, during a moist period there, with lands in the west where those arts had been developed.

THE SITUATION IN EUROPE AND OTHER NORTHERN REGIONS

The latter part of the warmest post-glacial times also saw great advances of civilization in the west, even as far north as northern Europe. The New Stone Age was succeeded by the Bronze Age. And the general warmth and evident freedom from storminess allowed great development of travel and cultural and trading contacts by land and sea. The map in fig. 46 shows how far the forest and grassland limits at 2000 BC extended northward and through the heart of the continents beyond their present positions all across Eurasia and North America. In the north and in the continental

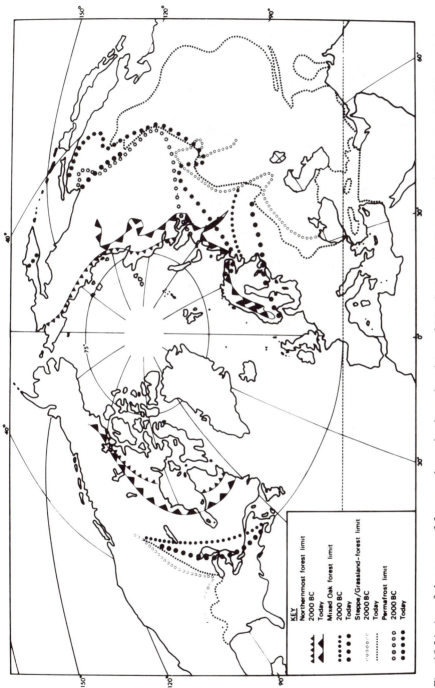

Fig. 46 Limits of the zones of forest and grassland on the plains of North America, Europe and Asia around 2000 BC and now.

Fig. 47 A stone circle (the Ring of Brodgar) in the Orkney Islands, which can be shown to have been built as a lunar observatory. (Reproduced from 'The place of astronomy in the ancient world', *Philosophical Transactions of the Royal Society*, Series A, vol. 276, no. 1257, p. 152, by kind permission of the Royal Society and Professor A. Thom.)

interiors the permafrost – the permanently frozen subsoil – was more restricted too than it now is. This probably meant that there was less of the cold swamps, where the surface melt water of summer cannot easily drain away, and which limit the spread of trees in the Arctic fringe today. The glaciers and lingering snow in summer in the Alps were less advanced too, allowing passage over the passes and gold mining at quite high levels. And the freedom from stormy winds and seas in the north is indicated by the forest cover, which in the British Isles extended to the exposed Atlantic coasts in Cornwall and the northwest Highlands of Scotland and even in the Orkney Islands (though the range of tree species was very limited there). And in Iceland some valleys that are now filled with ice supported trees.

The spread of megalithic monuments from the Mediterranean to Britanny, Cornwall, Wales, the Outer Hebrides and Orkney surely indicates considerable seafaring links along this route, even though one of the most splendid of these monuments was inland at Stonehenge on the open chalk plateau of southern England. The apparent construction of many of these stone circles as solar, or astronomical, observatories (figs. 47, 48) suggests – particularly in the case of the Hebrides and Orkney – that the skies were less frequently clouded over than they are today. This is a

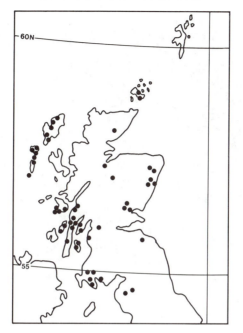

Fig. 48 Distribution of stone circles dating from about 2000–1500 BC in northern Britain, which have been shown by Dr A. Thom to have been astronomical (mostly solar) observatories. Many of these sites are in what are now the cloudiest districts of the British Isles, the sun being visible only between 22 and 30 per cent of the time.

suggestion that is entirely consistent with the reconstruction of the prevailing wind circulation, with a more northern position of the anticyclones (cf. fig. 44), accompanying the warm climate regime. The recent discovery that some of the megalithic tombs and circles at Carrowmore in Ireland are the earliest examples so far found anywhere, dating from between 4500 and 3700 BC (corrected radiocarbon dates) does not alter this picture. Whether or not they also had astronomical associations, reconstructions of the climatic patterns prevailing indicate already from well before those times regimes with frequent anticyclones and more frequently clear skies than now in this part of the world.

The seafaring links which we have referred to must have been mainly concerned with trade. It is well known that in the last two millennia before Christ tin from Cornwall was traded south to the Mediterranean for the manufacture of bronze and that finished metal objects were brought north. The Bronze Age probably first reached the British Isles with the 'Beaker folk' across the North Sea. But already around 2000 BC the megalithic monuments culture had spread in the reverse direction, from Scotland and Orkney to Denmark and southern Sweden (Skåne), apparently with the

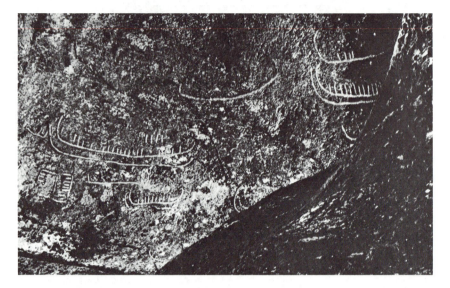

Fig. 49 Rock engravings of boats from the Bronze Age at Revheim near Stavanger, Norway. (Kindly supplied by the Arkeologisk Museum, Stavanger.)

Fig. 50 Ancient skiers: rock pictures in Norway and near the White Sea, attributed to Bronze Age or late Stone Age time. (From G. Berg, 'The origin and development of skis throughout the ages', *Finds of Skis from Prehistoric Time in Swedish Bogs and Marshes*, Stockholm, Generalstabens Litografiska Anstalts Forlag, 1950.)

136

Fig. 51 Stumps of large pine trees which grew about 4000–4500 years ago, found at 650 m above sea level on the Cairngorm Mountains in central Scotland. This height is near, or slightly above, the limit of dwarfed and stunted trees today.

Baltic amber trade. Norway was a notably poor country at that time, remaining outside these developments; but there, too, conditions were evidently favourable for the development of sea traffic, as is witnessed by the numerous rock drawings (fig. 49) of boats – rowed by up to fifty men – as well as the animals with which the population was familiar. Reindeer are most prominent among these, but there are also horses and cows as well as sleds and people on skis (fig. 50). Clearly, one should not imagine that there was no snow covering the Scandinavian fells in the winters of the warmest post-glacial times. Similar pictures are found near the borders of Lake Onega and the White Sea in Russia.

A gauge to the relative warmth of those times is found not only in the relics which establish the tree line (fig. 51) but also in the extension of Bronze Age cultivation on the hills of southwest England (Dartmoor) to over 450 m (1500 ft) above sea level, compared with the absolute limit of 300 m in the same district today. During the third millennium BC late Neolithic and early Bronze Age people occupied sites, perhaps only seasonally, on the 600–800 m high plateau of the Pennines in northern England. This occupation seems to have ended when it was overtaken by peat formation in the succeeding millennium. The heights were abandoned, apart from a temporary resettlement of levels between the valley swamps and the peat mosses of the Pennine summit plateau some time between 1000 and 500 BC. Also in the hills of Wales there were occupation sites in late Stone Age and Bronze Age times; but there, and in the Pennines, the former forests were declining and the herding and foraging of animals may have so damaged what tree growth there still was on the uplands as to play a significant part in instigating the spread of the blanket of peat, which ended the occupation.

8

TIMES OF DISTURBANCE AND DECLINE IN THE ANCIENT WORLD

THE TURNING POINT IN POST-GLACIAL CLIMATE DEVELOPMENT

The most general conclusion from the evidence so far examined – by pollen analysis, former lake levels and so on – regarding the moisture history of Australia is that somewhere about 4000 to 2000 or (more likely) 2500 BC there was a turning point. Before that time throughout the Australian region many places were wetter, and more consistently wetter, than they are today and rainfall had been increasing both in the tropical and temperate regions. Since that time there has been a marginally cooling and drying trend, with rather wide fluctuations superposed on it. These are the views of Professor D. Walker of the School of Pacific Studies at the Australian National University, Canberra, stated in 1978. The same turning point seems to be indicated in the other great deserts of the present world, as is apparent from the last chapter. And much the same epoch marks the turning point in the history of prevailing temperatures as registered by the height of the upper tree line in every part of the world where this has been examined, from northern and central Europe and the Yellowstone Park in Northern America to Japan and the mountains of New Guinea, to New Zealand, and the Andes in South America. The same is confirmed by the history of the northern limit of forests and by the glaciers on the mountains in temperate and lower latitudes. Only in parts of northern Canada and northern Greenland, where there was most residual ice still melting in mid post-glacial times, was the climax of warmth delayed significantly – in some areas until 2000 BC or after. Thus, the evidence of a global event is clear: the climax of our interglacial. The moisture maximum in subtropical and tropical latitudes and the temperature maximum shown by the vegetation in middle and higher latitudes, and on the mountains everywhere, must be seen as related aspects of the warmest postglacial time.

It is obvious now that the ancient civilization could not for ever continue in the Indus valley and that the other cultures in Asia and north Africa had to shift and change, as the moisture supply became more restricted and the crop and vegetation limits were drawn in. But what exactly happened and when? And how and where did man's activities hasten the process of desertification? Can we fill in any of the details?

Various peoples in various times have had legends of a Golden Age in some earlier time. The notion occurs in the literature of Classical Greece and Rome and of other peoples. Often it refers to an idealized state of society, but occasionally there are references to a lost landscape, the best known being the biblical tale of the Garden of Eden. It may well be that some of these myths enshrine dim folk memories of some of the changes with which this book is concerned. The times of highest civilization and of their decline and breakdown were, of course, not generally synchronous in different regions. But there does seem to have been a very wide-ranging reduction of occupation of the north African and Arabian desert lands around 3000 BC and on into the following centuries and another phase of very widespread disturbance in the steppes and deserts of Hither Asia and north Africa around 1200 BC. In so far as these events may have been caused by climatic change, the pattern of change and its effects were perhaps no more monolithic and uniform than in the decline in the late Middle Ages for which much fuller evidence exists and which will be described in Chapter 11.

RECORDS OF THE DOWN-TURN OF CLIMATE

The most detailed record we have so far, which goes all through the times with which this chapter is concerned, is the series of yearly tree ring widths since 3431 BC in bristlecone pines growing near the upper tree line on the White Mountains in California, shown here in fig. 52. The ring widths at this height are, of course, mainly a response to the level and duration of the summer temperatures. While the curve in fig. 52 cannot be taken as applicable in detail to the climatic history of Europe or elsewhere in the Old World, the hundred-year average measurements over the centuries since AD 1100 for which we have more or less reliable derived temperature values for England do appear to be significantly correlated with the temperatures in Europe. The differences which we know of between this Californian record and Europe, where various types of evidence have been used, are:

1 There seem to have been one or perhaps more further recurrences in Europe for periods of the order of a century or so of temperatures approaching the warmest post-glacial level between about 1100 and 800 BC.
2 The warmth in Roman times in Europe seems to have continued, and perhaps reached its maximum and greatest consistency, in the fourth century AD, whereas the breakdown appears 100–150 years earlier in the Californian tree ring record.

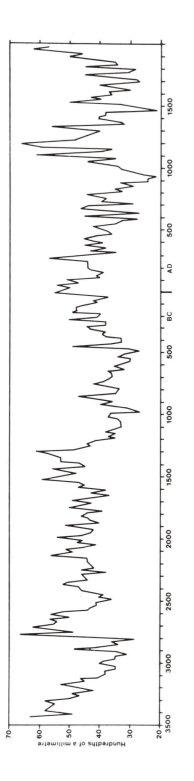

Fig. 52 Ring-widths (twenty-year averages) in growth rings of bristlecone pine trees near the upper tree line in the White Mountains, California from 3431 BC. The variations at this height may be taken as indicating variations of summer warmth and/or its seasonal duration. (From data kindly supplied by Professor V. C. La Marche at the Laboratory of Tree Ring Research, University of Arizona, Tucson.)

3 Again in the Middle Ages the warmth continued about a century later in Europe, at least in northern and western Europe, and perhaps culminated in the thirteenth century AD. This warm period in Europe seems also to have begun earlier, in the 900s.

4 The cold climate which followed the medieval warmth had a severe phase in Europe, as in California, in the 1400s but was severest in and around the 1590s and 1690s.

Fig. 53 compares the generalized history of the height of the upper tree line in Europe with that in California. Here, the parallelism seems closer. It becomes clear that the major peaks of warmth shown by the Californian tree rings (fig. 52) and the medieval and modern temperatures in Europe (fig. 30, p. 84) did not last long enough to restore the limits of the forest on the mountains and towards the northern coasts to their most advanced post-glacial positions. In many places, of course, restoration of the ancient forest cover had been rendered impossible by soil deterioration and the development of peat in the meantime.

Other 'proxy' indicators which give us a record, albeit less complete, of the history of the climate since 3000 BC include radiocarbon dated moraines and other traces of former glacier advances, also radiocarbon dated pollen analyses of the deposits in peat-bogs and lake beds (so long as we are careful to avoid areas where the vegetation was disturbed, or managed, by man), and the study of the yearly layering of deposits such as that of the Crimean lake illustrated in fig. 34 (p. 91) and the ice-caps in Greenland and Antarctica.

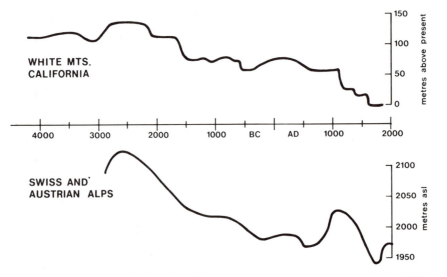

Fig. 53 Changes in the height of the upper tree line in two areas in the White Mountains, California and in the Alps in Switzerland and Austria. (From work by V. C. La Marche and V. Markgraf.)

142

From all this material we can trace the course of the decline from the warmest post-glacial times in different regions of the world. The prevalence of average lapse rates of temperature with height between 0.6 and 0.7 °C per 100 m in nearly all parts of the world, and with most uniformity at heights above the lowest 1000 m above sea level, gives us a yardstick with which to gauge the long-term changes of temperature to which the lowering of the upper tree line corresponds. The changes seen in fig. 52 indicate that by about 500–200 BC the long-term average temperature level in Europe was about 1 °C lower, and in the southwestern part of North America about 0.5 °C lower, than it had been in the warmest post-glacial period. (In the latest millennium, after the Middle Ages, a further decline is registered.) These figures are modest compared with other assessments of some of the more abrupt changes in the last two millennia before Christ (and again in the late Middle Ages). It is clear, however, from fig. 52 that the growth rings of the trees in California indicate a number of very sharp shorter-term changes, in both directions, affecting the average temperature level over periods from a few years up to a few centuries, such as have been deduced from the indications of past glacier advances and retreats in Europe and North America, and that these have exceeded the magnitude of the more persistent changes registered by the upper tree line. Some of these abrupt changes were certainly associated with epochs of enhanced volcanic activity and frequent volcanic dust veils in the high atmosphere. The changes over a few centuries seen in fig. 52, as around 1300–900 BC, before and after AD 1200, and again since 1800, are more than twice as great as the differences in average ring thickness from millennium to millennium: they probably corresponded to changes of prevailing temperature by rather over 1 °C there and rather over 2 °C in Europe. In northern Canada, and probably elsewhere in high latitudes, the changes were greater still.

The evidence of sharp cooling in California, which continued in progress from 3300 BC or earlier until about 2800 BC (fig. 52), presumably corresponds to the Piora Oscillation with its glacier advances in Europe. This was also a sharp event, but it is as yet doubtful whether it continued quite so long. It is inconceivable that either Europe and Asia or North America were totally unaffected by any of the pronounced cooling phases evidenced at any time on the other side of the hemisphere, and this seems to be borne out by what is known of the record in the respective sectors. But it is also clear that some episodes were much more strongly developed on one side of the hemisphere or the other and that there were differences of timing of the major developments, even though these showed a substantial overlap. Presumably, in the case of these earliest developments of what geologists ominously call the Neoglacial an extensive mass of thick pack-ice was built up with its 'centre of gravity' on one side of the Arctic Ocean, while the ocean currents maintained a strong enough transport of warm water to keep the opposite side of the polar basin largely ice-free.

In New Mexico and Arizona, in what is now the southwestern United States, the first agriculture had been spreading northwards in the warm period between about 4500 and 4000 BC, presumably supported by a reliable northward penetration of the summer rains. There is so far no evidence of agriculture continuing there during the next two thousand years after about that date. So it was presumably knocked out by a drier regime that set in there at the time of the earlier cooling episode before 3000 BC; and the climatic recovery, which seems to have been sharp in California around 2800 BC, did not last long enough to be exploited by movement of the doubtless sparse population.

Farther north, the prairie landscape had been extending northeastwards at the expense of the forest for some thousands of years and reached its farthest limit near the Mississippi River and the south end of Lake Michigan probably between 6000 and 5000 BC. This development[1] was presumably associated with maximum development of the westerly winds and of the extent of the rain-shadow effect of the Rocky Mountains. It was still early post-glacial times in North America, with much of the old ice-sheet present in Canada and the thermal gradient produced by the warming in the south must have generated strong winds. Since 5000 BC there has been a slow, fluctuating retreat of the natural prairie limit westwards. The retreat has been most pronounced at times when the belt of westerlies weakened or shifted farther north, allowing a moister climate to return to the great plains of the Middle West. These fluctuations of the wind circulation and climate have continued to have important effects on the history of the region and its peoples right to our own day, as will be noticed in later chapters of this book.

During the third millennium BC the history of the composition of the woodlands in Europe indicates a gradual recovery from the sharp changes associated with the elm decline, though neither the elm nor the lime ever fully recovered. The year rings from this period in European oak trees also indicate that the year-to-year and decade-to-decade variability was significantly greater than in recent times and presumably greater than in the equable climate of the 'Atlantic' regime before 3500 BC. This variability seems to have calmed down again during at least part of the Bronze Age. Around 2300 BC, approximately dated by reference to the counted year-layers in former lake bed sediments, the famous Swedish varve chronology, there was a pronounced change to greater wetness in the peat-bogs of Sweden leading to intensified growth of the peat. This may well be the counterpart in the north to the shift to a drier regime in southeastern Europe of which we see evidence around 2200–2100 BC in fig. 34. Such an opposition between the rainfall tendencies in the north and northwest of the continent as against the southeast is normal, a feature produced by the prevailing size of the anticyclones and breadth of the cyclonic depression zone. We also know from ancient Egyptian records inscribed on stone tablets, or 'steles', that there were great famines around 2180–2130 BC and again between about 2000 and 1950 BC and yet again some two hundred years after that: and it is made clear that these were associated with an abnormal prevalence of southerly winds from the desert and low level of the River Nile – i.e. failures of the yearly flood. The periods concerned, in the twenty-second and eighteenth centuries BC, were, moreover, times when Egypt was invaded by peoples coming from the east, bringing the Old and Middle Kingdoms to an end. It is legitimate to wonder whether the invaders themselves had been unsettled at those times by droughts in their former homelands. So the periodic variations demonstrated by the lake sediment in the Crimea were also serious elsewhere, including farther

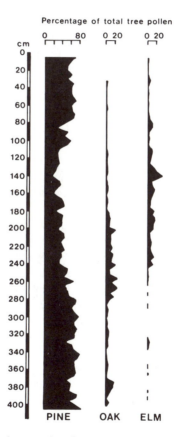

Fig. 54 Variations through post-glacial time in the occurrence of pollens of pine and of more warmth-demanding trees in a mire over 10,000 ft up (3120 m above sea level) in Kashmir, near 34 °N 75 °E. The vertical scale is a scale of depth in the bog and can be read as a time-scale, with the latest deposit at the top. The 80 cm level is thought to be from about 2500 years ago. (From work by G. Singh and Vishnu-Mittre.)

south in Africa, though the drier and wetter periods were probably some-what out of phase in those different latitudes.

It seems likely that it was some development associated with these vari-ations involving the movement of the rain-bringing cyclonic activity in Africa and the Near East that caused some of the disruptions noted in the Indus valley in the last chapter. Unfortunately precise dates are lacking.

A pollen diagram from Kashmir (fig. 54), where the forests were not apparently affected by human activity, shows the decline of the more warmth-demanding trees and return of the pine about this time, although precise dates are lacking. Thus, we find in that latitude the same overall temperature history as in the north.

The 'Neoglacial' development[2] began to make itself felt in earnest between about 1500 and 1300 BC with advances of the glaciers in Alaska and in the Alps and perhaps the first beginnings of renewal in the United States Rockies (Colorado), in Scandinavia and New Zealand. Around 1300 BC a sharp cooling phase arrived in California, as witnessed by the bristlecone pine record in fig. 52. And this time there was no full and lasting recovery at any time since. Most – and perhaps all – of the glaciers present today in the United States Rockies south of the Canadian border are believed to have formed since 1500 BC. In Central and South America, too, there is evidence of sharp cooling about 1500 BC.

Investigations of the sea bed on the coast of Maine, in the northeastern United States, indicate that Gulf Stream water regularly followed the coast as far north as that all through the warmest post-glacial times until roughly 1500 BC. Since then, it has moved out into the Atlantic (as the warm saline North Atlantic Drift) from more southerly points on the American seaboard and has never renewed its dominance so far north.

Changes in Europe are signalled about the same time or soon after by regrowth phases in the peat-bogs and by fluctuations of the levels of the lakes in and around the Alps, which affected the human settlements at their margins. The many bogs in Ireland and elsewhere in northwest Europe show fluctuations at a variety of dates, which are sometimes purely locally determined by poor drainage and cycles of growth of the plant life of the bog and collapse of the morass with its water content, built up from time to time above the surrounding surface. But the renewals of growth after periods of drying out of the peat surface are commonest in Ireland around 1500 and 800 BC and AD 500. In Sweden regrowth or 'recurrence surfaces' are marked about 1200 and 600 BC and AD 400 and 1200.

EFFECTS ON EUROPEAN LAKE SETTLEMENTS AND MINING IN THE MOUNTAINS

In central Europe distinctive settlements had been built on piles in the edges of the lakes in the warmest post-glacial times between about 4000 and 2400 BC or after, for instance at the Burgäschisee and Thayngen in Switzerland and Federsee in the Alpenvorland. These sites seem to have been abandoned later after catastrophic flooding episodes. C. E. P. Brooks wrote that some of these floods may have been connected with the 'great eruption of Bronze Age peoples from the Hungarian plain, which probably occurred soon after 1300 BC, and carried the Phrygians into Asia'[3] (actually Asia Minor). This was about the same time as the Hittites were abandoning the Anatolian plateau. There was also an infiltration of peoples from the north into Italy, who doubtless formed part of the ancestry of the Etruscans and of the first Romans. The lakes had a low phase, and the European peat-bogs seem to have dried out, in a warmer interval broadly around 1000 BC. New lake settlements were built after that and farming activity was resumed even above the Alpine forest limit. But the whole of this renewal collapsed in the wetter, colder climate which ruled after 800 BC. Towards, or about, 500 BC the great Lake Constance (Bodensee) rose rapidly by about 10 m, and all the lake settlements succumbed to a new disaster. Within the whole Alpine region population seems to have fallen to a minimum, largely confined to the warmest valleys.

As noted already by H. Gams in 1937 in his history of the Alpine forests,[4] the deterioration of climate at the close of the Halstatt time and of the wealthy society built up around the salt-mines and trade of that area near Salzburg around 800–700 BC seems to have been catastrophic. In about the same centuries when the lakes flooded the surrounding settlements, the glaciers were advancing and brought to an end the previously flourishing high-level mining – for instance, for gold in the Hohe Tauern – and stopped the traffic over the Alpine passes. The upper limit of the forests fell sharply, and their composition altered. The oaks and other broad-leafed trees lost a great deal of ground to the firs and pines, and, as we now know, the period after 1200 BC was one of the main stages in the infiltration of spruce, coming from the southeast.

EFFECTS IN NORTHERN LANDS

Also in both north and south Norway the glaciers were advancing, and by 800–700 BC they reached positions almost as far forward as in the worst period of recent centuries. Similarly, this was one of the main periods of advance of the spruce from the east into Scandinavia, the time when it became general in southern Finland and, crossing central Sweden, migrated into Hedmark in eastern Norway. It is generally thought that this is the period when the legend of *Ragnarök* originated, the twilight of the northern gods, presumably a folk tale of the end of a former way of life in the north. We may reasonably guess, as the great Swedish meteorologist Tor Bergeron believed, that this was when the events recorded in writing long afterwards by Snorri Sturluson (in his *Edda* poem written about AD 1220) occurred – the dreadful *Fimbulvinter:* 'the snow drives from all quarters with a biting wind; three such winters follow one another and there is no summer in between'. We cannot exclude the possibility, however, that the events thus remembered accompanied the earlier marked cooling of climates, of which we have seen evidence elsewhere, in the previous millennium. If so, and if the account as quoted from Snorri were taken literally, it is intriguing to notice that such might well have been the effects in the north of the exceptionally dense veil of volcanic material from the huge eruption of Santorin in the Aegean about 1450 BC, which overwhelmed Minoan Crete and the Aegean islands associated with it. The quantity of rock blown up into the atmosphere, much of it as submicroscopic particles , has been estimated as five or more times as much as that produced by the eruption of Krakatau in AD 1883. The latter was the greatest eruption within the last hundred years, and like other comparable eruptions in recent centuries it seems to have lowered the global temperature by about 0.5 °C for a year or two. The effect is likely to have been several times greater in high latitudes, and would doubtless be greater still with a substantially greater eruption. Snorri's account ends with a great conflagration, perhaps horrifying forest fires, burning the dried out remains of the more warmth-loving trees which had died at their former northern limit.

The last-named suggestion is one which ties in interestingly with the conclusion from pollen analysis studies, notably by Dr Harvey Nichols in Boulder, Colorado, that have surveyed the history of the forests in the Canadian north. It seems that all across Canada around 1500 BC the forest rapidly, and finally, retreated 200–400 km from the northern limit that it had achieved from 4000 BC onwards. The withdrawal was accompanied by extremely widespread forest fires, presumably started by lightning strikes on the dead wood, and within about a century the whole zone had been converted to tundra. In that Arctic continental region it is concluded that this and the further cooling stage around 500 BC altogether lowered the prevailing summer temperatures by as much as 3–4 °C.

EFFECTS IN THE EASTERN MEDITERRANEAN AND HITHER ASIA

The stage in these changes of the world scene reached around 1200 BC found the lands around the eastern Mediterranean and in the Near East in a state of general disturbance and in some areas of turmoil. And as early as about 1500 BC the Aryan peoples had poured out from Iran to settle in northwest India. The Minoan civilization in Crete had been overwhelmed by, or had suffered severely from, the enormous volcanic eruption of Santorin on the island of Thira in the Aegean a couple of centuries previously – the

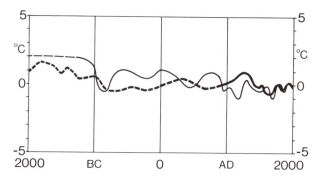

Fig. 55 Temperatures prevailing since 2000 BC in England (bold line) and China (thin line) compared. (Adapted from a diagram by Chu Ko-chen, using Manley's derivation of temperatures in England.)

date is usually taken as about 1470–1450 BC – and had been effectively replaced by the empire based on Mycenae in the southern part of mainland Greece. About 1200 BC this society collapsed, and a century or more later the Dorian tribes were able to move into Greece from the north, into a virtually empty countryside. Greece seems to have continued only sparsely populated until 850 BC. Similarly, around 1200 BC the Hittites seem to have largely abandoned their former homeland on the plateau of Anatolia, in modern Turkey. Egypt found itself threatened by invasion of peoples from the northeast and bands of Hittite and Syrian raiders as well as other migrants coming from Libya. All these migrations point to drought in the lands of origin as the probable cause: this view was ably put by the American professor of archaeology, Rhys Carpenter, in the 1960s and has since been strongly supported by a variety of meteorological and palaeoclimatic researches. In the case of Crete, we have the report of Herodotus that after the Trojan War the island was so beset by famine and pestilence that it became virtually uninhabited – conditions which certainly point to drought.

In a comparison between the late period of old Babylon between 1800 and 1650 BC with the later Babylonia of 600–400 BC J. Neumann, of the Hebrew University of Jerusalem, found that the mean date of the beginning of the barley harvest had become later by more than a month, shifting from late March to early May.

Agriculture and stock rearing, and the use of metals, had begun to spread from the south into the forest zone of European Russia some time after 2000 BC, but the Neolithic culture continued around the upper Dniepr and the Volga and survived into the first millennium AD in the extreme north. There were, however, cultural changes in the Volga region around 1500–1200 BC and in western Siberia along the Yenesei and Ob rivers between 1200 and 700 BC.

EFFECTS IN CHINA

Also in China, where the chronology probably differed most from Europe or North America (see fig. 55), between about 1100 and 800 BC the warmth of the most genial post-glacial times came to an end, never to be fully restored. As early as 1500 BC the ancient Chinese Yang-shao and Lung-shan cultures had been overwhelmed by a horse-riding people, invading from central Asia and presumably signalling an early stage of increasing difficulty there (probably aridity increasing or colder winters or both). The change now, a few centuries later, was certainly accompanied by enhanced variability, which included great droughts. The bamboo retreated from northern parts of its domain, the dates of the rice and fruit harvests became later and freezings of the Han river in central China (33 °N) were recorded: a climate similar to the present one. According to Chu Ko-chen, there are indications that the climate became once more slightly warmer than today after about 800 BC and still more after 500 BC, making it possible to grow two crops of millet a year in the southern part of Shantung province (36 °N) in eastern China, though a colder regime returned by around 200 BC. Much of this period was, however, a time of confusion in China known as the Warring States Period (from 480 to 222 BC). It may well be that the times of colder and warmer conditions within that millennium in China are not yet precisely enough determined.

COMPARISONS WITH LOW AND HIGH LATITUDES

Also, in the lower latitudes, at least in Africa and Hither Asia, there seem to have been several centuries of colder conditions around the middle of the last millennium before Christ. Among the evidence of this are a number of references in the Old Testament books of the Bible, as, for instance, the remark of the prophet Jeremiah about 600 BC (chapter 18): 'Therefore, thus says the Lord . . . What, shall the snows of Lebanon melt from those wild peaks, shall they be dried up at their sources those icy torrents that flow down from it? . . . but my people have forgotten me.'[5] The context suggests that such things were unthinkable. There has been more doubt as to whether this period was also cold in Central America, but archaeological studies of cultural development in the Valley of Mexico indicate that there were three distinct periods of maximum human settlement there in the period before the Europeans arrived: approximately 500–100 BC, more briefly around AD 800 and then during three or four centuries of the Aztec period, particularly towards its end, i.e. in the century up to 1520. These periods coincide with the main periods of colder climate in Europe and much of North America, notably the United States east of the Rockies and the central Canadian Arctic; they were probably moister than most times in Mexico, if the summer rains did not move so far north. Indeed, there may be evidence of greater moisture at these times in that the settlements were able to spread out increasingly from the lakeshore plain to the foothills zone, where the soil is thin and the climate today is too dry for most kinds of agriculture. The period about 600–300 BC saw the first such population expansion. It was very rapid. The total numbers are estimated to have grown five- to nine-fold.

Taking an overall view of the changes of climate established during the millennia with which this chapter is concerned, it is clear that they have to do with the end of the warmest post-glacial times. It would be presumptuous to suppose that human activities had much to do with it. There were several abrupt cooling stages and sometimes the recoveries, though these were either shorter-lived or incomplete, were quite rapid too. The difference of mean temperature between the warmest and coldest individual centuries between 3500 and 500 BC in central Europe may well have amounted to 2 °C or perhaps a little more. The changes were greater than this in high latitudes, to judge from the Canadian and Scandinavian evidence, as also in the case of the winters in China. Some cooling is also evident in Babylonia (Mesopotamia) and northern India, though in subtropical and lower latitudes it is increasing drought that is the most obvious feature and gravely affected the civilizations there. With so many and great changes in the environment, is it any wonder that there were effects on the human societies of the time, or that in later Classical times there were memories enshrined in talk of a former Golden Age?

It is also clear that the severity of different stages in the climatic development differed in different parts of the world. On the whole, it seems that in North America the cooling trend in the second millennium BC, from about 1500 BC onwards, was the sharpest – though perhaps not in the far north where it was not many centuries since the warmest times had begun. In Europe and in most of the rest of the hemisphere, some shock around 1200 BC seems to have been more important; and it was through the middle of the next millennium that the sharpest change and the severest weather came.

In keeping with the late opening up of the northeast Canadian Arctic and northern Greenland, the first evidence of Eskimo arrivals there is dated around 2500 BC, whereas various peoples had moved through and settled in Alaska many thousands of years earlier. Cultural changes followed at times which correspond closely to the climatic chronology which we have recognized elsewhere. From 1500 to about 1100 BC activity declined in northern Greenland, and there was a general southward movement of the early peoples traceable also along the shores of Hudson Bay and Labrador. After a certain recovery tendency around 1100–700 BC, the north Greenland settlements were again abandoned and the general southward movement was repeated.

DETAILS FROM NORTHWEST EUROPE

It is worth while finally to look further at these culminating stages of the development in northwest Europe, where there is evidence of some climatically important details.

Already by 2000 BC the forest had been retreating from the exposed coasts of northwest Scotland and from most of the highest places which it had reached in Scotland and northern England. Farther south in Ireland, Wales and Cornwall woodland still extended to the Atlantic shores and higher on the hills than any present woods, until the Bronze Age. In southern England human activity had been considerable, and was disturbing the natural vegetation cover, particularly near the chalk uplands, ever since Neolithic times as early as the third millennium BC. Oak and hazel were abundant at the foot of the hills and on the slopes, and were used for firewood. The Neolithic farmers grew crops and raised sheep and themselves lived on the plateau of the chalk downs, which was already grassland. In the second millennium BC the tops were abandoned except for grazing and burials and for their convenience as travel routes: this may well mean that in the drier periods of the Bronze Age the springs were lower and water supply was difficult on the higher ground. The open areas on the chalk hills, which had been cleared by man's activities and then abandoned, or partly abandoned, became colonized by beeches, the last of the big trees to immigrate into Britain.

That stage was followed between about 800 and 400 BC by a period of such unmatched wetness in the west that close dating by multiple radiocarbon tests shows that in the great bog at Tregaron in west Wales nearly one metre thickness of peat was added during these four centuries. (This is as much as the entire thickness added in the succeeding two thousand years.) In eastern England such indications as we have are rather of dryness in at least the earlier part of this time. Farming extended to the flat lowlands of Holderness near the east coast of Yorkshire around 1000 BC and after, though the region ultimately became marshy again. The situation in which an abandoned boat, dated about 750 BC, was found near Brigg in Lincolnshire suggests that the flow of the eastern rivers may have been sluggish as late as the eighth century BC and possibly after. These observations point to an unequalled predominance of westerly winds in the centuries concerned. This was the time when the glaciers in the Alps and in Norway were advancing, perhaps more rapidly than at any other stage since the warmest post-glacial times; they came forward from their minimum extent to produce moraines almost as far forward as those of the coldest recent centuries (and in a few places overstepping this later advance).

Meteorological reconstruction of the probable characteristics of the wind circulation, from what is known of the thermal pattern over the northern hemisphere about 500 BC, suggests that a cooling Arctic had pushed the cyclonic activity south over northern Europe. This probably meant a strong development of cold westerlies across the Atlantic from Canada and Greenland to Britain and central Europe in winter and prevalence of cyclonic northwesterly winds in summer over the British Isles and across Europe to the Mediterranean. An impression, mentioned in some of the earlier literature, that sea level was high on the coast of eastern England around this time is almost certainly mistaken, since it was a time of sharp climatic cooling and glacier growth in most parts of the world. The impression may have been caused by what was more likely the mark of exceptional high waters and flooding from North Sea storm surges, when the winds veered to the northwest and north behind some of the most intense cyclonic depressions.

That the period was one which included some outstanding storminess is indicated by activity of the sand-dunes around the coasts of northwest Europe from south Wales to Denmark. Certain spits of land on the east coast of Scotland in the Firth of Forth appear to have formed as sand-dunes, or sandbanks, in northerly storms and have been dated to about 500 BC. A slight lowering of the sea level due to the build-up of the glaciers may have contributed by exposing greater expanses of sand in the estuaries and along the shores. Ancient harbour works, dating from about 500 BC and now generally submerged, at Naples and in the Adriatic suggest a mean sea level about one metre below present. Although the Mediterranean region has too much tectonic instability to allow much confidence, this figure is a reasonable one in relation to what we know of the climatic change and its duration. It suggests that the water level may have dropped by 2 m or rather more from 2000 to 500 BC. What does seem certain is that there was a tendency for world sea level to rise progressively during the time of the Roman Empire, finally reaching a high stand around AD 400 comparable with, or slightly above, present.

The Dutch coast seems to have lain generally about 50 km east of its present position between 3000 and 2000 BC, a finding that is quite consistent with the supposed highest world sea level around that time. The coastline was transferred westwards, and the North Sea receded, with new coastal sand-dune barriers beginning to form in the second millennium BC; but the main shift came in a series of blowing sand incidents, identified by sand layers radiocarbon dated by tests on incorporated objects, between about 600 and 150 BC. Also in Denmark, study of a peat-bog in Jutland has shown the intrusion of layers of blown sand between about 600 BC and the time of Christ.

Landscape studies inland in the south of England also indicate that the cold – and in the west wet – climate regime lingered on. Wooden track-ways were laid at some time after 900 BC across the Somerset levels, apparently in an effort to keep open accustomed routes across the area when it was becoming increasingly marshy. And towards the end of that time, around 350 BC, boats were used instead. In the next century, around 250 BC, lake villages were built on piles at Glastonbury and Meare, perhaps taking advantage of the value for defence of such a position.

About 500 BC the climate became much wetter than before in eastern districts of England as well; so that all parts of England, Wales and Ireland were then affected by the notably wet regime. This must imply that the winds were no longer so very predominantly from the west and that cyclonic activity frequently passed near, or over, the southern parts of the British Isles, extending farther south and east than it had done in the immediately preceding centuries. The regime apparently came to resemble more the one which we shall see affected Europe in the fifteenth century AD. And the ancient ridge routes which had already been established across England – the Cotswold ridgeway from near Bristol to Lincoln, the Icknield Way from near Stonehenge to Norfolk, and the route which much later came to be known as the Pilgrims' Way from Winchester to Canterbury – doubtless acquired an extra merit in avoiding not only the thicker forest but also the often swampy lowland areas.

THE TIME OF BIRTH OF GREAT RELIGIONS

Before we leave this period of history, it may be of interest to notice that it was in the last millennium with which this chapter has been concerned that some of the great religions and philosophies of life and the world evolved. This should not be taken to suggest that the climatic events of the time in any way affected their founders' thinking or that they were even conscious of it. Any individual living in this or that region must in any case experience in at least some years weather that is out of character for the times, and it may even happen that his whole life is lived in a region whose experience is untypical. We have already seen in the case of the rise of Harappa and the Indus civilization how the prolonged experience of a whole region in central Asia to the north seems to have differed from that of northwest India at that time. The incidence of climatic, like cultural, changes usually has a definable geographical pattern. When there is some quite general, even global, climatic trend, there are usually some regions that have a contrary experience. Nevertheless, the impact in the monsoon regions and near the arid fringe of times of globally increased variability of the weather from year to year, like the impact of periods of very extensive climatic anomaly such as a drought affecting much of one latitude zone, may provide conditions favourable to the spread of a new religion by its enthusiastic missionaries and/or armed supporters, perhaps most of all through the breakdown of the old way of life and its ordered customs. There is evidence of such a breakdown, through drought, in parts of the Mediterranean world about the time of the spread of Islam in the first millennium after Christ.

The climatic record in China in the last millennium BC certainly seems to differ materially from that in Europe (fig. 55) and probably most other European lands, where there were times of great climatic stress, perhaps most of all around 1200–800 and 600–200 BC, which probably had to do with some great folk migrations in those times. These movements and the resulting ferment probably affected all the peoples of Europe and Asia either directly or indirectly. Buddha (563–483 BC) and Confucius (551–479 BC) each offered solutions to the universal problem of suffering in human experience. Confucius taught that all men are brothers and should sustain each other. The Buddha commended meditation to seek Nirvana, ultimately to reach a state of reconciliation to the terms of our existence and a serene view of pain and suffering. The period from about 600 to 536 BC saw the captivity of the Jews in exile in Babylon, accompanied by renewal of their spiritual leadership and exhortations to get back to the laws that should govern life in the community, which had been laid down seven centuries earlier during another migration. And in Greece the middle and later centuries of the millennium were the times of the great philosophers, whose teachings influenced Christianity and all later European thought, leading on to the development of modern science and democratic debate.

9

ROMAN TIMES AND AFTER

THE MEDITERRANEAN WORLD IN ROMAN TIMES

Rome was reputedly founded in 753 BC. About the same date Greeks were already establishing colonies in Sicily and southern Italy and by 600 BC at Marseilles. Until about the second century before Christ Rome seems to have been outshone by the seafaring, trading, agricultural, and colonizing activities of the Phoenicians, Carthaginians and Greeks. There was a Phoenician circumnavigation of Africa about 600 BC. And the first Roman historians, before 200 BC, wrote in Greek. The peoples of the eastern and southern Mediterranean were perhaps helped by the colder climatic regime of those centuries, which brought more winter rains and therewith a period of greater fertility to Greece and to the northern fringe of Africa, making possible the Carthaginian and later Roman croplands there. (It seems also likely that the water table was still higher than now, and the oases in the deserts more extensive, as a legacy of the moister climates before 3000 BC.) Rhys Carpenter has pointed out too that the Greeks of early classical times had gone over to warmer clothing and pitched gable roofs on their houses in contrast to the flat roofs and semi-nudity of the earlier cultures of Mycenae and Minoan Crete. Historical reports (e.g. in Livy) tell us of at least a few severe winters in Rome in those times, when the River Tiber froze and snow lay for many days, and that beech trees grew there around 300 BC, whereas by the time of Pliny in the first century AD the climate seemed to be too warm for them: the beech was regarded by the Romans in Pliny's day as a mountain tree.

Around 310–300 BC an exploration of the coasts of western Europe by Pytheas was extended by way of the Hebrides to northern Scotland, where heavy seas '80 cubits high' were encountered, and six days sail farther north to an island at latitude 66½ °N called Thule, which seems identifiable as Iceland, especially since a strange substance, described as consisting of 'earth and air suspended', presumably volcanic pumice, was observed. One day's sail beyond this island the sea was found to be frozen. An unexplained part of the report is the mention that the island had inhabitants, who lived on the wild berries and honey and some evidently wild grass grain, which they threshed indoors in barns because of the continual rains. Who were these folk? Can it be that earlier seafarers had come from Europe in the better climate of earlier times? Or were they Eskimos, and did Eskimos perhaps even reach the Orkney islands at some time in the last millennium BC, as the design of the small, igloo-like, partly subterranean stone houses at Skara Brae might suggest?

As late as between about 120 and 114 BC, according to the account given by Strabo, there was a great storm, or series of storms, in the North Sea with sea floods, which pushed back the coast of Denmark and Germany. This, so-called, Cymbrian (or Kymbrian) flood caused a southward migration of the Celtic and Teutonic peoples who had been living in the areas affected. Thus, over about a thousand years to this time, while a rather sharp cooling of the Arctic was going on, of which there is positive evidence at least in Canada, Greenland and Scandinavia, and while the mountain glaciers were advancing in all the regions investigated, there were population movements southwards from northern and central Europe. But after this, for a few hundred years there is little mention of major climatic disturbances or dislocations of society of that order.

Julius Caesar's long wait in the summer of 54 BC while persistent west to northwest winds delayed his expedition across the Channel to England was certainly nothing that would be unusual today. Similar conditions had persisted in the summer of 55 BC, and William of Normandy was similarly held up in AD 1066 until a favourable wind came in early October.

Roman horticultural writers in Pliny's time, and in the previous century, drew attention to the fact that the vine and the olive could then be cultivated farther north in Italy than had been the custom in earlier centuries. This agrees with the general indications of various kinds of fossil or proxy climatic data that there was a continued tendency towards recovery of warmth in Europe through Roman times, and of increasing dryness, until about AD 400. A gradual, global warming up to AD 400 would, of course, be consistent with the evidence of rising sea level referred to in the last chapter (p. 153). This background to the heyday of Roman rule may account for the growth of a widespread supposition in modern times that the continual ups and downs of the weather were never more than a minor nuisance and have had nothing to do with the course of history either then or since.

We see, in particular, that despite the well-known political sources of unrest with which the Roman empire was concerned, Christ – unlike the Buddha and Confucius – seems to have been born in a relatively benign period as regards the tendency of the climate. There were evidently considerable similarities to the climate of our own times, except for the continuance of a somewhat moister regime in north Africa and the Near East. This, no doubt aided for a long while by residual soil moisture and

Fig. 56 Remains of Petra, a thriving city between 300 BC and AD 100, in the Jordan desert. (Photograph published by R. G. Veryard in his article 'The changing climate', *Discovery*, vol. 23, p. 8, 1962.)

vegetation from an earlier time, made possible the extensive African crop-lands, the granary of the Roman empire, and the thriving settlements at places like Petra (fig. 56) that have since been conquered by the Syrian and Jordanian desert.

There is a weather diary kept by Ptolemy (Claudius Ptolemaeus) of Alexandria in about AD 120. It shows some remarkable differences from today's climate there in the occurrence of rain in every month of the year except August, of thunder in all the summer months (as well as at other seasons), and in that days of great heat were commonest in July and August: today the continual north and northwest winds off the sea in those months lower the temperature.

Another sign of the easier climate of Roman times may be seen in the building between AD 101 and 106 of a bridge with many stone piers across the River Danube at the Iron Gate (between modern Yugoslavia and the Transylvanian highlands in Rumania). The bridge (fig. 57) was designed by Appolodorus of Damascus for the Emperor Trajan for passage of the Roman armies and administration into Dacia. It stood for about 170 years, at a point where in any recent century such a construction would surely have been carried away by ice in some heavier than usual ice winter. In the end it is said to have been destroyed by the Dacians after Roman rule had been withdrawn.

The continued northward spread of vine cultivation is registered by an edict of the Emperor Domitian late in the first century AD prohibiting vineyards in the western and northern provinces of the empire, beyond the Alps. This edict was revoked by Probus about AD 280, and it was in fact the Romans who introduced vine growing to England and Germany. Around AD 300 evidence of the importation of wine into Britain ceases, and it seems at least possible that the province had become self-supporting in wine. This is not to suggest that the entire northward march of the vine from southern Italy over a few centuries was due to a shift of climate of that magnitude, the case was rather that easier times and fewer difficulties experienced by the cultivators in Italy led to a realization that the vine might in fact be cultivated a good deal farther north.

LINKS WITH THE EAST: TIMES OF TRADE AND OF MIGRATIONS

For centuries during Roman times, from about 150 BC until AD 300 or a few decades later, caravans of camels used the Great Silk Road across Asia to trade luxuries from China. But by the fourth century AD, as we know from changes of level of the Caspian Sea and studies of the intermittent rivers and lakes and abandoned settlements in Sinkiang and central Asia, drought developed on such a scale as to stop the traffic along this route (fig. 58). Other serious stages of the drought occurred between about AD 300 and 800, and particularly around those dates, as can be established from old shorelines of the inland seas and old harbour installations which indicate a very low level of the Caspian culminating around those times. The suggestion, made by Ellsworth Huntington in his book *The Pulse of*

Fig. 57a The Roman bridge, built AD 101–6, across the River Danube at the Iron Gate, a wooden structure on stone piers, as illustrated on Trajan's Column in Rome. (Reproduced by courtesy of the Trustees of the British Museum.)

Fig. 57b The River Danube at the Iron Gate in the 1960s. (Picture by courtesy of the Yugoslav National Tourist Office, London.)

160

Fig. 58 Carved niches in the cliffs of Jiaohe in the Turpan Depression in the Tien Shan mountains in central Asia (Sinkiang), relics – perhaps a former Buddhist shrine – of a city on the Great Silk Road which flourished from about AD 200 and was abandoned in the centuries before AD 1000. (Photograph by Bruce Dale, copyright National Geographic Society, Washington, DC, kindly supplied for this book.)

Asia in 1907, that it was the drying up of pastures used by the nomads in central Asia that set off a chain reaction of barbarian tribes and un-settled peoples migrating westwards into Europe, where they ultimately undermined the Roman empire, looks a sensible one in the light of this evidence.

This time of migrations of peoples – the *Völkerwanderungen* time, as it is known in central Europe – during the long decline of the Roman empire is characterized, like that in the last millennium before Christ, by migrations predominantly in one direction. But, whereas in the previous case the direction was from north to south, this time it was from east to west. Both cases suggest that some global cause was at work. In the former case, it seems clear that there was increased difficulty in maintaining life in the Arctic and in regions where the climate was becoming more disturbed because of a spreading out of Arctic cold air. This time the trouble was plainly not in the north but in the east, in the heart of the Eurasian continent. And, as the winters are always harsh there, the critical change was more likely to have to do with drought – of which we have, in fact, unmistakable evidence.

Fig. 59 displays what we can tell of the variations of climate in Europe over nearly two thousand years, from the fourth century BC to AD 1300, from documentary reports. The variations of frequency of floods and wet years in Italy seem broadly to parallel indications from elsewhere along the northern fringe of the Mediterranean region as well as the variations which we have noted affecting the Caspian Sea and moisture in central Asia. And there is no doubt that the drier periods caused great difficulties in those regions.

CRITICAL DISEASE EPIDEMICS

There was a history of plague epidemics of various kinds in the later Roman times, which may have been not unconnected with increasing warmth and general dryness. In AD 144–6 and 171–4 the population in parts of Egypt was reduced by a third, and in 166 a plague brought from Macedonia reached Rome and spread to much of the empire. There were worse epidemics in AD 251–68 in Italy and in Africa, with deaths in Rome reported to have reached five thousand a day. Worst of all was the bubonic plague which seems to have come from Egypt or the upper Nile valley, or possibly Ethiopia, in 542–3, in the Emperor Justinian's reign, and spread far and wide over the Roman world and beyond, reaching Persia and the Indies and the ports of Europe. As with the Black Death in the Middle Ages, there was terrible mortality with the first outbreak, but other, less severe or less widespread recurrences were to follow. Half the population of the Byzantine empire and of Europe is believed to have died of it between 542 and 565, perhaps one hundred million deaths in all. It may have been late recurrences of this same pestilence that reached Ireland in 664 and 682, when it had effectively died out in its earlier centres. If climate was indeed involved, the significance of the times of drought in its initial spread was doubtless in connection with the difficulty of hygiene under such conditions. The incidence of the disease in the Middle Ages seems to have been worst where it was concentrated in locally warm and moist habitats in cities and in other concentrations of population and contacts along the routes of travel. This was probably also the case with Justinian's plague.

SEA LEVEL AND COASTAL CHANGES IN NORTHERN EUROPE

The slow rise of world sea level, amounting in all probably to one metre or less, that seems to have been going on over the warmer centuries in Roman times, not only submerged the earlier harbour installations in the Mediterranean but by AD 400 produced a notable incursion of the sea

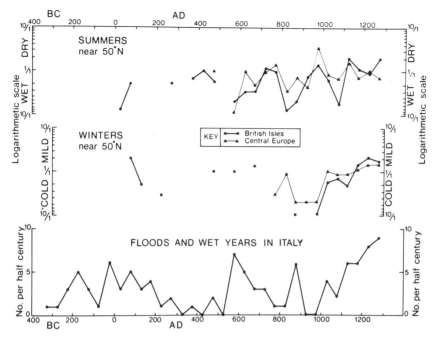

Fig. 59 A survey of the climatic record in Europe as documented between 400 BC and the Middle Ages. Frequencies of reports of wet or dry summers and cold or mild winters in the British Isles and Germany, and of wet years in Italy, by half centuries. The prevailing character of the reports of the summers and winters is indicated by ratios.

from the Wash into the English fenland and maintained estuaries and inlets that were navigable by small craft on the continental shore of the North Sea from Flanders to Jutland (fig. 60). This is a circumstance which may have helped the Anglian and Saxon migrants launching out across the North Sea from their previous continental homelands. The transgression of the sea over the previous coastline of Flanders and the Netherlands between about AD 250 and 275 had caused a depopulation of the coastal plain there.[1] The existence of pre-Norman conquest salterns – saltpans, or 'sandacres', over which the tide washed and from which the salt-saturated sand was then taken – outside the later sea-dykes in the English fenland, on the Lincolnshire coast, may or may not point to a period of slightly lowered sea level between the late Roman and the medieval high water periods. There is other evidence to suggest this between the seventh and tenth centuries. But many later saltpans are known in the area, also on the sea-banks, standing up to 3 m above the present mean sea level. Investigations in the Netherlands have established that the previous activity of blowing sand and shifting dunes on that coast was followed after about

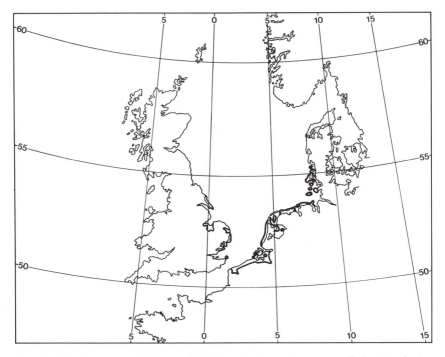

Fig. 60 Sketch map to indicate (bold lines) where the coasts of the North Sea around AD 400 and 1000–1100 differed significantly from today's coasts (thin lines). If we are looking for climatic causes, or triggers, of historical events during the first millennium AD, it is clear that drowning of coastlands by the sea affected only rather small, localized regions, whereas the droughts that prevailed in the continental heartland from Kazakhstan or the Ukraine eastwards across Asia affected a huge, if sparsely populated, area. And it may well have been a remote consequence of the loss of pastures in Asia, caused by those droughts, which brought by a chain reaction Slavonic tribes wandering westwards across the European plain and establishing settlements as far west as the border of Denmark.

100 BC by eleven centuries of stabilization, with forest ultimately colonizing the dunes. This must mean that, among other changes, the water table was higher than before and storminess seems to have been reduced. There were only minor and localized intrusions of blown sand in the period of colder climate, which we shall describe, in the middle and later part of the first millennium AD. The coastal forest was cut down by man in the Middle Ages, and it seems certain that this allowed the effects of the subsequent period of renewed dune activity to spread farther eastwards than would otherwise have been the case. Close study by Sylvia Hallam over many years of the history of human settlement near the coast of the Wash in eastern England (*Antiquity*, vol. 35, pp. 152–6, 1961) has indicated that sea level was rising from some centuries before up to a maximum attained

in the last century BC. There was then some recession of the water until about AD 200, followed by a major high stand and incursion of the sea around AD 300–400. Sea level was again rather lower in the seventh and eighth centuries and possibly later, but seems to have been again high in the late thirteenth to fifteenth centuries. The present writer's opinion is that the impression of a high level of the sea as late as the fifteenth century may in reality owe a good deal to storm surges – i.e. to recurrent sea floods as storminess increased.

THE CLIMATIC SEQUENCE IN EUROPE THROUGH THE FIRST MILLENNIUM AD

Let us now return to consider the climatic sequence over central and the more northern parts of Europe during the first millennium AD. Variations in the amount of documentary information are allowed for in fig. 59 by using ratios of the frequencies of reports of dry and wet summers, mild and cold winters. Even so, the numbers of reports are generally too low to provide reliable ratios, and for some periods no estimates could be made. There is, however, reason for some confidence deriving from the rather close parallelism of the curves for the British Isles and for central Europe. Again, too, general support can be found in pollen studies and glacier histories, etc. An increasingly warm, dry tendency of the summer climate is indicated up to AD 400; there were some cold winters, but these seem to have been insufficiently severe to have any lasting effects. There are hints in the diagram of some periods of generally rather colder and more disturbed climate later in the millennium, particularly in the sixth century and at certain times between about AD 750 and 900, accompanied in the sixth and ninth centuries by wetness in at least northern parts of the Mediterranean and in northern, western and central Europe also. It is from the sixth and ninth centuries also that most of the storms and sea floods around the coasts of the North Sea in this millennium, of which reports still survive, are dated. And in, or about, AD 520 it is alleged that a whole county, Cantref y Gwaelod, was lost on the west coast of Wales in Cardigan Bay, when the sea breached a dyke in a storm. Another indication of a cooler and more disturbed climate around this time may be seen in the widespread abandonment of land and cultivation in the relatively low-lying Jaeren coastal region of southwest Norway.[2] A few very remarkable winters were reported in this time. That in 763–4 is the earliest winter to be documented from many parts of Europe, with enormous snowfalls and great losses of the olive and fig trees in southern Europe: there was ice on the sea in the Dardanelles. Another such winter in 859–60 produced ice strong enough to bear laden waggons on the edge of the Adriatic near Venice. Finally, in 1010–11 there was ice not only on the Bosphorus but even on the Nile.

It used to be thought that these extreme winters, like the run of disastrously wet years in the 580s in Europe, should be considered as isolated events and that there was no development of a significantly colder climatic regime at any time in the millennium which we are considering. Recent work in the Alps, most notably by Röthlisberger and Schneebeli in the Geographical Institute of the University of Zurich,[3] and in Norway and north Sweden by Wibjörn Karlén, suggests that this view needs revision. Radiocarbon dating of the old moraines marking former glacier termini in the valley bottom in Val de Bagnes, in southwestern Switzerland, revealed that positions reached by the glaciers coming down from the heights on either side around AD 600–700 and perhaps again as late as 850 were as far forward as those registered in the well-known Little Ice Age period between 1550 and 1850.[4] These glaciers clearly cut an old Roman route across the mountains from Italy which passed down through this valley. Moreover, tree ring studies on larches which grew near the upper tree line near Zermatt (fig. 61) indicate what appears to be a gradual build-up of warmth of the climate with only small variations from year to year in the late AD 300s, followed by rather sharp variations between about 400 and 415 and a marked cold period thereafter. So, if this dating is reliable, the Roman administration faced further difficulties besides the growing threat of barbarian migrations at the time of the collapse of the empire in the west.

The same period has been established as one of glacier advances and upper tree line depression in other parts of the Alps. Glaciers in north Norway also advanced to prominent maxima between AD 450 and 850, and some studies[5] have found indications of glacier maxima in both Baffin Island and Alaska in this period. We have mentioned the indications of storminess on the seas and coasts of northwest Europe in the sixth and ninth centuries. There was one other bout of storminess, between AD 400 and 440, which accounted for a quarter of all the sea floods known from that millennium, with coastal changes in the south of England and losses of life on the Dutch coast.

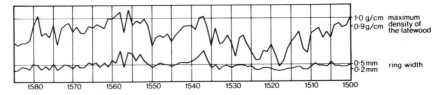

Fig. 61 Mean ring width and density of the late-summer wood in larches that grew near the upper tree line near Zermatt, Switzerland. (These measurements are essentially indicators of summer temperature.) The dating is not absolute but by radiocarbon test, and is counted in years before AD 1950 (i.e. 1550 on the scale is taken as AD 400). The turn towards colder years after about 400 is plainly indicated. (From measurements at the Swiss Forestry Research Institute, reported by F. Röthlisberger.)

166

From the point of view of meteorological research it is noteworthy that this cold climate regime which we recognize in middle latitudes in part of the first millennium AD seems to have had a somewhat different pattern, and followed a different course of development, notably in the subtropics and also in parts of the high latitudes, from the well-known one in the latest millennium. This is obvious in the isotope record from northwest Greenland seen in fig. 36 (p. 93), where the colder regime of the AD 400s appears as a very minor development, though somewhat prolonged, and is followed by relative warmth as early as the 600s, which continued and built up to a maximum in the twelfth century AD. Warmth in that quarter, however, at a time when most other places in northern latitudes were cold may simply be attributed to recurrent anticyclones near and over northern Greenland, repeatedly giving southerly winds over the whole of western Greenland and the regions around the Davis Strait. Moreover, the cold regimes around AD 400–900 and 1400–1900 both seem to show some differences from the regime in the middle of the last millennium before Christ, with its very persistent wetness (and presumed westerly winds) in western Europe. There is no doubt, however, that colder climates developed, at least in middle latitudes of the northern hemisphere, in the middle centuries of each of the last three millennia. Indeed, some degree of parallelism with the developments around 3500 and 1500 BC also may be recognized in the turn towards colder weather and glacier advances in middle latitudes: it may even be that a somewhat similar tendency has shown itself in the middle of every millennium in this part of the world.[6]

There is an interesting example of how the climatic developments around the end of the Roman empire in the west may have affected local history in East Anglia at West Stow, near Bury St Edmunds, Suffolk. A Saxon village was established, evidently with Roman permission, on the edge of what is now marshy land in a shallow valley about AD 400 in the last years of Roman rule. Perhaps it was less marshy then in the apparently drier climate that had prevailed in this latitude, and farther south, in Europe for a century or two. At any rate, wheat, barley, oats, rye and flax were grown for a time. But in the seventh century the site was abandoned. It may be that better sites became available after the Romans left, but the timing of the abandonment suggests that the site may have become too marshy following the wet years of the sixth century and after, especially in the 580s and early 600s. The archaeologists' report suggests that the marsh was encroaching.

The early monastic institution at Glastonbury in Somerset in about this period drained the marshes there: the monks became regarded as the leading experts of the time in land drainage. There are hints of some preoccupation among the Saxon population of England in those times with draining marshland in river valleys, and the monks were given such land to drain elsewhere.

THE SEQUENCE IN THE MEDITERRANEAN AND FARTHER SOUTH

The character of the same millennium farther south, in the Mediterranean, north Africa and far to the east into Asia, is most marked by the periods of drought, which seem to have had two maxima, around AD 300–400 and 800. The Caspian Sea fell to low levels at these times. In Italy, and perhaps elsewhere in the northern Mediterranean, the driest periods were evidently separated (cf. fig. 59) by intrusion of the wet, cold influence affecting northern and central Europe around AD 600 and again in the later 800s. But over wide areas farther south and east the dryness seems to have persisted. This was when Ephesus, Antioch and Palmyra decayed: there and in southern Italy and Greece people were migrating to the coasts and leaving a depopulated hinterland. And in Arabia, places where agriculture had been carried on with the aid of elaborate irrigation works, which had survived earlier periods of desiccation, were abandoned around AD 600. According to Rhys Carpenter, the seventh century AD was the climax of this. So it appears that the rapid spread of Islam took place at a time when there were stresses due to widespread drought over the areas affected.

The evidence suggests that most parts of the northern hemisphere south of about 35 °N continued as warm as, or warmer than, before through these centuries. The yearly floods of the Nile, supplied by the summer monsoon rains over Ethiopia, were low; but the winter flow of the river, which depends on the rains near the equator, was high, as was the level of Lake Rudolf in eastern equatorial Africa. This probably means that the equatorial rains had a restricted seasonal migration north and south at that time. They seem to have supplied more water to equatorial Africa, and therefore to the White Nile, than in the years of drought in the Sahel and Ethiopia in the 1970s: possibly the seasonal migration of the rain system was even more restricted in the centuries which we are considering here than it was during the 1970s.

This, or something like it, was the setting in which a Christian kingdom, cut off from its cultural links, had the vigour to survive through these centuries in the Nile valley in Nubia, in what is northern Sudan today, and build churches adorned with art of Byzantine tradition, actually until the fourteenth century AD. It was only after that that these buildings were buried in sand.

On the other side of the Atlantic the Mayan civilization, whose temple-pyramids are now lost in the region of dense tropical rain-forest and warm, moist, enervating climate in the southern lowlands of the Yucatan peninsula, had its high period from about AD 300 to 800. Its realm spread from latitude 14 or 15 to 25 °N in Central America. Is it just a coincidence that the time of its fullest flowering coincides so precisely with the period of droughts in the zone from the Mediterranean to central Asia or was the climate also in those latitudes in Central America drier than before or since?

THE TROPICAL FOREST LANDS: CENTRAL AMERICA AND SOUTHEAST ASIA

Admittedly, there had been a long formative period in the Maya lands before the Classical period began, but after AD 800 the decline was rapid. Pollen studies in lake sediments and bogs on the edge of the highlands of Guatemala near 17 °N certainly show a quick change of the surrounding vegetation from grassland to deciduous forest about AD 850–900, but a previous change in the reverse direction is dated as early as about 900 BC. It is clear that the highly organized Mayan civilization developed just in that part of Central America where the effort to keep back the forest would nowadays be greatest. The beginnings seem to have been established in the drier, or better drained, highlands; but the civilization spread to the moist Guatemalan lowlands, where a network of drainage canals was constructed. Ultimately, the population, which rose to two or three million, may have grown too big to be supported in this way, or perhaps the effort to maintain the social organization to operate the irrigation and drainage system and keep back the forest became too much. Perhaps it could only be managed in a dry period. The interpretation is made difficult by strife among the peoples of the region about the causes of which we know nothing.

With evidence of so much human activity, the suggestion that climatic changes also played a significant part in the Mayan development and its final collapse has been little regarded in recent years. It has nevertheless been suggested from pollen analysis in the 1970s that in the early phases of the Mayans' civilization they noticed a natural change weakening the forest in the lowlands and realized that by burning they could at last conquer it and extend their agriculture to the lowlands in what are now Guatemala and El Salvador. And others have suggested on the basis of archaeological evidence that towards the end of the Classical Maya time, about 800, the climate in the valley of Mexico and in Yucatan became so dry that there was concern over water supply and soil moisture and the drought may have been prolonged to the point when it became necessary to abandon the driest regions. However that may be, it seems certain that soon afterwards Yucatan and the regions to the south in Guatemala and Honduras became so much wetter that the forest spread rapidly. And in the period of cultural decline towards AD 900, after, it is thought, revolts of the peasants against masters who worked them too hard, people continued to visit the ceremonial places and live in shacks among the crumbling buildings and changing landscape.

C. E. P. Brooks pointed long ago to the similarity of the Mayan sequence with the history in the next centuries of the Khmer empire, in southeast Asia between the Mekong river and the frontier of Thailand, whose capital. Angkor, at latitude 14 °N, was founded in AD 860 and flourished for four or five centuries before disappearing in ruins into the increasing jungle.[7] With such vegetation changes there is clearly a strong case for believing that some climatic shift was at work, but in spite of the increasing interest in climatic research in recent years we still lack knowledge of the details that might firmly resolve the background to the rise and fall of the civilizations in Central America and southeast Asia.

The difference of timing of the dry – or at least drier – periods in southern Mexico and Cambodia (Kampuchea) may reasonably be related to the difference of a few degrees of latitude and, presumably, some changes in the incidence of the southeast Asian monsoon. This history of the Khmer civilization in Indo-China, however, belongs to the period with which the next chapter deals.

10

THROUGH VIKING TIMES TO THE HIGH MIDDLE AGES

ASYMMETRY OF THE MEDIEVAL WARMTH OVER THE NORTHERN HEMISPHERE

As indicated in the last chapter, there seem to have been some regions of the world – particularly in low latitudes and in the Antarctic, possibly also around the north Pacific and in parts of the Arctic – where the rather greater warmth of the climate established around AD 300–400 continued, with variations but more or less unbroken, for several centuries longer and in some cases right through to AD 1000–1200. In Europe and much of North America, as well as in the European Arctic, there clearly was a break. But by the late tenth to twelfth centuries most of the world for which we have evidence seems to have been enjoying a renewal of warmth, which at times during those centuries may have approached the level of the warmest millennia of post-glacial times.

China and Japan evidently missed this warm phase. A warm period can be discerned in the historical records in those countries from about AD 650 to 850, more or less covering the time when Europe had its colder break. But in the eleventh and twelfth centuries the data collected by the late Dr Chu Kochen make it clear that the climate of China took a much colder turn, with frequent references to snow and ice in the winters and snows a month later in spring than in the present century. The plum trees were disappearing in north China; frosts killed the mandarin trees in the coastal province near Shanghai and the lychees in parts of the south. In Japan the long records of the dates of the cherry blossom in the royal gardens at Kyoto indicate on average the earliest springs in the ninth century and the latest springs of the whole record in the twelfth century, when the mean date was a fortnight later than it had been three hundred years earlier. There are hints that this was a cold time generally in and around the wide expanse of the North Pacific Ocean. If so, part of the explanation of the medieval warmth in Europe and North America, extending into the Arctic in the Atlantic sector and in at least a good deal of the continental sectors on either side, must be that there was a persistent tilt of the whole

171

circumpolar vortex (and of the climatic zones which it defines) away from the Atlantic and towards the Pacific sector, which was rather frequently affected by outbreaks of polar air.

In this chapter we shall concentrate our study on the Atlantic side of the hemisphere and the lands where the warmth of the high Middle Ages was most marked, since it happens that these are the areas where both the climatic and the human historical record are at present most accessible.

THE MEDIEVAL SEQUENCE IN NORTHERN EUROPE AND THE NORTHERN ATLANTIC

The reconstituted western empire of Charles the Great did not coincide with a particularly favourable climatic period. Nor did it last very long. The campaigns by which it was established between about 770 and 800 seem to have been in a time with more than usual tendency to cold winters; the other seasons, although perhaps more often dry than wet, revealed both drought years and some years when floods created difficulties. There is a suggestion in this that it may have been one of those times when 'blocking of the westerlies' by anticyclones in this or that longitude in 45–65 °N was frequent, with a consequent disposition to extreme seasons of various, even opposite, sorts depending on where the stationary anticyclone lay: but further evidence is required before we can be sure of this.

Where there is no reasonable doubt is that over the next three to four centuries, as reports indicating the character of the seasons in Europe become more numerous, we see that the climate was warming up (cf. figs. 30 and 59), until there came a time when cultivation limits were higher on the hills than they have ever been since. Trees seem also to have been spreading back towards the heights. Certainly the upper tree line in parts of central Europe (cf. fig. 53) was 100–200 m higher than it became by the seventeenth century. The isotope record from the Greenland ice-sheet (fig. 36) shows us that the climate had already been in a relatively warm phase in the far north since AD 600, though the warmth there too was becoming more sustained and was increasing. On the heights in California the tree ring record (fig. 52) indicates that there was a sharp maximum of warmth, much as in Europe, between AD 1100 and 1300.

The variations shown by the more than one-thousand-years'-long record of the tree rings in European oaks from the lowlands of Germany are harder to interpret climatically, because both temperature and rainfall come into it. The records from different areas agree in producing the extreme narrowest and the extreme widest ring series both within the times covered by this chapter. The extremely narrow rings prevailing in the tenth century, especially between about 910 and 930 and again in the 990s, must surely indicate prolonged and repeated drought. One cannot suggest that any general coolness of the summers was responsible; the sparse documentary records point more to some of the summers being notably hot. The impression on present data is rather that the tenth century saw a remarkable amount of anticyclonic weather over Britain, Germany and southern Scandinavia, giving low rainfall, rather warm summers and rather cold winters. The latter point seems to be confirmed by the numerous bone skates revealed by the archaeological investigations in York from the Anglo-Scandinavian period in that city. The other extreme of the German oak chronologies occurred between about the years 1052 and 1160, when the decade average ring widths were 35–80 per cent wider than in the tenth century. We may deduce, if not excessive wetness (apart from isolated years), at least more moisture than in the 900s and general warmth of the growing seasons. Of this warmth we shall see further evidence in the following pages.

There is no mistaking the fact that there was a general opening out of the European world in the period we are considering in this chapter. How much of it was directly dependent on the more genial climatic regime which developed?

There had been European seafarers occasionally wandering out over the northern Atlantic long before Viking times. Prominent among them were Irish monks apparently seeking peaceful shores on which to establish a foothold far from the troubled times of cultural decay and barbarian migrations in Europe in the fifth and sixth centuries and after. It has been suggested that the annual migrations of the wild geese to and from Iceland and the Arctic gave them confidence that there was land to find in the north. One must suppose that there is some substratum of fact in the legendary voyage of St Brendan at some time between around AD 520 and 550 and that he got far enough in the direction of Greenland to encounter icebergs. Certainly Dicuil, an Irish monk writing AD 825,[1] assures us that

> there are many other islands in the ocean . . . which can be reached in two days and two nights direct sailing from the northernmost parts of the British Isles with full sails and a fair wind. . . . Some of these islands are very small . . . separated from one another by narrow sounds. On these islands hermits who have sailed from our Scotia [i.e. Ireland] have lived for about a hundred years. But, even as they have been . . . uninhabited from the world's beginning, so now because of Norse pirates, they are empty of anchorites, but full of innumerable sheep and a great many different kinds of seafowl.

The islands here described are by general agreement the Faeroes, which were therefore settled by Irish monks as early as about AD 700–25. (I have used the translation given by Gwyn Jones in *A History of the Vikings*, Oxford University Press, 1968.) But they left around 800, when the Vikings first appeared. The Vikings' first recorded exploration to Iceland (under Floki Vilgerdason) was not until about 860, though two earlier Scandinavian voyages had been blown there accidentally a few years before. The Norse settlement on the island seems to have begun during the 860s. But they found that Irish monks had preceded them. Dicuil reports one visit as early as the 790s. The Irish account records that the sea was frozen one day's sail north from Iceland, and Floki's party observed one of the big fjords of northwest Iceland (Arnarfjord) choked with ice. But after that time there is little mention of ice – only brief and, according to Lauge Koch,[2] doubtful reports of it in 1010–12, 1015, 1106, 1118 and 1145 – on the seas near Iceland until the 1190s, when it reappeared in some strength between Iceland and Greenland, and in July and August of the year 1203 it was at the coast of Iceland.

It seems likely that the beginning of the era of Scandinavian sea-going explorations, as of the rough story of Viking raids which harried the coasts of Europe from the 790s onwards, came with the mastery of sail by the northern peoples. Even then, they had no lodestone or compass until centuries later. But the spread of their voyages north into the Arctic and west to Greenland, and ultimately to Newfoundland and apparently into the Canadian Arctic north of Baffin Island, surely owed a great deal to the long period of retreat of the sea ice and probably a relative immunity from severe storms. Ottar, or Othere, whose home was in north Norway, told King Alfred in England of an exploration he had made about AD 870–80 beyond the customary range of the whalers of those days, evidently to the White Sea. And Harald Hardråde who was king of Norway and England is reported by Adam of Bremen to have explored 'the expanse of the Northern Ocean' some time between 1040 and 1065 with a fleet of ships, beyond the limits of land (Spitsbergen or Novaya Zemlya?) to a point where he reached ice up to 3 m thick and 'there lay before their eyes at length the darksome bounds of a failing world'. The medieval Icelandic sailing directions covered voyages, reckoned to take four days, north to Svalbard 'in the polar gulf', which it seems from the sailing time must have meant the east Greenland coast between 70 and 72 °N (not the Spitsbergen archipelago, to which the name is now applied). This coast was discovered in 1194; and seals, walrus and whales were hunted there already before the year 1200. Very soon, however, the increasing ice evidently put a stop to this, and the same coast seems to have been rediscovered in an easier year about 1285; but by 1342 the ice was so much increased that the old sailing route from Iceland to Greenland at the 65th parallel of latitude had to be abandoned for one farther south. Later, communication with Greenland was lost altogether.

The North American coast, Vinland (or Wineland) to the Norsemen, like Iceland and Greenland (where the first Norse settlement was established in the 980s) before it, was discovered by accident, by ships being blown off course, about AD 1000. The site of only one settlement, at L'Anse aux Meadows in northern Newfoundland, has so far been discovered, though another farther south is also referred to in the sagas. It seems, in any case, that the settlement and the America voyages were discontinued after a few years, and it appears that difficulties with the native inhabitants rather than weather or sea ice were the cause. Further accounts indicate that crossings from the Old Norse settlements in west Greenland to Markland (Labrador) were resumed much later, in the fourteenth century (one as late as 1347), when the climate and ice conditions had deteriorated and communications with Europe had almost ceased, to collect timber for building.

That the waters off west Greenland in the heyday of the Norse settlements were at least as warm as in the warmest periods of the present century is indicated by the abundance of cod which the inhabitants caught, the bones of which are found in their middens. We may probably safely conclude that an even greater warm anomaly occurred in the quiet waters within the fjords of southern Greenland west of Cape Farewell from another circumstance, a rare case where the limits of tolerance of man himself may yield reliable information on past temperatures. For it is recorded in the *Landnámabók*, a book written in Iceland about 1125 cataloguing the settlement of Iceland a couple of centuries earlier and describing the Old Norse settlement of Greenland between AD 985 and 1000, that one of the first Greenland settlers, Thorkel Farserk, a cousin of Erik the Red who founded the colony, having no serviceable boat at hand, swam out across Hvalseyjarfjord to fetch a full-grown sheep from the island of Hvalsey and carry it home to entertain his cousin. The distance was well over two miles. Dr L. G. C. E. Pugh of the Medical Research Laboratories, Hampstead, has given his opinion, from studies of the endurance of Channel swimmers and others undertaking similar exploits, that 10 °C would be about the lowest temperature at which a strong person, even if fat, not specially trained for long-distance swimming, could swim the distance mentioned. As the average temperatures in the fjords of that coast in August in modern times have seldom exceeded 6 °C (+3 to +6 °C being more typical), it seems that the water must have been at least 4 °C warmer than this limit in the year in which Thorkel swam it and brought home his sheep.

Other items point to a similarly great departure of the temperatures ashore in that area: for Old Norse burials took place deep in ground which has since been permanently frozen. It is harder, however, to be sure of the climatic implications of another report from the time of the old Greenland colony. Lauge Koch cites a medieval report that in 1188 or 1189 – i.e. at a time when the climate in the area may already have begun to be colder and the sea ice to reach somewhat farther down the coast towards south Greenland – a ship, the *Stangfolden,* on passage from Norway to Iceland came to be wrecked off the east coast of Greenland. Some years later, about 1200, the dead bodies of seven of the ship's company were found in a rocky cave near that coast, among them the clergyman Ingemond who had left a written report in runic letters on their fate beside him. Ingemond's brother, also wrecked about the same time, is reported to have succeeded, with two other men, in crossing the southern part of the inland ice, only to perish when near the main Norse settlement in Greenland, the so-called East Settlement (actually their southernmost settlement), a little west of Cape Farewell. This suggests that the inland ice in that neighbourhood was not thought of as such a hostile environment that one would not venture on it in an emergency, but nevertheless the going would be easier in the absence of melting and a crossing would doubtless require some days of reasonably good weather without strong winds.

By about AD 1250 the *King's Mirror* (*Konungs Skuggsjá*), a Norwegian work of that time, reports that

> as soon as the great ocean has been traversed there is such a great superfluity of ice on the sea that nothing like it is known anywhere else in the whole world and it lies so far out from the land that there is no less than four or more days journey thereunto on the ice, but this ice lies more to the NE or N outside the land than to the S and SW or W.

A further passage about Greenland around 1250 in the same work reports that 'men have often tried to go up into the country and climb the highest mountains to look about and see whether there was any land free from ice and habitable'. A number of reports indicate that in this period of the early stages of the climatic deterioration the Norse Greenlanders were induced once more to roam more widely afield in search of materials and hunting food, including penetration farther north than before to the west of Greenland, reaching Baffin Bay and making contact with the Eskimos who were tending to move south.

Having to this extent taken the measure of the early medieval warm period at the limits of the Arctic region reached by the contemporary Europeans, let us now look at the evidence from other regions. The northern limits of the cultivation of grains show a corresponding expansion of range during the centuries with which this chapter is concerned. Grain was grown in Iceland from the time of the first Norse settlers there, apparently fairly continuously, until its abandonment in the late sixteenth century. There was also undoubtedly more scrub birch woodland there in the early days of the settlement than at any time since, though the settlers themselves seem to have been largely responsible for its destruction. Its area is believed to have been reduced from perhaps a fifth of the country to 1 per cent by the thirteenth century. Investigation by Dr G. S. Boulton, with colleagues from the University of East Anglia and from Iceland, of a farmhouse site at Kvisker in southeast Iceland that has been occupied for a thousand years revealed that the oldest of the successive houses on the site, dated before the volcanic ash layer of AD 1090, was the biggest and richest.

Fig. 62 One of the ancient farms, Svinafell, in southern Iceland, established in the earliest settlement times on a south slope. A great glacier can be seen now filling the valley close to the site of the farm. (Kindly supplied by Dr Sigurdur Thorarinsson of Reykjavik and reproduced by permission.)

Its midden contained relics of diverse and luxurious foods, including (imported) oysters. And the forest surrounding the farmed land there produced birch stumps of a good size, never attained since. From pollen analysis it appears that the farmer at Kvisker gave up growing oats about AD 1200 and reduced the amount of barley grown by about a half. In the next century much of the ground was covered by river gravels and part of it by a glacier (see fig. 62).

THE PEAK OF MEDIEVAL WARMTH IN EUROPE

In Norway some kind of corn, probably barley, was grown as far north as Malangen (69½ °N) in north Norway, at least from Ottar's time (around 880) until the eleventh century, and wheat in Trøndelag, the district about Trondheim, where pollen studies and other records again indicate that it came to an end sharply in the later Middle Ages. Professor Andreas Holmsen[3] reports that it was just between about AD 800 and 1000 that the area of forest clearance and settled farming in Norway, which had long remained more or less static, spread 100–200 m farther up the valleys and on to the higher ground. Most of this ground was lost again after AD 1300.

Fig. 63 Relics (ridge and furrow) of medieval tilled fields between 350 and 400 m (1150–1300 ft) above sea level on the heights of Dartmoor in southwest England beside the abandoned settlement of Houndtor which lies just to the left of the picture. The Greator rocks are seen in the picture. (Photograph, copyright by G. Beresford, who kindly supplied it for this book.)

Fig. 64 Ridge and furrow, the result of thirteenth-century tillage, seen on the fells on a south-facing slope above Redesdale, Northumberland at 300–320 m above sea level.

178

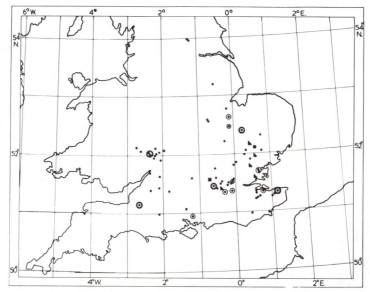

LEGEND

. Vineyard, usually 1–2 acres or size not known.

▲ Vineyard, 5–10 acres.

■ Vineyard, over 10 acres.

○ Denotes evidence of continuous operation for 30–100 years.

⊙ Denotes evidence of continuous operation for over 100 years.

Fig. 65 The distribution of known medieval vineyard sites in England.

In many parts of Britain, also, tillage was extended to greater heights than for some long time previously or since, on Dartmoor in the south-west (fig. 63) to about 400 m (1300 ft) and in Northumberland, near the Scottish border (fig. 64) to 320 m (1050 ft). In AD 1300 one grange at 300 m (roughly 1000 ft) above sea level, belonging to Kelso Abbey, in the south of Scotland had over 100 hectares of tillage, 1400 sheep and sixteen cottages for shepherds and their families. An approximate gauge of the temperatures prevailing in the summer half of the year in England and central Europe, serving as a check upon the figures derived by the method used in fig. 30 and explained on pp. 84–5, may be obtained by consideration of the limits of vine cultivation in the Middle Ages and comparing the present climates of those sites with the modern limits of wine production. Fig. 65 is a map of the distribution of known medieval vineyards in England. The comparison indicates that the average summer temperatures were probably between 0.7 and 1.0 °C warmer than the twentieth-century average in England and 1.0–1.4 °C warmer in central Europe. (The quality

179

Fig. 66 The medieval English vineyard site at Tewkesbury, Gloucestershire. The ground slopes gently northwards to a ditch in the middle ground of the picture. Surely a frost hollow site, which suggests that the medieval cultivators were not much troubled with late frosts in May after blossom time.

of the English medieval wine is indicated by the efforts of the French trade at that time to have them closed down under a treaty). In England particularly it seems that there must have been less liability to frost in May in the period between 1100 and 1300. (Fig. 66 is interesting in this connection.)

Thus, it seems that the great period of building of cathedrals in the Middle Ages, in what Kenneth Clark[4] has called the first great awakening in European civilization, and the sustained outburst of energy of the European peoples, which produced among other things the more controversial activities of the Crusades, coincided with an identifiable maximum of warmth of the climate in Europe. Hugh Trevor-Roper[5] makes no comment on the climate but notes the time around AD 1250 as the turning point:

> the highest point of the European Middle Ages. . . . Up to that date we see – from about 1050 onwards – only advance . . . growth of population, agricultural revolution, technological advance. The frontiers are pushed forward in all directions. . . . Already in the middle of the thirteenth century the territorial expansion had been halted . . . in 1242 the eastward advance of the Teutonic knights . . . was

180

held up by the ruler of the Russian Slavs. . . . By 1300 all that remained of the Eastern Empire of Christendom was a few shrinking relics of Greece.

The warm phase, which had already passed its peak in Greenland in the twelfth century, seems to have broadly continued in Europe until 1300 or 1310 though with a marked increase in the incidence of severe storms in the North Sea and the Channel and with flooding disasters on the low-lying coasts. The warmth may even have reached its maximum at this late stage: for there are documentary records to tell us that it was in the 1280s that the tillage reached so high on the Pennines and Northumbrian moors that there were complaints from the sheep farmers that too little land was left for grazing. Such a peak of warmth in the last stages before Europe itself was affected by the down-turn of temperatures in the Arctic would be meteorologically consistent with the development of a strong thrust forward of the Arctic regime in the longitudes of Greenland and Iceland, distorting the pattern of the circumpolar vortex with a sharp trough there and a recurrent warm ridge over western Europe. Something like this pattern seems to have recurred at times in the middle and later parts of the fourteenth century, bringing notable droughts in Europe after an extremely wet phase which had marked the first break in the early part of that century. (It is likely that some of the troubles about this time with the massive buildings – cathedrals, churches and castles, with collapsing towers and cracking walls and arches (fig. 67) – were not so much due to faults of design as to soil moisture changes and consequent settling.)

The occurrence in medieval York of the bug *Heterogaster urticae* (F.), whose typical habitat today is on stinging nettles in sunny locations in the south of England, discovered by the city of York archaeological investigations to have been present there both in the Middle Ages and in Roman times, presumably indicates prevailing temperatures higher than today's. Another revelation from insect studies is the abundance also in medieval York of a beetle *Aglenus brunneus* (Gyll.) whose habitat preferences indicate high temperatures generated in decaying vegetable refuse. Both these discoveries hint at rather high prevailing temperature of the urban environment itself in the tightly built-up medieval city centre.

There are many indications that in eastern Europe, as in Greenland and Iceland, a colder, more disturbed climate set in already in the 1200s. And, indeed, as far west as the Alps, some trouble was caused by advancing glaciers during the thirteenth century. During some part of the warmest period, perhaps in the tenth and early eleventh centuries, there seems to have been concern about drought in the Alps: for a water supply duct, the Oberriederin, was laid from high up near the Aletsch glacier to the valley below, and similar water supply installations were engineered in the Saastal

Fig. 67 An arch deformed by subsidence in Carlisle Cathedral. No movement seems to have occurred after about 1300–50. Compaction of the site through drying out of the soil in the previous centuries has been suspected as the cause of the damage seen (see pp. 197–8).

(also in Switzerland) and in the Dolomites, only to be overwhelmed by the advancing glaciers between 1200 and 1350.

The ancient gold-mines in the Hohe Tauern in Austria and other high-level mines in central Europe, abandoned before the time of Christ, were opened up and worked again in the warmth of the high Middle Ages, only to be abandoned again later. Underground water began to cause difficulties about 1300: at Goslar it was reported in 1360 that water had been increasing in the mines in the Harz Mountains for more than fifty years. In Bohemia the same difficulty led to some mines being abandoned as early as 1321. In the Alps some of the mine entrances were again closed by the glaciers.

THE CONTEMPORARY SCENE IN THE MEDITERRANEAN, EASTERN EUROPE AND ASIA

In the Mediterranean, as also in the region of the Caspian Sea and on into central Asia, the period of warmth in high latitudes in the Middle Ages seems to have been a time of greater moisture than the present century. Lake levels were high, the Caspian Sea as much as 8 m above its present level during much of the time between the ninth and fourteenth centuries.

Fig. 68 (a) The medieval bridge (Ponte dell'Ammiraglio – Bridge of the Admiral) at Palermo, Sicily, built in 1113 to span a much larger river than now exists there. The River Oreto, which has now been diverted, as seen in (b), was used by ships up to this bridge when it was first built. (Photographs kindly supplied by General Fea of the Servizio Meteorologico, Aeronautica Militare Italiano, Rome.)

183

Two of the rivers of Sicily, the Erminio and the San Leonardo, were described as navigable in the twelfth century – something which would now be impossible even for the vessels of those times. Bridges were built, as across the Oreto at Palermo in Sicily (fig. 68), of a size not required by the present rivers. (The famous Pont d'Avignon finally built across the lower Rhone in southern France in 1177–85 at a difficult point, where roads converge but the current is always strong and the Romans had been unable to bridge the river, suffered many collapses of parts of the bridge in the following years but was not finally abandoned half destroyed until 1680.) There was also in the high Middle Ages more general flow of the streams in Greece and in the wadis of north Africa and Arabia. Fig. 45 (p. 131) indicates a more adequate rainfall in medieval times also in the dry area of northwest India.

These features seem likely to be explained partly by a displacement of the anticyclone belt of the desert zone during the warm epoch north of its present usual position to an axis from the Azores to Germany or Scandinavia as in some of our modern fine summers. Such partly meridional wind circulation patterns, with a cold trough deformation of the circumpolar vortex, commonly thrust cold surface air south over eastern Europe and western or even central Asia, and from there it would be deflected by the mountains westward and southward towards the Mediterranean. This is an eastern position for such a development in the circumpolar vortex, requiring a longer wave-length (or spacing of the troughs and ridges) than commonly prevails in the upper wind flow from the more or less fixed disturbances over North America caused by the Rocky Mountains. Such a longer wave-length would be likely to occur at a time when the main flow of the winds was displaced towards higher latitudes and particularly when, as in the thirteenth century, Arctic cooling strengthened the thermal gradient and the winds.

Our knowledge of the past variations of lake levels – archaeologically determined in the case of the Caspian Sea – indicates that the barbarian movements out of Asia which troubled the Roman empire over a long period can be associated with times of drought in central and western Asia around AD 300, which also returned around 800. By contrast, the great outbreak of Mongolian tribesmen in the thirteenth century seems to have occurred in a moist period, when the Caspian Sea was rising. The sudden outburst of energy of the peoples of inner Asia, which brought Genghis Khan and his Mongol hordes within the space of twenty years, between 1205 and 1225, deep into European Russia, to the Indus and to the gates of Peking, could reasonably be supposed to have had its origin in a build-up of population in the arid heart of Asia in times when the pastures were in better than usual shape. But its suddenness, and the coincidence of its timing with what we know of the cooling in high latitudes from the isotope record in northern Greenland and the great advance of the Arctic sea ice towards Iceland, raises a suspicion that some more sudden event connected with the cooling may have triggered it off. This could have been some invasion of the heart of Asia by colder Arctic air than before, the effects of which would be particularly noticeable if it happened in summer. This is speculation, but China had long been experiencing a cold regime and some scientists have thought that this anomaly gradually spread westwards until it enveloped Europe in the Little Ice Age of later centuries.

There was clearly some difference between the sectors of the northern hemisphere with which these paragraphs have been concerned and the situation over east Asia, where the climatic zones seem to have been pushed south over a long period of which the twelfth century marked the climax. The swing to the southeast of the isotherms and of the flow lines of the circumpolar vortex from a northward displacement (or ridge) over the Indian sector to a southward displacement (or trough) over east Asia is a pattern which seems liable to have introduced an anticyclonic tendency over Thailand and northern Indo-China, reducing rainfall there. This meteorological speculation suggests an explanation of temporarily easier – i.e. drier – conditions favouring the Khmer empire of Angkor in Cambodia (Kampuchea) in the region, which after 1300 returned to jungle.

EFFECTS ON SEA LEVEL AND LOW-LYING COASTS

Our survey of the European scene during the warmer centuries of the Middle Ages would not be complete without mention of the things that suggest a slightly higher stand of the sea level, which may have been gradually rising globally during that warm time as glaciers melted – and particularly in the area around the southern North Sea where the land-sinking due to the folding of the Earth's crust in that basin was presumably going on then as now. Fig. 60 draws attention to the greater intrusions of the sea in Belgium, where Brugge (Bruges) was a major port, and in East Anglia, where a shallow fjord with several branches led inland toward Norwich. The English fenland south of the Wash provided an extensive watery landscape of shallow brackish channels and low islands, fringed by reeds and brushwood, in which the island of Ely was so cut off that the Anglo-Danish inhabitants were able to hold out for seven to ten years after the Norman conquest of the rest of England. And the coastal plain of the Netherlands and Belgium had a fluctuating population in the eleventh and twelfth centuries, as the state of flooding varied, leading finally to a more general emigration to Germany.[6]

THE SEQUENCE IN NORTH AMERICA AND SOME COMPARISONS

In North America east of the Rocky Mountains there is evidence that the prevailing temperatures followed a sequence very similar to that in Europe and that there were interesting and important changes in the moisture climate. Only in northern Labrador and the neighbouring Ungava region is there no sign so far of a medieval interruption in the cooling off that began 3000–3500 years ago and put the forest into retreat before the advancing tundra. In northern Quebec and in the North-West Territories west of Hudson Bay, the extensive pollen-analytical researches co-ordinated by Dr Harvey Nichols of the University of Colorado Institute of Arctic and Alpine Research indicate some recovery of the forest, associated with warming of the summers, from about AD 500 to some time about 1000–1200 or 1250. Farther south, in the Middle West of the United States, the archaeological studies of Baerreis and Bryson at the University of Wisconsin have indicated that the Indian people of the Mill Creek culture grew corn (maize) in northwestern Iowa before the year 1200, in an area which today is somewhat marginal as regards enough rainfall for the crop. Elk and deer, both woodland animals, which they evidently hunted, together accounted for most of the flesh in their diet before about 1100; in the twelfth century the proportion of these among the bones in the middens rapidly declined and was overtaken by bison, an animal of the open plains. The abundance of bison bones increased towards the west where the climates are drier, in the 'rain-shadow' of the Rocky Moutains. But from about AD 700 onwards the climates of the whole region seem to have become moister than before, the prairie giving way to landscapes with more trees, until an abrupt reversal about the year 1200. Farming peoples were spreading their occupation northwestward on the plains, moving northward into Wisconsin and on up the Mississippi and other valleys into Minnesota as early as the eighth century. They maintained a thriving culture until 1200, when their sudden disappearance coincides with evidence of drought and vegetation change. Such a change in the region concerned is readily explained by increased sway of the westerly winds, intensifying and extending the rain-shadow of the mountains, as the thermal gradient increased with the cooling of the Arctic then setting in. We have referred to the evidence of this on Greenland and Iceland waters.

The climatic history reviewed in this chapter has led one historian[7] to summarize the matter by saying: 'intriguingly, the profile of long-run average temperature in England shows a crude but clear congruence with that of material welfare broadly conceived'. And he goes on 'The medieval expansion, the crises of the fourteenth and late sixteenth centuries, and the revivals of the fifteenth (to early sixteenth), eighteenth and nineteenth centuries, broadly correspond with movements in the trend line of temperature.' Yet, he argues that climatic change has little explanatory value and that one cannot assert that the course of European history would have been much different if the climate had not changed. The period covered by the next chapter will give us an opportunity to examine this contention a little more closely.

11

DECLINE AGAIN IN THE LATE MIDDLE AGES

THE DOWN-TURN OF CLIMATE IN THE ARCTIC

The deterioration in their situation which announced itself to the Old Norse Greenlanders in 1197–1203 by the increase of ice encroaching on the seas that were used for their links with Iceland and with Europe, at first in occasional years but later on seeming permanent, clearly had to do with a cooling of the Arctic (see fig. 36, p. 93).

Already during the twelfth century the Eskimos of the Dorset culture, once (about 700 BC) widespread across the eastern Canadian Arctic, who had returned to high latitudes after AD 800–900, had been moving south. Archaeology suggests that this was partly because another Eskimo culture, developed near Thule in northwest Greenland, was more successful in hunting the resources of the far north; but it is probable also that increasing ice and dwindling seal and walrus populations were making the competition more difficult. And so it was around 1200–50 that Norsemen and Eskimos first came into contact in Greenland. At first some trading went on between them. But about 1350 the smaller of the two Norse centres in Greenland, with only about seventy-five farms, the Vesterbygd ('West Settlement'), which was the more northerly of the two areas occupied in west Greenland, was wiped out either by conflict or disease, possibly the plague. (Some cattle and sheep were found wandering unattended by any human owners when a ship visited the area from the other settlement.)

The larger Østerbygd ('East Settlement'), where there were about 225 farms, survived until about the year 1500, though in evident decline: the average stature of the grown-up men buried in the graveyard at Herjolfsnes in the fifteenth century was only 164 cm (5 ft 5 in.) compared with about 177 cm (5 ft 10 in.) in the early period of the settlement. By about 1342 it is recorded that the old sailing route along the 65th parallel of latitude between Iceland and Greenland was finally changed to a route farther south because of the increase of ice. After the wreck off Norway of one of the ships used in the late medieval royal monopoly trade in 1369 regular communication between Europe and the Greenland colony ceased. Some

187

ships bound for Iceland arrived in Greenland in later years after being blown off course, and there is indirect evidence of occasional visits by traders and freebooters from England and elsewhere in the fifteenth century. In 1492 Pope Alexander VI wrote of his anxiety over the situation in that outpost of Christendom:

> the church of Garda is situated at the ends of the Earth in Greenland, and the people dwelling there are accustomed to live on dried fish and milk for lack of bread, wine and oil . . . shipping to that county is very infrequent because of the extensive freezing of the waters – no ship having put in to shore, it is believed, for eighty years – or, if voyages happened to be made, it could have been, it is thought, only in the month of August . . . and it is also said that no bishop or priest has been in residence for eighty years or thereabouts.[1]

In fact, the Herjolfsnes graveyard preserved bodies and clothing in the subsequently permanently frozen ground, the dresses including European models of about the year 1500. But ships from Hamburg beaten off course to Greenland about 1540 found only one dead Norse body and no inhabitants alive. From that time on only whalers, or explorers such as Hudson in 1607, occasionally happened to get through the ice belt to this or that point on Greenland's deserted Arctic shores, until in the 1720s the Danish-Norwegian state once more founded posts, again in southwest Greenland. There were no settlements in east Greenland before the nineteenth century.

It has been suggested that the explorations in the fifteenth century which led the fishermen from Bristol ever farther west across the Atlantic, until as early as the 1470s or 1480s they may been fishing on the Newfoundland Banks,[2] may have started because the fish stocks of the higher latitudes in the northeast Atlantic had deserted their former grounds as a result of the increasing spread of the Arctic cold water. The situation was doubtless aggravated by Hanseatic competition in Iceland-Greenland waters. However that may be, it is clear that the searches of the English sixteenth-century seafarers such as Chancellor in 1553 and the Dutch expedition under Willem Barents in the 1590s to find a North-East Passage, and of Frobisher in the 1570s, of Davis in the 1580s and soon Hudson, to find a North-West Passage through the Arctic to the Indies were undertaken at a peculiarly unfavourable time. The same was true of Hudson's attempt in 1607 to reach the North Pole and still in 1827 of Edward Parry's attempt, and of the renewed efforts around that time to seek out a North-West Passage, as well as the voyage of Sir James Clark Ross in 1831, which succeeded in reaching the north magnetic pole.[3] The severer Arctic climate had then ruled for hundreds of years, though there were still some openings in the polar pack-ice controlled by the wind pattern; the whalers had found some of these in their operations near northeast Greenland, and this had produced a misplaced belief in the existence of an ice-free sea in the central Arctic.

In Iceland the old Norse society and its economy suffered a severe decline which set in first about AD 1200 and could be said to have continued over almost six centuries. The population of the country fell from about 77,500, as indicated by the tax records in 1095, to around 72,000 in 1311. By 1703 it was nearly down to 50,000, and after some severe years of ice and volcanic eruptions in the 1780s it was only about 38,000. The people's average stature also seems to have declined, much as in Greenland, from 173 cm (5 ft 8 in,) to 167 cm (5 ft 6 in,) from the tenth to the eighteenth century. It is clear from the surviving records that years when the Arctic sea ice was close to the Iceland coast for long months (usually between January or March and any time from June to August) played a big part in this. In such years the spring and summer were so cold that there was little hay and thousands of sheep died, especially all over the northern and eastern part of the country. The shellfish of the seashore were also destroyed by the ice. Gradually all attempts at grain growing were given up. The glaciers were advancing. And there were some volcanic disasters besides, when whole areas of the island were covered by volcanic ash or lava flows, the pastures were ruined by the fluorine or sulphurous content of the ash and the sheep and cattle were killed by it. One of the worst cases was the great eruption in Öraefi in the south of Iceland in 1362.

It cannot be denied that the trade monopoly claimed by the Danish-Norwegian crown through most of these centuries must also have had some effect, its restrictions probably contributing to the country's difficulties, but it seems that the main causes of decline were the natural disasters – Iceland's 'thousand years struggle against ice and fire', as Sigurdur Thorarinsson's 1956 article called it.[4]

That there were some easier times as well as periods of great severity during these centuries can be clearly discerned despite the scarcity of records at certain times. For example, the widespread use of polar bear skins in the late Middle Ages for carpeting the church floors in Iceland indicates a large supply of the bears, and therefore presumably of the ice which brought them, in the fourteenth century. A hundred years later the skins were getting scarce, and many were old and in poor condition, but there was some increase in the sixteenth century before this item became restricted by the trade monopoly of the monarchy in Denmark. This information seems to confirm the inference from the direct reports of the sea ice which survive that there was much ice from the late 1200s through the fourteenth century, and then some improvement before the drastic increase of ice in the late 1500s and after. The times of most ice and coldest climate in Iceland seem to have started suddenly in 1197–8 and 1203 and reached culminating phases around 1300, from about 1580 to 1700, especially the 1690s, and again in the late eighteenth and nineteenth centuries.

Fig. 69 Two pictures of North Sea storm waves assailing the sea defences of the small island of Heligoland in a northerly storm, Beaufort force 10, on 10 October 1926. The island is but a remnant of its former size. (Photographs F. A. Schensky, reproduced by kind permission of his daughter Miss L. Schensky of Schleswig.)

HOW EUROPE WAS FIRST AFFECTED: STORMS

The first symptoms of the change already affecting Greenland and Iceland which may have been noticed by the inhabitants of Europe, particularly around the North Sea, were the increased incidence and severity of wind storms and sea floods in the thirteenth century. Some of the latter caused appalling loss of life, comparable with the worst disasters in Bangladesh and China in recent times. In at least four sea floods of the Dutch and German coasts in the thirteenth century the death roll was estimated at around 100,000 or more; in the worst case the estimate was 306,000. As a result of the floods of 1240 and 1362 it was reported that sixty parishes accounting for over half the agricultural income of the (at that time) Danish diocese of Slesvig (Schleswig) had been 'swallowed by the salt sea'. In some of these storm floods the Zuyder Zee in the Netherlands was formed, and enlarged, and it was not drained until the present century. Islands, and other inlets, were formed by losses of land on the German and Danish North Sea coasts. Other islands were destroyed by the stormy seas. The island of Heligoland (50 km out in the German Bight), which is believed to have measured over 60 km across in the year 800, had been reduced to 25 km by about 1300, perhaps half of it being lost in a storm in that year. Today it measures only about 1.5 km on its longest axis (fig. 69). In England the great ports of Ravenspur or Ravensburgh (east of Hull) and Dunwich (on the Suffolk coast in East Anglia) were lost in successive stages in the sea storms of these centuries. Deaths of 100,000 or more people in flood-ings of the continental shore of the North Sea were again reported in storms in 1421, 1446 and 1570. In the 1570 storm great cities were flooded, and

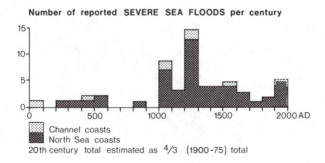

Fig. 70 The distribution by centuries of reports of severe storm floods which caused much loss of life or land on the coasts of the North Sea and English Channel. The data and sources of data on which this diagram is based are fully tabulated in the author's book *Climate: Present, Past and Future*, vol. 2, London, Methuen, 1977, pp. 120–6. Any apparent mistakes by earlier collectors of the data from the distant past producing repetitive reports of the same incident have been cut out in the counts of numbers of storms for this diagram.

Fig. 71 Some results of the storms of blowing sand in the late Middle Ages: (a) (top) The lagoon at Kenfig on the coast of south Wales, near Port Talbot, formed by sand-dune movements, reputedly around 1316, which closed the medieval port there. Further movements between 1344 and 1480 finally buried the old Roman coast road and with a storm in 1573 carried a line of high sand-dunes 3 km inland. (b) The coast edged with a belt of great sand-dunes protecting the flatland of Morfa Harlech in northwest Wales. These dunes lie more than 1 km seaward of the former port of Harlech, in use until about 1385, which they closed.

Fig. 71 (c) The Sands of Forvie: the 30 m high sand-dune which covers the medieval township of Forvie, on the east coast of Aberdeenshire, Scotland, which was obliterated by a great southerly storm in August 1413.

the deaths were estimated at 400,000. And in 1634 there were again great losses of land from the Danish and German coast and the off-lying islands.

Fig. 70 shows the distribution over historical time of known reports of severe sea floods in this part of the world. In the southern North Sea on the Netherlands coast the occurrence of devastating storm surges was greatest in the early 1400s and late 1600s;[5] the late 1500s were remarkable for a few storms of outstanding range and severity, most of all the storm of 1–2 November 1570 when the flooding affected the coasts from France to northwest Germany. In reading the diagram allowance must be made for the reduced chance of reports having been made and surviving from early times, but it seems safe to conclude that there were real maxima of storm flood occurrences for the region as a whole in the eleventh and in the thirteenth centuries AD, and in the southern North Sea at the times mentioned above. There is also a suggestion of more severe floods in, and soon after, late Roman times and again in our own century than at other periods. This distribution suggests that storm floods on the low-lying coasts of the North Sea have been most troublesome: (a) when the sea level may have been somewhat raised after long periods of warm climate and glacier melting; and (b) when a cooling Arctic has produced a strengthened thermal gradient in latitudes between about 50 and 65 °N, leading to increased storm frequency and severity over this zone. In the thirteenth century, and perhaps again in recent decades, both these conditions were present. One must conclude from the much more restricted range and loss of life in modern storms that the dykes which have been built along the coasts of the North Sea, and continually improved, in later centuries are among man's greatest successes in defence against natural disasters.

Another accompaniment of some of the severe storms of the northeast Atlantic and North Sea region in the late Middle Ages and after was the overwhelming of a number of coastal places by blown sand (fig. 71). There was a long epidemic of such disasters on the sandy coasts of northwest Europe from Brittany to the Hebrides and Denmark, starting about the thirteenth century and continuing to about 1800. As examples, the little medieval port of Harlech on the west coast of Wales was permanently obliterated by a line of great sand dunes around 1400, within at most a few decades of the other cases pictured in fig. 71. In the seventeenth century a great storm destroyed the fine natural harbour at Saksun on the northwest side of the Faeroe Islands by filling it with sand, and another overwhelmed an area – now known as the Culbin Sands – of perhaps 60 km^2 of fine farmland, including nine farms and a mansion house, in northeast Scotland. In the sandy terrain of the Breckland in East Anglia and in similar country in the Netherlands even places inland were affected by frequent blowing sand in this period.

It is interesting that the case pictured in fig. 71c on the east coast of northern Scotland took place with a southerly storm, a circumstance which lowers the level of water in the North Sea. Moreover, the date reported was within a few days of a date when the astronomically calculated tide was only 4–7 cm short of the extreme of the nineteen-year cycle, and this was itself only one cycle short of a roughly 2000-year extreme. This coincidence may point to a combination of factors which led to the shifting of so much sand as to destroy a coastal township in a single severe storm. An exceptionally low tide seems likely to have occurred, laying bare a wholly abnormal expanse of sand to be scoured by the wind. It is, of course, possible that previous storms and high tides had played a part in preparing the situation through moving sand towards the shore by wave action and leaving uneven accumulations of it. At all events it is noteworthy that the epochs of widespread sand-dune activity on northwest Europe's coasts both in the last millennium before Christ and in the late Middle Ages were not only times of relatively cold, or cooling, stormy climate in this latitude but were also more or less centred around long-term maxima of the range of the tides.

COOLING AND WETNESS IN EARLY
FOURTEENTH-CENTURY EUROPE

The cooling trend, which should be seen as the basic element in the climatic deterioration with which this chapter is concerned, began to affect Europe directly soon after 1300. The generalized temperature curve presented in fig. 30 (p. 84) seems to be verified by the history of the vineyards in England and central Europe and of the upper limit of trees on the hills from the Vosges in the west to the Erzgebirge on the borders of Czechoslovakia, Germany and Poland. This smooth curve, however, masks the real shocks.

The change which broke the medieval warm regime must have appeared devastatingly sudden. It came first in the regions mentioned with the extraordinary run of wet summers, and mostly wet springs and autumns, between 1313 or 1314 and 1317. And it continued with little intermission at least to the early part of 1321. Moreover, this followed closely upon one of the really notable periods in the Middle Ages of mostly warm, dry summers, from 1284 up to 1311. (The first decade of the new century was a time when many had the confidence to start new vineyards in England.) The year 1315 (see fig. 72a), when the grain failed to ripen all across Europe, was probably the worst of the evil sequence which followed. The cumulative effect produced famine in many parts of the continent so dire that there were deaths from hunger and disease on a very great scale, and incidents of cannibalism were reported even in the countries of western Europe. Great numbers of sheep and cattle also died in the 'murrains' or epidemics of disease which swept the sodden and often flooded landscape. Thereafter the fourteenth century seems to have brought wild, and rather long-lasting, variations of weather in western and central Europe, the later 1320s and 1330s and also the 1380s with mostly warm, dry (often seriously droughty) summers and a few other decades, notably the 1360s, predominantly wet. In eastern Europe there seem to have been troubles with heat and drought in the summers throughout the century. The type of variability of the climate in western Europe here described, which affected the winters also, continued in the fifteenth century and spread to eastern Europe as well. The 1430s produced a very remarkable sequence of severe winters, or winters which at least included long severe spells, in central and western Europe, including 1431–2 (fig. 72b) and every winter from 1433–4 to 1437–8. Within the last thousand years only the 1690s seem to have produced so many cold winters or severe spells within the span of one decade. Furthermore, the winters of 1407–8 and 1422–3 had been of historic severity, permitting traffic over the ice across the Baltic and with wolves reported to have passed over the ice on the easternmost part of the North Sea from Norway to Denmark.

A graphical 'history' of the wetness of the Bolton Fell Moss peat-bog on the England–Scotland border near Carlisle, produced by a variety of

195

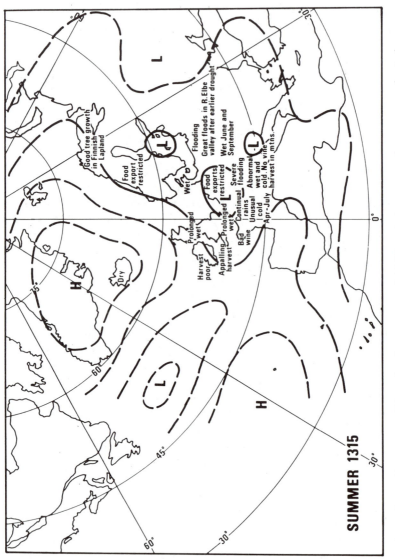

Fig. 72a Reported weather and crop conditions in the summer of 1315 and the proposed pressure and wind pattern which seems to be implied.

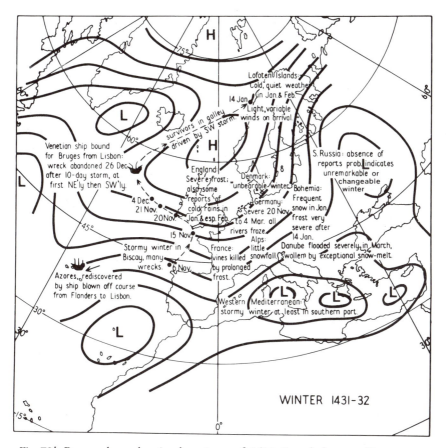

Fig. 72b Reported weather in the winter of 1431–2 and the prevailing pressure and wind pattern which seems to be implied.

researches, is shown in fig. 73. This seems to agree with the temperature and rainfall sequences presented elsewhere in this book and therefore may be regarded as supporting evidence of them. (There is an apparent discrepancy in the wetness indicated in the tenth century, but wetness in the northwest corner of England could be consistent with the pattern we have supposed at that time with westerly winds there and anticyclonic situations producing droughts in the southeastern half of England and in Germany.) The Bolton Moss curve certainly supports Trevelyan's contention[6] that the rivers of England were generally deeper and bigger in the fifteenth century than they are now (and, perhaps, earlier in the high Middle Ages). What is abundantly clear from fig. 73 is that there was a very great change in the prevalence of soil moisture, at least in northwest England, about 1300. The change seems in fact to have occurred much more widely, in view of

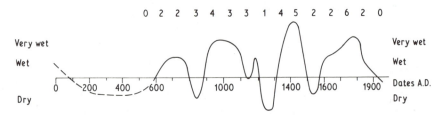

Numbers of data points each century indicating wet or very wet conditions

Fig. 73 The record of long-term variations of the surface wetness of Bolton Fell Moss peat-bog northeast of Carlisle, derived by Dr Keith Barber of Southampton University (from pollen and macrofossil analysis, soil chemistry investigations and records of land-use history). (From data kindly supplied by Dr Barber.)

the frequency of regrowth phases reported in the peat-bogs of Sweden and elsewhere in northern and western Europe about this date. Corresponding difficulties, caused by increasing wetness, were noted in the last chapter in the mines in the Harz and other mountains in central Europe.

Perhaps the most remarkable aspect, devastating in all its effects, of European climates in most of the decades studied in the fourteenth and fifteenth centuries AD was the extraordinary frequency of easterly winds which seem to have largely dominated latitudes between about 50–55 ° and 60–65 °N in the summers and winters alike.[7] This we deduce from the weather maps for individual seasons which we have reconstructed in similar manner to the ones here illustrated in figs. 72a and 72b. It certainly applies to the summers in the decades starting in 1310, 1330, 1340, 1420 and 1430, and to the winters in the 1420s and 1430s. The reconstructions were made possible by the availability in the literature of enough reports from around Europe of those seasons of dramatic weather, enough even to supply some support of each other. The decade maps were produced by averaging the maps of the individual years of the decades referred to. Among the most interesting are the maps for the summers of the decade 1310–1319, when there were famines and economic difficulties, and of the 1340s because of the extraordinary wetness of those summers all over western and central Europe followed by the heat of 1348 when the plague, the Black Death, arrived. Equally, the winters of the 1430s, which produced a remarkable number of spells of severe weather, produce an interesting decade map.

The climatic effects which marked those decades can nowhere have been stranger to our ideas of normality than in Norfolk – and probably in eastern England generally – where the usually dry climate owes most to the shelter from the prevailing westerly winds and their moisture provided by the hill ridges of southern and western England and the mountains of Wales and the Pennines. This shelter was replaced in such times by continual supplies of moisture carried by cold north, northeast and east winds from the North Sea and the Baltic. Norfolk particularly, but also much of East Anglia and northeast England, doubtless became much wetter places than we know today. This seems to be confirmed by the frequency of legal disputes in the local courts in those years in pursuit of people who failed to keep their drains running.

Another part of the map seemed also to be supported by a change of the rainfall experienced. In parts of the Alpine region screened from rainfall from the north by the high mountain ridges, notably in northern Italy (Val d'Aosta) and even in the upper Rhône valley and its tributaries (as in Saas), networks of water channels, essentially narrow wooden aqueducts, were built in this period to bring water from the streams that emerged from the glaciers of the high Alps to irrigate the summer pastures on the valley sides. One of these constructions even brought water from high up in the Val d'Ayas 25 km along the high cliffs of Mount Zerbion and over a pass to deliver it on the south side.

Changes in the geographical distribution of rainfall or snowfall, where mountain shelter is involved, can, as in these cases, provide a sensitive detector of changes of the prevailing winds – a distinctive indicator tool in reconstructing climate patterns. The harshness of the climatic effects of these changes in northern Europe, in Scotland and Scandinavia, in the late Middle Ages, brought about by the apparent frequency of northerly and northeasterly winds, is attested by the reports of harvest failures and populations reduced to making bread from the bark of birch trees, and the abandonment of the poorer and more exposed upland farm villages in those countries, and in northern England and Norfolk and the east Midlands besides.

A TIME OF DISEASES

The prevailing wetness during parts of the fourteenth century and, perhaps still more, in the fifteenth century undoubtedly made this an unhealthy time. There were many troubles with the diseases of mankind, animals and crops.

It seems established that in England the average expectation of life decreased by about ten years from the late thirteenth century (when it was apparently about forty-eight) to the period 1376–1400.[8] One of the most horrifying of the diseases of the period – and most clearly associated with the weather – was ergotism, or St Anthony's fire, produced by the ergot blight (*Claviceps purpurea*) which blackened the kernels of the rye in damp harvests. Even a minute proportion of the poisoned grains, baked in bread, would cause the disease. The course of the epidemics was such that the whole population of a village would suffer convulsions, hallucinations, gangrene rotting the extremities of the body, and death. In the chronic stage of the disease, the extremities developed first an icy feeling, then a burning sensation; the limbs then went dark as if burnt, shrivelled, and finally dropped off. Even domestic animals caught it and died. And pregnant women miscarried.

More often mentioned than this disease from the blighted corn in connection with the collapse of confidence and of the economic and cultural structure of Europe's medieval society has been the great bubonic plague, the 'Black Death' which arrived in 1348–50, and its subsequent recurrences. It is estimated that in different districts of Europe from one-eighth to two-thirds of the population died. The consequences in terms of harvests not gathered in, of labour shortage and rising costs, have been much written about. The death rate was heavy in the cities and ports and along much-frequented routes of trade and pilgrimage. Overall probably more than a third of the population of Europe succumbed to the pestilence. Interestingly, the Black Death seems to have originated in China, or in central Asia, in a region where bubonic plague is endemic, during or immediately after exceptional rains and flooding in 1332: this flooding was itself one of the greatest weather disasters ever known, alleged to have taken seven million human lives in the great river valleys of China, and destroying not only the human settlements and their sewage arrangements but also the habitats of wildlife, including of course the rats, over a wide region. Thus, there was a complex of factors in which climate was deeply involved, rather than the Black Death and economic troubles alone or the intellectual questionings of the time, which brought the end of the old medieval era.

DESERTION OF FARMS AND VILLAGE SETTLEMENTS

The fact that the climatic change played a part, independent of the debilitating effects of disease on the population and on the economy, can be seen in the failures of the northern vineyards in England and on the continent, in the retreat of corn-growing too from its former northern limits and of all cultivation from the heights, and in the depopulation of villages and farms. It is recorded in *Nonarium Inquisitiones*, a valuation of agricultural production in the year 1341, a few years before the arrival of the Black Death, that there were large numbers of villages with uncultivated land in every part of England, mostly said to be due to shrinkage of population since the famine years earlier in the century but also to soil exhaustion and shortages of seed corn and ploughing teams.[9]

This abandonment of former settlements was going on all over northern and central Europe and on the higher ground even in the south. The sites of many thousands of deserted medieval hamlets and villages have been

Fig. 74 All that remains of Whatborough, a village on the highest ground in Leicestershire, in the East Midlands of England. It is recorded that Whatborough declined in size between 1430 and 1446 and was deserted by 1495. (Reproduced by kind permission of the photographic air surveys branch of the Ministry of Defence, London.)

identified within the area of pre-war Germany alone.[10] In Germany and England the phenomenon became prominent in connection with the famines of the decade about 1315, but had begun even earlier, and was already reaching its first peak in the twenty years before the Black Death. Of over eighty deserted village sites for which population figures can be deduced from tax records in two counties in central England only about 10 per cent were attributable to the Black Death, but all had suffered severe losses of population in the famine times between 1311 and the 1320s. And those that did disappear around 1350 were generally the same places that had declined most (on average by two-thirds) in the years of famine earlier in the century. The fact that some villages disappeared and others survived in neighbouring positions in various parts of the country has caused many to doubt the climatic explanation, but it seems that these differences of fortune can often be explained by differences of soil and exposure.[11] The coincidence of timing of the waves of desertion over much of Europe points to a widespread, and presumably external, cause such as the behaviour of the climate. Moreover, the period when most desertions took place in England, between about 1430 and 1485, coincides with a fairly well-documented time of frequent cold winters and wretched summers, the latter particularly in the 1450s and later 1460s (fig. 74). But the climatic influence is hinted at most clearly in that it seems to have been Norway that, apart from Iceland and perhaps eastern Europe, was worst hit.

THE SEQUENCE IN THE NORTH OF EUROPE: NORWAY, DENMARK, SCOTLAND

We know a good deal about Norway in the Middle Ages and after, thanks to the wealth of information on taxes, occupations and properties in the 'church books' (kirkebøker) and the pioneer researches by Professor Andreas Holmsen of Oslo.[12] These have borne fruit in numerous indications of the interplay between climatic and environmental history and social history in northern Europe, and inspired the Deserted Farms Research Project (Ødegårdsprosjekt)[13] in which all the northern countries, including Finland and Iceland, have collaborated over many years.

The abandonment of farms began first in north Norway already before 1200, accompanied by an expansion of the areas used by Lapp hunters and a drift of the Norwegian population south and towards the coastal fisheries. At the end of the Viking period there must have been about a thousand farms in Hålogaland in north Norway, and they grew barley, oats and rye. By the 1430s, in and near the rich fishery districts in the Lofoten islands up to 95 per cent of the farms had been abandoned, and elsewhere about 60 per cent. At the coast, in fact, the numbers of the population and their economy expanded between 1350 and 1500, and it seems possible that, for so long, the increased cold water outflow from the Arctic near Greenland was compensated by a strengthening of the inflow of the warm Atlantic water with its fish stocks on the Norwegian side. (But later on, in the seventeenth century, the Norwegian fishery too seems to have been affected by the climatic deterioration.)

West Norway was the next to be affected, with some decline of population during the thirteenth century and reduction of the taxes in the 1330s and 1340s on account of lowered farm yields and losses caused by natural disasters such as rock-falls. Owing to the nature of the country there were big variations from district to district and from farm to farm. The decline was on the whole sharpest in the sheltered districts in the inner parts of the fjords and in Trøndelag, the district about Trondheim, which had been richest earlier in the Middle Ages. Wheat had been grown there. Particularly interesting is the case of the marginally situated upland farming village of Hoset (fig. 75), 350 m above sea level, east of Trondheim near the Swedish frontier. The place has been the object of interdisciplinary studies by Professors Sandnes and Hafsten and colleagues, notably Dr Helge Salvesen.

Archaeological work and pollen analysis show that a small area of the forest was cleared for cultivation, including cereals, about the fourth century AD. The farmers may have been attracted by the possibilities of iron production in the neighbourhood. Twice, or perhaps three times, since then the area has been abandoned and reconquered by the forest, each time in periods of colder climate. This is not surprising because in periods of prevailing southwesterly winds Trøndelag enjoys the shelter of the great mountains of southern Norway (and some additional warming of the south and southwest winds by an effect like that of the Alpine foehn wind[14]), but whenever winds from the northwest and north become prominent the district is directly exposed to these winds from the Arctic seas. Hoset may have been abandoned first for a time about the sixth to ninth centuries AD. There was, however, a climax of cereal cultivation there in the high Middle Ages, and the two later abandonments were precisely in the periods of sharpest climate stress in 1435 and 1698. Full-scale farming was not resumed there until about 1930.

In the most sheltered part of Norway, the central and southeastern part (Østlandet), the medieval expansion continued right up to the Black Death. It would be correspondingly easy to attribute all that followed to the disaster

Fig. 75 Hoset: a farm village at latitude 63° 24′ N 11° 10′ E, east of Trondheim, Norway, 350 m (approximately 1150 ft above sea level). The position is so marginal for agriculture that it was twice abandoned in periods of climatic deterioration in or about 1435 and the 1690s and reconquered by the forest. The first period of cultivation there in earlier times seems also to have gone into decline and possibly been abandoned some time between AD 500 and 900.

of the plague. The incidence of the disease itself was very patchy, the death rate amounted to 90 per cent of the population in the great Hallingdal valley, with its through route, and about two-thirds along the pilgrim route to Trondheim through southern Sweden, while blood group research suggests that the more remote parts of Telemark in central south Norway were never touched. But it is noteworthy that there was no real recovery in Norway for about two hundred years. The farms on the higher ground stood empty for that long, partly because any surviving occupants had been able to take up vacant farms on richer land in the valleys. But by 1387 production and tax yields were only from (in some districts) as little as 12 per cent to barely 70 per cent of what they had been around 1300. Even on the bishop's land near Oslo only oats were grown. And in the 1460s it was becoming recognized that the change seemed permanent. As late as the year 1665 the total Norwegian grain harvest is reported to have been only 67–70 per cent of what it had been about the year 1300, and in west Norway the medieval production was not exceeded until around the middle of the eighteenth century.[15]

In parts of Denmark, particularly Jutland, near the North Sea, the situation seems to have been not much better, with many farms deserted, corn growing given up and those farmhouses that were still maintained were shared by several families.[16] English visitors to a Danish royal wedding in 1406 reported seeing much sodden uncultivated ground and that wheat was grown nowhere. There was, in fact, a gradation across the country with much less stress in the more sheltered districts of the islands of Fyn and Sjaelland farther east.[17]

It is clear that the changes registered in agriculture and husbandry in various parts of Europe in the late Middle Ages were influenced by impact of the climate as well as by the disastrous depopulation brought by the Black Death. The growing season everywhere shortened, perhaps typically by three weeks or more, its accumulated warmth decreased and the frequency of harvest failures increased – the dreaded 'green years' when the crops fail to ripen – in the north. Wheat has a rather higher requirement of summer warmth than barley or oats and thrives best in regions where the yearly rainfall is less than 90 cm; but it can be successfully carted wet for drying indoors, whereas the other cereals soon overheat. Rye withstands severe winters better than other cereals and is the most productive grain on poor soils. As the climate deteriorated, barley, oats and rye were therefore to be preferred to wheat except in the warmer parts of Europe. On the other hand, there were many places where cereal growing ceased to be profitable and was given up in favour of sheep rearing to meet the increasing demand for wool.

In the Highlands of Scotland, it seems, the long history of clan warfare and of the Highlanders raiding cattle from the Lowlands, as also in this period the cattle raids from the Southern Uplands across the border into

England, may be explained by the stress of a deteriorating climate upon the settlements which had been established far up the glens in the 'golden age' of the twelfth and thirteenth centuries. As early as the 1070s and 1080s, when King Malcolm III and his queen (who became St Margaret) held their court in Dunfermline, Scotland was a haven for innumerable English exiles from Norman rule. But as in Iceland and in north and west Norway, much of the country is exposed to a drastic change whenever northwest and north winds become more frequent at the expense of the benign southwesterlies. Internal troubles of various kinds, not all connected with the incursions of the English king and his forces, began about 1300. The fifteenth century historian Boece wrote that in 1396 all the north of Scotland was engulfed in clan warfare. There was more of it in 1411 and fairly clearly the fifteenth century was the peak period for such troubles. It was in 1433 that the estate of the Earls of Mar who had ruled the central area of the Highlands collapsed, and poverty rapidly worsened in the region. In that decade of the 1430s in the Scottish Highlands, as in Sweden, bread had to be made from the bark of trees for want of grain. And in the accompanying unrest, in 1436, King James I of Scotland was murdered when hunting on the edge of the Highland region near Perth. It was then that it was decided that at no place north of Edinburgh Castle could the king's safety be guaranteed, and so Edinburgh became the capital of the country. In the same decade, the severity of which in other parts of Europe we have already noted, dearth and famine were recorded for the first time in the annals of Dunfermline. And W. G. Hoskins estimates that the famine in England in 1437–9[18] was second only to that in 1315–17.

The physical background to these developments in the history of Scotland shows itself in the fact that the upper limit of cultivation on the Lammermuir Hills[19] southeast of Edinburgh, which had been as high as 425 m (nearly 1400 ft) above sea level at one point in the mid-thirteenth century, fell in stages until by 1600 it was 200 m lower. Over the period from 1300 to 1500 on the hills of continental Europe, from the Vosges in the west, through middle and southern Germany to Czechoslovakia, the upper tree line fell by 100–200 m. And after 1300–1430 the upper limit of vineyard cultivation in Baden in southwest Germany was brought down by 220 m. These height changes tend to verify the approximate magnitude of the change of summer temperatures as derived in fig. 30 (p. 84). We also have a register of the climate of this whole period in the yearly growth rings of larches near the upper tree line near Berchtesgaden in the German Alps:[20] between 1330 and 1490 the rings were of unusually variable width, but from 1490 to 1560 there was a period of good growth. Decline followed and from 1590 the growth rings have on the overall average only had half the width of the 1490–1560 period, though 1770–1810 and 1850–1950 appear as relatively good growth periods.

CENTRAL, SOUTHERN AND EASTERN EUROPE

The changing climate with its enhanced short-term fluctuations, including some runs of three to five years, or even more, of wet, flood-ridden seasons, of droughts and either severe or very mild winters, made itself felt also farther south in Europe after 1300. The wheatlands and the vineyards of northern France shared in the harvest failures and the resulting famine and deaths by the million in the decade beginning in 1310. Ladurie[21] has shown how the dates of the southern French wine harvests beginning in 1349 (but only forming a continuous series from about 1550) can be used as an index[22] of the climate. And K. Müller[23] derived a similarly informative index, from the early Middle Ages to our own times, from the percentage of the wine harvests in south Germany which were reported as good in different periods. Although the early records are fragmentary, the German record shows a decline from figures ranging between 30 and 70 per cent before 1300 to figures never above 53 per cent and at times under 20 per cent between 1400 and 1700.

In the widespread famines of the 1420s and 1430s there were reports of cannibalism in eastern Europe, as there had been also in the west in the 1310s. The repeated famines gave rise to an emigration from Russia westwards into Germany, (It would be useful to have an estimate of the size of this population movement.) And in the severe winters in the 1430s the wolves were active in many parts of Europe, from Smolensk in the east to England in the west. (In England, but not in Scotland or Ireland, this may have been the last time that wolves were reported.)

It was not only in the Highlands of Scotland that there was turmoil in the fifteenth century period of climatic stress. In Denmark and in what is now the southern province of Sweden (Skåne) the deepening crisis in agriculture led to a drift to the towns and by the end of the century apparently to a more general emigration affecting the towns as well. In Bohemia the 1420s and 1430s saw the Hussite risings; and, although these were basically concerned with religious and political ideas of democracy and independence, we may suppose that the times of bad weather and harvest failures made many people rootless and more readily persuaded to join the conflict. Something of the same influences may have applied in England, where the Wars of the Roses dragged on from 1455 to 1485: Trevelyan[24] mentions that, although the common people were probably little concerned about the dynastic causes of these campaigns, the effect of starvation and the run-down state of the country on soldiers returned from France and the Hundred Years War probably encouraged them to enlist. In many, perhaps most, parts of Europe – in England, Sweden and south Germany, for example – it was in the fifteenth century that the main abandonment of the small, unsuccessful settlements occurred. In England John Rous of Warwick, writing in 1485, listed fifty-eight sites, mostly in that one county, which had become depopulated in his life-time.[25] There was much agitation about the conversion of previously tilled land to other use, usually sheep rearing (with the shepherds using any abandoned houses, which had not fallen or been pulled down, for shelter). And the landlords who organized the conversion and enclosed the land became a focus of hostility. In Germany the rising civic pride and splendour in the merchant cities in the fifteenth century seems to have been linked to some extent with the drift to the towns and the protection which they offered against the lawless state of the countryside, in which peasant revolts grew worse until the general rebellion in 1525.

In European Russia a greater proportion of the apparently increasing climatic troubles after 1300 seem to have been due to summer droughts than farther west. This trend seems to be confirmed by the general decline of ring widths shown by the timbers used in the successive surfacings of the streets of medieval Novgorod. There seems also to have been an increasing incidence of severe winters. And the impression given by the chronicles of the monasteries[26] is that the results were of a severity in terms of famine and loss of life, and indeed of the frequency of such events, unmatched in western Europe except in a few decades such as the 1310s, 1430s and 1690s.

In southern Europe, although we have so far disappointingly little direct evidence of the climate in the fifteenth century, grain prices and vintage dates alike suggest that there were no severe effects in the 1430s nor from other parts of the period between about 1420 and 1480 which produced so many harsh seasons farther north. (Fig. 33a, p. 88, suggests that the southwest peninsula of England escaped similarly.) Preliminary meteorological analysis of the 1430s indicates an extraordinary predominance of blocking anticyclones over northern Europe. The southerly winds at the western limit of the anticyclones could well explain the impression that this period was one of some recovery in Iceland. If this analysis is right, the fifteenth century probably saw an abnormal amount of cyclonic activity in parts of the Mediterranean, giving more rainfall than is now normal there but few extremes of temperature. It is greatly to be hoped that the documentary archives of the Spanish and Italian cathedrals will some day be systematically studied for what they may contain in the way of direct information on the climatic history of the Mediterranean region.

DEVELOPMENTS IN AFRICA AND INDIA

Farther south again, in the desert regions of north Africa the writings of the great Arab geographers indicate that there was more moisture than now all through the high Middle Ages and after, from the eleventh to the fourteenth centuries. This probably applied to Arabia too. There are descriptions of journeys across the Saharan region[27] from the north African fringe to Ghana and Mali and to the Kufra oasis (24–25 °N 22 °E), in the eastern desert. The desert did not extend north of latitude 27 °N. Crossing of the uninhabited region took two months, but even there, on a journey in 1352, it was reported that a large number of wild cattle often approached the caravan. By that date it seems, however, that a drying tendency had set in, since it was also remarked that the rearing of beef cattle had been given up in the Kufra region. Formerly, great herds had found pasture there in regions which had become desert. From the thirteenth century until the fifteenth there was a Mali empire, which at its height between 1307 and 1332 is said to have covered most of west Africa. In 1325 the Mali sultan built a royal palace in Timbuktu and a tower for the mosque. After the temporary loss of Timbuktu, Mali power was restored there in 1353 and continued until it was abandoned to the Tuareg nomads in 1433. Although it can never be safe to deduce climatic changes from human political history, in this extreme region the events described most probably confirm that drying out of the desert region was proceeding and causing increasing difficulty. In the meantime, we do know from the pollen analysis researches of J. Maley of the Université des Sciences et Techniques du Languedoc at Montpellier in France that in the Lake Chad Basin there was a maximum occurrence of the pollens of the plants of the Sudan-Guinean monsoon zone flora between about AD 700 and 1200 and that these and other water-demanding plants declined rapidly over the period 1300–1500. A curious feature of the period between the moisture optimum around AD 700–1200 and the greater difficulties experienced in this area in the Little Ice Age is that there were successive waves of human migration southwards at two-hundred year intervals, in the thirteenth, fifteenth and seventeenth centuries. In these regions we may hope for further and more direct information from the Arabic libraries, which are reported to contain records of at least the more important years of drought, and from the continuing studies and dating of the former levels of Lake Chad and other African lakes.

The position is rather similar regarding the climatic sequence in the Indian subcontinent. K. S. Lal[28] has described the sources of information on famines and population in India during the Middle Ages and after. Although the data on famines and behaviour of the monsoon in this early period have not been analysed yet, the population estimates are interesting, since they once again produce a sequence which (apart from the under-lying long-term increase) roughly parallels our estimates of the temperature trend in higher latitudes. According to Lal, the best estimates of the total population of the subcontinent rise to a maximum, around 200 to 300 millions, about AD 1000, already fall slightly to 190 to 200 millions about AD 1200 and to 170 millions in 1388, followed by a sharper fall to a minimum, around 120 millions, between about 1525 and 1550. Around 1600 a population of about 130 to 140 millions is suggested. When all allowance is made for the effects of wars and massacres, it seems likely that famines and disease must be the main explanation for the fall of population on such a scale as is indicated in the late Middle Ages.

THE SEQUENCE IN NORTH AMERICA: HOW THE PRE-EUROPEAN CULTURES WERE AFFECTED

When we turn our attention to North America, the researches carried out by the Institute for Environmental Studies in the University of Wisconsin at Madison under Professor R. A. Bryson indicate drastic population shifts, the timing and the nature of which point strongly to a meteorological explanation. About AD 1000 the Amero-Indian people had been growing corn all across the high plains from the base of the Rockies, through eastern Colorado and western Nebraska; and farther east there were substantial settlements in the river valleys, where oaks and cottonwoods grew. One of these places, now known as Cahokia, in southern Illinois just east of St Louis, is estimated to have had a population of 40,000 people. From pollen analysis studies, and from counting the bones of different animals found in the refuse dumps (or kitchen middens) of these farming and hunting communities. Bryson and Baerreis have found that the scene underwent a rapid change after AD 1200. There was least change in the valleys close to the watercourses, but the oaks disappeared in most of the places where they had grown, and the overall numbers of trees declined in favour of the plants of the prairie. And among these the shorter grasses gained at the expense of the bigger, more moisture-demanding types. At one site investigated in northwestern Iowa the increase of grass pollen from a negligible proportion to about 70 per cent of the not-tree pollens took only forty-five years or less. Correspondingly, the forest animals, the deer, gave way to bison in the people's diet. These are signs of a significant decrease of rainfall. Moreover, this suggestion accords with the idea of increased dominance of the west winds, generated by the increased north to south gradient of temperature at a time of cooling of the Arctic. This would extend the rain-shadow effect of the Rocky Mountains farther east than before and intensify the dryness within it. The picture is completed by wholesale abandonment of the settlements after about AD 1200. At first, it seems the smaller villages in the driest areas were deserted and people tended to congregate in the bigger places in the river valleys. But ultimately even the biggest of them, Cahokia, was abandoned, seemingly about 1300; and when the first European (French) traders arrived in the area in the eighteenth century they found only scattered, small Indian settlements. As Bryson and his collaborators have demonstrated,[29] the rainfall pattern over the United States in one of those summer months in modern times that have more than usual development of the westerly winds typically produces a long eastward-pointing 'finger' of severe rainfall deficiency, exceeding 50 per cent, an extension of the rain-shadow of the Rocky Mountains. And this feature is so placed that the main concentration of the village sites of the Mill Creek culture in the northern Middle West and Cahokia itself lay close to its axis. In the always drier parts of the plains nearer the Rockies the change in the thirteenth century was plainly catastrophic: all the small village sites there were soon abandoned.

Bryson estimates that the period of extreme dryness lasted two hundred years and coincided with the strong development of the circumpolar vortex which carried the westerlies mostly far to the north in the European sector, accounting for the warm periods of the thirteenth and fourteenth centuries there but also for the vigorous development of the cyclonic rains in Europe around 1315 when the westerlies came farther south. Much farther south, over the southern plains in northwestern Texas and adjacent parts of Oklahoma, the same investigation indicates that there would be an increase of rainfall – as appears to have occurred in the Mediterranean in the high Middle Ages – and it may have been substantial. It is presumed that it was towards these regions that the former population of the northern plains went. Certainly, archaeology indicates that the numbers inhabiting the so-called Panhandle region of Texas rapidly increased around AD 1200. Karlstrom and his associates in the US Geological Survey and in the University of Northern Arizona at Flagstaff have found that the Indian populations on the Colorado plateaux and neighbouring parts of northern Arizona and New Mexico experienced changes corresponding to those over the northern plains.[30] The economy was based on maize, squashes (i.e. plants of the pumpkin family) and beans, and some wild plants, supplemented by hunted game. The population had been increasing and spreading over the area from AD 550 or thereabouts, until between 800 and 1150 almost every habitable part of the plateaux was occupied. It was these people who created the great cliff dwellings of the Mesa Verde and built the many-storeyed stone villages and towns of Pueblo Bonito and the Chaco Canyon in the tenth to thirteenth centuries. They also built water control channels, roads and signalling stations. But after 1150 many areas, especially in the higher parts, were deserted in favour of positions along the bigger stream courses and more control channels for water for irrigation and domestic use were provided. After 1300 the former homelands were almost entirely deserted and the population had moved south and southwest, along the Rio Grande and to the Hopi Mesas area in central Arizona. The associated environmental changes, particularly in terms of moisture availability and water table, were demonstrated by pollen studies and tree ring work, which was also used for the dating.

12

THE LITTLE ICE AGE
Background to the history of the sixteenth and seventeenth centuries

THE SIXTEENTH CENTURY

During the sixteenth century we reach the period from which a great many more documentary reports of the weather survive. This is particularly true for Europe, where the reports are increasingly specific, verifiable and often precisely dated. But also around this time documentary reports begin to be available for other parts of the world. And the middle and later seventeenth century provides the earliest instrument observation records. These, like the evidence of the glaciers in many parts of the world and of the Arctic sea ice, introduce us to a colder climate than that of the twentieth century. In England the late seventeenth-century thermometer record indicates annual mean temperatures about 0.9 °C (1.6 °F) lower than in the period 1920–60. Over the years 1690–9 the deficit was 1.5 °C (2.7 °F).

The temperatures which we derive from the sixteenth-century material available for England (fig. 30, p. 84) and other parts of Europe, like the indications of tree ring width in California (fig. 52, p. 141) and the palaeotemperatures indicated by isotopic studies of the calcite in a cave in New Zealand,[1] point to generally rather warmer conditions between about 1500 and 1550 than in the previous century. We cannot yet say whether this was (in terms of the wind circulation patterns) any sort of counterpart to the temporary recovery in part of the fifteenth century registered by isotope measurements on the north Greenland ice (fig. 36, p. 93) and which seems to have affected Iceland too. The warmth of the early sixteenth century in Europe was probably produced by rather frequent anticyclones affecting the zone near latitudes 45–50 °N and westerly winds over northern Europe, whereas the previous century – like the period from 1550 to after 1700 – was characterized by a remarkable frequency of anticyclones north of 60 °N and winds from between northeast and southeast over Europe south of that latitude.

Despite the mostly genial character of the period 1500–50, there were at least three winters in England with enough severe weather to freeze over the Thames in London – it froze more easily in the days before the tributary streams were put into pipes and the new bridges allowed the tides to reach so far up the river – and the summers of the 1530s on the continent alternated in quality so strongly that graphs of the tree ring record from the oaks in Germany and the vintage dates recorded in France and Switzerland produce a regular saw-tooth zig-zag appearance.[2] (This *Sägesignatur* is a prime example of the more or less biennial, or alternate years, cycle which is present and at times prominent in many series of climatic data.) These observations suggest that the warmth of the 1500–50 period in Europe did not quite match that of 1900 to the 1950s, though the difference was probably not great. From examination of weather diaries covering the years 1508–31 from two places in Bavaria (Eichstatt and Ingolstadt) Professor Flohn found no significant difference of the winter temperatures from the level of 1880–1930, but the summers were on average slightly (7–8 per cent) wetter and by implication less warm.

In the middle of the sixteenth century a remarkably sharp change occurred. And over the next hundred and fifty years or more the evidence points to the coldest regime – though accompanied by notably great variations from year to year and from one group of a few years to the next – at any time since the last major ice age ended ten thousand years or so ago. It is the only time for which evidence from all parts of the world indicates a colder regime than now. This may reasonably be regarded as the broad climax of the Little Ice Age, though we can distinguish some severer years and decades within it and others that were less so. From another point of view it would be reasonable to regard the whole period between about 1420, or even 1190, up to 1850 or 1900 as belonging to the Little Ice Age development.

THE CHANGES IN CENTRAL EUROPE:
1500s TO 1800s

In a weather diary kept at Zurich from 1546 to 1576 the relative frequency of snow among the snowy and rainy days of winter was 44 per cent up to 1563 and 63 per cent from 1564 onwards. From this and similar statistical studies of other data Flohn has concluded that the mean winter temperature from 1560 to 1599 in central Europe was about 1.3 °C lower than in 1880–1930 or the first half of the sixteenth century, while Tycho Brahe's observations in Denmark from 1582 to 1597 seem to imply winter temperatures 1.5 °C lower there than in about the same modern period used for comparison. We also know from the Danish observations, as well as from a survey of ships' experience on the seas between the Netherlands and southern Europe, that easterly winds became prominent. In Tycho Brahe's observation series in Denmark southeast was the commonest single direction over the year as a whole, and northwest winds were as common as southwest. No equally reliable assessment of the summer temperatures has been possible yet in the absence of instrument measurements, although work on the stable isotopes of the chemical elements present in tree rings may lead to such a result when the problems of interpretation are better understood. Meanwhile, we note that the proportion of good wine years between 1550 and 1620 in Baden in southwest Germany was rather under half the frequency between 1480 and 1550. In fig. 30 (p. 84) we showed that the summer temperatures derived for England in the late sixteenth and seventeenth centuries averaged 0.6 to 0.8 °C below those of 1900–50 or the earlier sixteenth century, and it is likely that the difference would be a little greater on the continent. It seems reasonable therefore to attribute much, if not all, of the sharp increases of grain prices seen in figs. 33a and b (p. 88) to the climatic change.

We can follow the course of the changes over three centuries in central Europe in some detail in fig. 76, and I am greatly indebted to the Swiss historian Dr Christian Pfister of Bern for allowing me to reproduce these results of his close analysis of the wealth of documentary reports from Switzerland. The graphs show a progressive cooling of the winters from the 1540s to the end of that century, which was repeated after the recovery to the 1620s and culminated in the very cold 1690s. Another recovery followed, but the winters of the 1750s to 1780s were again on average cold. The springs and summers largely pursue parallel courses, but some of the changes from decade to decade appear sharper than those that affected the winters. The springs of the 1690s and the summers of the 1570s and 1810–19 appear as outstandingly cold. The warm summers of the 1550s were dry in Switzerland, but there followed more than half a century of predominance of wet summers, notably so from the 1570s to the 1620s inclusive. There were recurrences of the wet character of the summers in Switzerland in the 1690s, in the 1720s and 1730s and from about 1760 to the 1780s. None of the other seasons of the year showed long periods of wetness during the Little Ice Age, except the autumns between about 1760 and the first decade of the nineteenth century and to a lesser extent the autumns of the 1550s, 1570s and 1590s and the period 1690 to the 1720s. The most noteworthy feature of the autumns is that they continued warm – in fact it was the Augusts, Septembers and Octobers which continued warm – until the 1560s, even rising to a maximum of warmth in that decade. There were other peaks of autumn warmth in the 1630s, 1660s and 1680s and in the 1770s, in all cases followed by swings to a much colder climate which usually included a sharp change to colder autumns as well. The coldness of the autumns in the 1690s, as with the other seasons, stands out.

It is not surprising that the generally cold wet years that predominated from about 1570 to 1600, and from 1690 to 1740, produced great advances of the glaciers in the Alps (fig. 77). (The report of a traveller, Sebastian Münster, in 1546 tells us that the Rhone Glacier in Switzerland was already

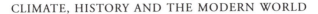

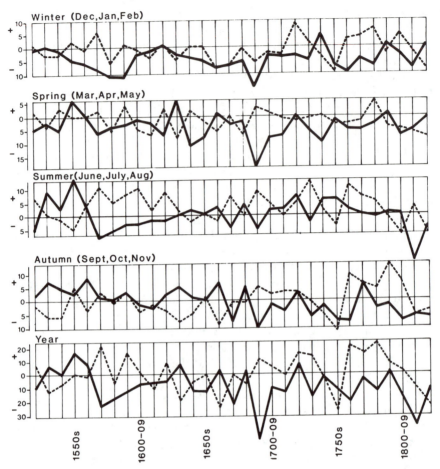

Fig. 76 Thermal index (bold line) and wetness index (broken line) for each season of the year near Bern and Zurich, Switzerland, averaged by decades from 1525–9 and 1530–9 to 1820–9. The indices are defined by the numbers of unmistakably warm (or wet) months minus the numbers of ummistakably cold (or dry) months (respectively) indicated by available documentary records. (Kindly supplied by Dr Christian Pfister of the Geographical Institute, University of Bern, Switzerland and reproduced by permission.)

as far forward as it was around 1900, reaching the broad valley bottom at the foot of the steep ascent to the Furka pass, though by no means reaching the size it had in the eighteenth century.) Pfister has examined the later eighteenth century recurrences of cold wet years in more detail, using the instrument readings of the climatic observation network set up in 1759 by the Economic Society of Bern.[3] There were strong short-term fluctuations: the years 1759–63 and 1778–84 had a warm tendency; but the period

214

Fig. 77 The Rhone glacier viewed from the same viewpoint:
(a) (top) in 1750; (b) in 1950.

1764–77 was notably cold, the summers being rainy on the Swiss lowlands and snowy in the Alps, the winters long and snowy, especially around 1770. The summers were too short to melt the snows on the alpine pastures, and the glaciers advanced strongly; and in 1769–71 the dearth of wheat, potatoes and milk produced famine. There was a brief repetition of conditions similar to these in 1812–17, with famine in 1816 and 1817.

Pfister has also been able to study weather diaries from the areas around Bern and Zurich from the year 1683 onwards and finds that the average number of days a year with snow covering the ground was about 70 in the first twenty years, 75 in the 1690s, though in the ten years 1705–14 it was as low (42 days) as in the decade (the 1920s) of mildest winters in the present century. More remarkably, the winters of 1684–5, 1730–1, 1769–70 and 1788–9 produced totals of about 110–112 days around Zurich and in 1784–5 over 150 days in Bern. It is thought that 1613–14 also had about 150 days. These figures are to be compared with the total of 86 days recorded in 1962–3, the longest winter in Switzerland of the last hundred years. By statistical methods Pfister has been able to derive temperatures from the weather observations in these diaries and finds that the mean winter temperatures in Zurich in 1683–1700 were 1.5 °C below the 1900–60 average (which agrees well with the departure derived for central England, from the data used in fig. 30, p. 84). The greatest deviation from modern times was, however, in the months of March, which averaged 2.2–2.7 °C (4.9 °F) colder than in the present century. March was a full winter month and in all the extreme winters mentioned in this paragraph had a complete snow cover throughout. In 1687 at Einsiedeln (882 m above sea level) this applied to April also, and in the three years 1699–1701 the snow cover lasted until 15 May, implying mean temperatures 4–5 °C below the modern average. The effects of these years on the Swiss farms were drastic. It seems that the grain crops suffered from attacks of a parasite, *Fusarium nivale,* which is active under snow cover in spring in Scandinavia and northern Germany but is not known in Switzerland today. And the stocks of hay for the animals ran out when the snow still lay in March and April, so that the cattle had to be fed on straw and pine branches and many cows were slaughtered.

Having gauged the situation in Europe where we have these details, let us now survey the situation around the world as we have done for earlier periods in the preceding chapters.

ICELAND AND THE ARCTIC FRINGE

Greenland, as we have reported (see also fig. 23, p. 60), was already cut off by the spreading of the Arctic sea ice. And by the 1580s the broad Denmark Strait between Iceland and Greenland was in several summers found entirely blocked by the pack-ice. In Iceland the effects were most

216

severely felt in all the northern districts and in the east and southeast of the island. Later, around the end of the seventeenth century, there are documentary records of the advancing glaciers overrunning farms. And from about 1480 onwards there had been disasters, entailing losses of farmland, through glacier bursts connected with volcanic activity under the ice and consequential river floods bringing torrents of sand and gravel with them. These *jökulhlaupar* characteristically lasted about one week, and are estimated to have brought a maximum river flow a thousand times that of the Thames.[4] During these times, therefore, there was a general drift from the farms towards the coast, and an increasing activity in the fishing, in the more sheltered southwest of the country. The overall decline of the population, which we have noted in the last chapter, suggests that there may also have been at least some emigration out of the country altogether. The sea ice was tending (albeit with many shorter-term fluctuations) to increase further and in the worst year, 1695, surrounded the country entirely so that no ships could come in for many months. We have referred in chapter 4 to the still greater spread of the Arctic cold water. In these circumstances the cod fishery, which had been the island's relief, ultimately also failed, even in the southwest of the country, for twenty years, from 1685 to 1704. The primitive equipment used by the Icelanders for their fishery in those times played a part in the failure, for foreign vessels operating 20 km off the south coast were able to obtain cod.[5] At the Faeroe Islands the fishery failed for thirty years, and it seems that the cold water was extensive in that direction. As late as 1756 the Arctic sea ice was again at the coasts of Iceland for thirty weeks.

During the period of these manifestations of strong cooling of the Arctic, and its spread to middle latitudes, there were two sectors of the far north, in Alaska and Lapland (northernmost Finland), where the tree ring records show that a more genial climate allowing good growth continued until 1580 or somewhat after. This can reasonably be attributed meteorologically to frequent anticyclones, with sunshine and some southerly winds, in those sectors (which are still particularly prone to blocking anticyclones today). These were doubtless the same anticyclones that were responsible for the frequent northerly and easterly winds over much of Europe and North America at that time.

GREAT STORMS AND COASTAL FLOODS IN EUROPE

As we have remarked in connection with fig. 23, the spread of the Arctic ice to Iceland and of the polar water to the region of the Faeroes meant that the surface of the North Atlantic between there and southeast Iceland became 5 °C colder than is usual today. Consequently, there was a greatly strengthened thermal gradient between latitudes 50 and 61 to 65 °N. This

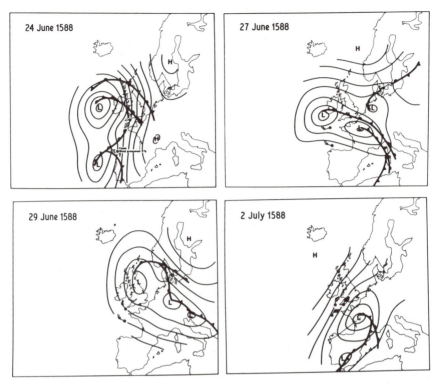

Fig. 78 A series of four daily weather maps from the summer of 1588, analysed with observations from ships of the Spanish Armada and the Danish astronomer Tycho Brahe.

seems to have been the basis for the development of occasional cyclonic wind storms over this part of the North Atlantic exceeding the severity of most of the worst storms of modern times. This is suggested by the many coastal disasters from sea floods – even at a time of slightly lowered sea level (as indicated by the first tide gauge, installed at Amsterdam in 1682) – and erosion and blowing sand. It most clearly indicated by meteorological analysis of the weather reports available from the Spanish Armada in 1588 (fig. 78). The analysis of the weather situations on sixty days during the Armada's expedition fixes the positions of the depression centres with sufficient accuracy to indicate that their rates of travel on at least six occasions during that one summer corresponded to jet stream winds at the limit of, or beyond, the maximum speeds expected from modern experience.

The development of great storms in this zone continued: from the vast North Sea floods and loss of life on the continental coasts in 1570, mentioned in the last chapter, to the permanent losses of land from the

Danish, German and Dutch coasts (and demolishing of the island of Nordstrand) in the storm of 21 October 1634; to the formation of the Culbin Sands in northeast Scotland in 1694, and the overwhelming of a four-thousand-year-old settlement site in the Hebrides with sand in 1697; to the great storm which passed across southern England on 7–8 December (New Style calendar) 1703 and was described in careful detail by Defoe.[6] The Eddystone lighthouse near Plymouth was blown down, as were houses in towns and countryside all across England to the east coast; the damage in London alone was estimated at £2 million; enormous numbers of trees were blown down; and many ships were blown up-river, or lifted beyond the usual reach of the tides, or wrecked on the coast and at sea; 8000 lives are said to have been lost. Despite the severity of these floods on the continental side in 1634, 1671, 1682 and 1686 and again at Christmas 1717, in each of which some thousands of people were drowned, and losses of land on many other occasions besides, it is unlikely – in a cold epoch of apparently more or less world-wide extent with glaciers generally in a state of growth – that the general sea level was as high as it had been in the Middle Ages around AD 1000 and between 1200 and 1400, though the difference may have been only of the order of 50 cm. The frequency of such floods between 1570 and about 1720 must be attributed to greater storminess.

EFFECTS IN SCOTLAND

A bizarre occurrence – serious for the individuals concerned – presumably resulting from the great southward spread of the polar water and ice was the arrival about the Orkney Islands a number of times between about 1690 and 1728, and once in the river Don near Aberdeen, of an Eskimo in his kayak. The situation in Scotland itself became serious. The recognition, based on the reports of sea ice and the fisheries, particularly the cod fishery, that the ocean surface between Iceland and the Faeroe Islands – only a few hundred kilometres to the north of Scotland – was probably 5 °C colder than it usually is today at last makes sense of the numerous reports by learned travellers of the time of permanent snow on the tops of the Cairngorms and elsewhere on the Scottish mountains. The cod, which thrives best in rather cold waters at between 4 and 7 °C, serves as a valuable indicator in this connection, because its kidneys fail at temperatures below 2 °C and it therefore cannot venture into colder seas. The cod fishery at the Faeroe Islands began to fail about 1615, and did so increasingly until, as we have noticed, there were no cod thereabouts for thirty years between 1675 and 1704. In the worst year, the same year 1695 in which the ice surrounded Iceland, cod became scarce also in Shetland waters and disappeared from the entire coast of Norway (except for a colony apparently surviving in the inner part of Trondheim fjord). It seems safe to infer that the Arctic cold water had spread across the surface of the whole Norwegian Sea. And although there was some immediate improvement the next year, the sea conditions seem to have been significantly colder than today until well after 1800.

The course of the development in Scotland and the periods of most severe climatic stress can be identified in the records of famines brought together in fig. 79. The information used in this diagram was mainly compiled from the economic records, annals and chronicles surveyed by Lythe and Smout.[7] Although most of the data relate to eastern Scotland, there are indications that the situation was worse in the north and in the poorer Highland districts in the west. The experience of recurrent famines in the later decades of the sixteenth century was at work in the movement of emigration from Scotland, then beginning, which was destined to became a well-known theme in the following centuries. Smout writes that 'the stimulus to leave Scotland was compounded of many factors, of which the general poverty and discomfort of the native land was the most obvious . . . Ulster and (later) America offered empty territory; Holland and England offered mercantile fleshpots; Russia, Sweden, Denmark, France and all the petty princedoms of Germany offered military opportunity' (p. 90). The Scottish mercenary soldier who figures in the writings of Sir Walter Scott was a familiar figure in the wars which troubled Europe in the seventeenth century, particularly in the service of the Swedish king in central Europe in the Thirty Years War: 'by 1660 the stream of military migration had fallen off. . . . Nevertheless even in 1700 there was hardly an army north of the Mediterranean without Scottish officers of some sort' (Smout, p. 92). But the most serious legacy of this time survives to our own day in the 'plantation' in 1612 of Scots farmers in the richer lands and more sheltered climate of Ulster in northeast Ireland after first evicting the native Irish. This seems to have been a device of King James VI at one stroke to stabilize the Irish political and religious situation in his favour and to relieve the impact of harvest failures in Scotland, by taking advantage of the power over Ireland that fell to him on his accession to the throne of England. In modern terms, it would surely be regarded as a model of how not to conduct international relations and a characteristic abuse of (near-)absolute power. It is estimated that by 1691 there were 100,000 Scots in Ulster, already about a tenth of the population of Scotland, and their numbers were soon to be swollen again by emigrants abandoning their Scottish homes in the disasters of the 1690s. Unlike the protégés from elsewhere who were introduced into Ireland in the seventeenth century, these were mostly humble folk who tilled the soil themselves.

If the Ulster plantation of 1612 was related in any way to the dearth in Scotland in that year, which doubtless awakened unhappy memories of the 1590s, it seems to have been an over-reaction. For more than sixty years dearths and famines were less frequent in Scotland than they had

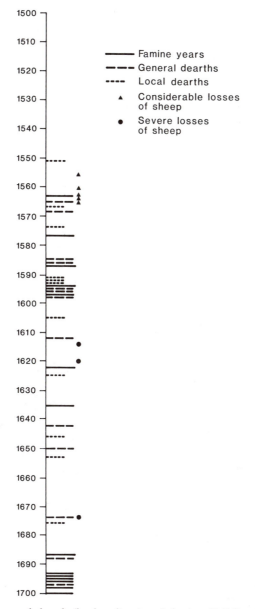

Fig. 79 Years of reported dearth (broken lines) and famine (full lines) in Scotland between 1550 and 1700. The information is mainly from eastern Scotland. Years with severe losses of stock (sheep and cattle), usually because of snow, are marked by dots.

been in the last forty years of the previous century. But from about 1670 the situation deteriorated again, with tremendous snows and frosts in that year and huge losses of sheep in the thirteen days of continously drifting snow in early March (by the modern calendar) 1674. Worse was to come in the last years of the century, when between 1693 and 1700 the harvests (largely oats) failed in seven years out of eight in all the upland parishes of Scotland. There are many accounts of those years parish by parish in the volumes of the *Statistical Account of Scotland* compiled by Sir John Sinclair a hundred years later. 'The poorer sort of people frequented the churchyard to pull a mass of nettles, and frequently fought over it . . . which they greedily fed upon . . .' (parish record of Duthil and Rothiemurchus in north central Scotland). Some were reported to have sold their children into slavery. In parishes all over the country from one-third to two-thirds of the population died – a greater disaster in many places than the Black Death – and great was the fear of being buried in a mass grave. Some whole villages and wide tracts of the countryside were depopulated at this time (fig. 80). And Andrew Fletcher of Saltoun in Midlothian appealed in the Scots parliament in Edinburgh in 1698 that the well-to-do should 'grudge themselves their luxuries' and recognize the nation's need, mentioning that 'from unwholesome food diseases are multiplied among the poor people' and that perhaps 20 per cent of the population of the country were reduced to begging from door to door. To the Jacobites these were the 'ill years of King William's reign', but to the rest of the population they probably made the union with England in 1707 seem inevitable.

A measure of the lowering of the general level of the temperatures prevailing in the northern and eastern Highlands of Scotland in those times is indicated by one or two reports of high-level tarns, or lochans, which had ice on them all the year round. Thus, there is a report in the Philosophical Transactions of the Royal Society dated 1675 of 'a little lake in Straglash [Strathglass] at Glencannich on land belonging to one Chisholm . . . in a bottom between the tops of a very high hill. . . . This lake never wants [lacks] ice on it in the middle, even in the hottest summer'. We also have the travellers' reports of permanent snow on the tops of the Cairngorms. These observations seem to require temperatures 1.5–2.0 °C below twentieth century values averaged over the year, a lowering twice to three times as great as that which has been substantiated in central England from actual thermometer readings, though not unreasonable in view of the apparent advance of the polar ocean water southeast of Iceland (see ch. 4, pp. 61, 205, and fig. 23).

Fig. 80 The site of a village, Daintoun or Upper Davidstown, in the southern uplands of Scotland, which was abandoned in the 1690s. The slope faces north and is 275 m (900 ft) above sea level. (a) (top) General view from the north, (b) at the site, looking northeast. The rectangular shapes of the footings of house-walls can be seen in the pictures. (Photographs kindly supplied by I. J. W. Pothecary.)

SCANDINAVIA AND FINLAND

As might be expected, the situation in Scotland was largely paralleled in Norway. In spite of the degree of recovery in the country in parts of the sixteenth and early seventeenth century, the total number of farms in 1665 was less than it had been around 1300, and there were more desertions later in the seventeenth century, among them the whole village at Hoset (as noticed in the last chapter). Over the next hundred years farms were in some cases overrun by the advancing glaciers and their land partly destroyed by avalanches, floods, rock-falls and landslides. On the Hardanger Vidda plateau small new glaciers were formed, one or two of which survive as dead ice today. Between 1936 and 1951 two sections of a rope fence, which had been erected on a mountainside near Olden in Nordfjord in west Norway by the farmer who owned the land from 1602 to 1624 to protect his sheep from wolves, were found to have been released from the snow and ice which had covered it in the intervening time. It seems that the climatic deterioration did not strike Norway as soon as Scotland and Iceland, although there were some harvest failures at the end of the sixteenth century. The country probably benefited from the influence of the

Fig. 81 Three nineteenth-century pictures showing the types of boat and the methods of drying cod used in the Norwegian coast fisheries over some hundreds of years: (a) A vessel with cargo in rough water on the west coast making for Bergen (picture published in *Norsk Penning Magazin* in 1836, but which apart from the flag could be in the sixteenth century).

Fig. 81 – continued (b) A rowing boat and fishing station with cod hanging up to dry (from *Norsk Skilling Magazin* 1868). (c) Split cod drying on the rocks near Kristiansund. (Photograph 1935 by Wilse, in Norsk Folkemuseum.) (The pictures at (a) and (b) were republished in Kari Lindbekk, *Lofoten og Vesterålens Historie 1500–1700*, Kommunene i Lofoten og Vesterålen 1978; copies of the pictures were kindly obtained and supplied for this book by Ivar Toflen and Øystein Bottolfsen of Stokmarknes in Vesterålen. The picture at (c) was similarly obtained and supplied by Professor Trygve Solhaug of Bergen.)

225

anticyclones and southerly winds of which we have noticed evidence in the tree rings in Lapland. The taking up again of long-abandoned farms on the higher ground in south Norway continued until about 1640. A reassessment of farm and land values carried out with great care in 1667 put up the taxes and land-rents, which were paid in kind, above what they had been in the previous century. But troubles and difficulties imposed by nature began soon after that. Reports of rock-falls, avalanches, landslides and flood damage led to pleas for reduction of taxes which were meticulously investigated and generally granted. By the late 1680s and 1690s these incidents had multiplied manifold, and the glaciers themselves were over-running farmland.[8] There was a great frequency of disasters in these categories in Norway between about 1690 and 1710, which continued with little abatement until the middle of the eighteenth century and then tailed off to reach a negligible level in the middle decades of the present century.

The worst years for the harvest in the Trondheim district of Norway often came three in a row, according to an eighteenth-century Norwegian historian, G. Schøning, writing in the first volume of the Trondheim Society's *Skrifter* (1761). He lists 1600–2, 1632–4, 1685–7, 1695–7 and 1740–2 as examples. In many of those years, however, the herring fishery was better than usual. Schøning wrote: 'the natural cause of this is without doubt that the self-same conditions which produce harvest failures with us, namely long-lasting harsh and stormy westerly and northerly winds . . . drive the great fish stocks of the Arctic Ocean [Barents Sea] in greater than usual numbers to our coasts'. This explanation was no doubt broadly right except in relation to the extreme situation in the 1690s when the fish seem to have been driven altogether farther south.

In north Norway the population fluctuated remarkably during the sixteenth and seventeenth centuries with the variations of the fisheries, as can be established from the taxation documents.[9] In Lofoten and Vesterålen there was a maximum of population about 1618, followed by a 20–30 per cent fall over the next thirty years, then an even greater maximum in the 1650s followed again by a nearly 30 per cent fall to the end of the century. It seems that these changes must have involved migration of fisherfolk in and out of the region from the south. The climate connection is partly obscured here by the effect of changes, controlled by war and peace in central Europe, in the amount of trading of fish – notably dried cod – for corn from the eastern Baltic lands (fig. 81). But it is clear that the fisheries were in poor shape in the latter part of the seventeenth century and that farms were once again being abandoned on a considerable scale in north Norway especially in the later decades of the seventeenth century.[10]

In Sweden most of the same pressures are registered as elsewhere in northern Europe, though less severely than in Norway or Russia. (There seems no mistaking that northerly storms in the Norwegian Sea and the

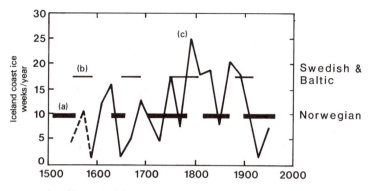

Fig. 82 Periods of herring fishery; (a) Norwegian coast; (b) Swedish coast (Baltic); (c) the incidence of Arctic sea ice at the coasts of Iceland. (Adapted from R. J. H. Beverton and A. J. Lee, 'Hydrographic fluctuations in the North Atlantic Ocean . . .', in C. G. Johnson and L. P. Smith (eds), *The Biological Signifcance of Climatic Changes in Britain*, pp. 79–107, London and New York, Academic Press for Institute of Biology, 1965.)

severity of the winters in Russia were prime aspects of the climatic deterioration.) In north Sweden there is little sign of the retreat of settlement and agriculture that was so widespread in other parts of Europe, probably because only the best land and the best sites for habitation had ever been occupied. And in Finland desertions hardly begin before the seventeenth century, although there was some migration of Finns to settle farther south in Sweden and Norway already in the sixteenth century.[11]

FISHERIES AND THE SEAFARING NATIONS OF NORTHERN EUROPE

The Baltic and North Sea–Norwegian Sea herring fisheries underwent sharp changes, largely alternating with each other in a way that had obvious climatic as well as historical significance. The variations are outlined in fig. 82. How far they represent migrations of the same fish stocks is not known, of course. All five of the Norwegian herring fishery periods, however, corresponded to minima in the occurrence of ice at the coasts of Iceland. In the sixteenth and seventeenth century cases the fish seem to have preferred the North Sea rather than the Norwegian coast. (The herring normally inhabit waters with temperatures between 3 and 13 °C.) Trevelyan wrote of the impact on English history:[12] 'The increase of deep-sea fishing was a feature of early Tudor times and helped to build up the maritime population and strength of the country. . . . The herring had recently moved from the Baltic into the North Sea' and in the words of the late-sixteenth-century English historian, Camden: 'These herrings, which in the times of our grandfathers swarmed only about Norway, now in our times

. . . swim in great shoals round our coasts every year.' Thus, in at least this aspect and perhaps in others, the Little Ice Age caused England to gain at the expense of her northern neighbours.

Similarly, the centuries we have reviewed witnessed the decline of Trondheim as a northern capital, the shift of the Norwegian court first to Oslo and the Akershus fortress and later to Copenhagen after the union of the northern kingdoms: finally in 1536 Norway ceased to exist as a separate country (until 1815). A parallel (but shorter distance) southward movement took place in Scotland and culminated in the union with England in 1707.

The prosperity of Holland in the first half of the seventeenth century also owed a good deal to the transference of the fisheries to the North Sea and Atlantic waters, as well as to an industrial revolution based on the exploitation of the windmill. The rise of Dutch sea power also had something to do with the chaos produced by the Thirty Years War in central Europe and the need to protect Dutch trading interests in troublous times. Later in the seventeenth century the prosperity declined somewhat, owing partly to the incidence of great storms and sea floods which broke the dykes as well as to poorer yields from both farming and fishing.

HARVESTS AND HEALTH IN ENGLAND

England did not altogether escape direct impacts from the development of the Little Ice Age climate, however. Hoskins's survey of English wheat harvests, mainly in the west, from 1480 to 1760[13] shows a few runs of terrible years, among which some in the 1550s and 1560s, 1594–7, 1692–8, as well as the years 1709, 1740 and 1756, stand out. There were notable runs of good harvests in the 1490s, 1537–48, 1685–90 and 1700–7, and a much greater proportion of good harvests from 1717 to the end of the survey.

For England the summers of 1555 and 1556 and the harvests they produced certainly came as a severe shock after the easier times that preceded them. Already in 1550, 1551 and 1554 the harvests had been mediocre or worse. Whether the outcome should be described as famine is debatable, but presumably malnutrition aggravated the influenza epidemic of 1557–8 in which whole families died.[14] A close study of the registers of births, marriages and deaths in a sample parish, Colyton (near Exeter) in southwest England, provides a survey of its population from about 1550 onwards.[15] There was a decline in the 1550s, when deaths exceeded the number of births for several years. Thereafter, there was fairly steady growth until the last plague epidemic in the area in the 1640s reduced the population by about 20 per cent in a single year. But afterwards, apart from a spurt of marriages in the 1650s, there was no real recovery for a long time. The yearly number of burials exceeded the births from the 1660s until about 1730. Only after the 1780s were the births substantially in excess. Delay of the marriage age and a general loss of fertility seem to have been the order of the day. From 1560 to 1645 the average age of the women of the parish at marriage had been 27, then until after 1700 it was 30 years. After 1720 it began to fall: the long-term average around 1800 was 25 and in the 1830s it had fallen to 23 years. Whatever the exact causes of these changes in this agricultural parish, with its small market town and involvement in the woollen industry, the numbers of the population and the expectation of life show an obvious and direct association with what we believe to have been the variations of prevailing temperature, apart from the plague of the 1640s.

THE VARIABILITY OF WEATHER IN THE LITTLE ICE AGE

The difficulties imposed by the climate in the Little Ice Age time were not only due to the lower temperatures, to which any generation could no doubt adapt, even if with some effects on health, fertility, length of life, etc. But, as the harvest results mentioned in the last two paragraphs have implied, there was an enhanced variability of the temperature level, which must have badly upset harvest expectations and posed a need for storage of reserves of foodstuffs beyond the resources of the community at that time. This was not just an occasionally very wide variability from year to year but, doubtless with more distressing effects, the wide differences between one group of up to six or eight years and the next. This is a characteristic which seems to have recurred in recent years. The well-known occurrence of very hot summer weather in the two summers of 1665 and 1666, when London experienced its last great epidemic of the plague which ended with the great fire that burnt the city in September 1666, occurred in the middle of the coldest century of the last millennium; this inevitably now arouses memories of the summers of 1975 and 1976. Similarly, the two winters with least Baltic ice as shown by the over four hundred years long record of ice closing the port of Riga occurred in 1651–2 and 1652–3; the winter of 1658–9 produced the opposite extreme – much as the great Baltic ice winters of 1962–3 and 1965–6 occurred only a few years before the ice-free winter of 1974–5. This tendency can be illustrated also by the listing of the most extreme winters and summers shown by the temperatures measured in England since 1659, in tables 2 and 3 below. Notice particularly the occurrences of opposite extremes within a few years of each other in the Little Ice Age, in the winters of the 1680s, 1690s and the 1790s, and in the summers of the 1670s and around 1720.

The temperature values quoted in tables 2 and 3 are from the series painstakingly homogenized by the late Professor Gordon Manley.[16]

Table 2 Average temperatures over December, January and February in the seven coldest and seven mildest winters in central England between 1659 and 1979 (long-term average for winter 1850–1950 4.0 °C)

Winter	1683–4	1739–40	1962–3	1813–14	1794–5	1694–5	1878–9
°C	−1.2	−0.4	−0.3	+0.4	+0.5	+0.7	+0.7
Winter	1868–9	1833–4	1974–5	1685–6	1795–6	1733–4	1934–5
°C	6.8	6.5	6.3	6.3	6.2	6.1	6.1

Some of the gentry who had taken over the former monastic estates in England after the Reformation were encouraged by some of the warmer summers of the late sixteenth to eighteenth centuries to try once more establishing vineyards, as the monks had done in the high Middle Ages, though protected by specially built walled gardens and not in the open field as of old. However, when Samuel Pepys went in July–August 1661 to see one of the grandest of them, the vineyard which the Cecils had established fifty years earlier at Hatfield House, he remarked only on the coldness of the day and the size of the gooseberries.

Table 3 Average temperatures over June, July and August in the fourteen hottest and fifteen coldest summers in central England between 1659 and 1979 (long-term average for summer 1850–1950 15.2 °C)

Summer	1826	1976	1846	1781	1911	1933	1947	
°C	17.6	17.5	17.1	17.0	17.0	17.0	17.0	
Summer	1868	1899	1676	1975	1666	1719	1762	
°C	16.9	16.9	16.8	16.8	16.7	16.7	16.7	
Summer	1725	1695	1816	1860	1823	1674	1675	
°C	13.1	13.2	13.4	13.5	13.6	13.7	13.7	
Summer	1694	1888	1922	1812	1862	1698	1890	1920
°C	13.7	13.7	13.7	13.8	13.8	14.0	14.0	14.0

NOTABLE WINTERS AND SUMMERS IN EUROPE

The period of history with which this chapter deals was, of course, the time of the great frosts which froze the rivers of Europe (fig. 83). The River Thames was frozen over in London at least 11 times in the seventeenth century, 20–22 times between 1564–5 and 1813–14. This phenomenon in itself was probably not of very great economic importance, particularly as it

Fig. 83 The frozen River Thames in London: (a) (top) in December 1676; (b) in February 1684. (The painting by Abraham Hondius in (a) is reproduced by courtesy of the Museum of London.)

231

came to be an expected norm and society was adjusted to it. Nevertheless, the careful records that were kept of when the Dutch canals were closed to traffic because of ice have made it possible to reconstruct the prevailing winter temperatures in the Netherlands back to 1634.[17] The series confirms the very low winter temperatures between 1670 and 1700, but suggests that in the Netherlands – more than in England – the coldness of the seventeenth century winters was fully matched for some decades around 1800. The long-term average winter temperatures for central England between 1670 and 1700 suggest that the normal yearly number of days with snow lying must have been 20–30 as against the 2–10 days which has characterized much of the present century. In the extreme cases of the seventeenth century we have a few reports of much greater totals: 60–70 days at Aldenham in 1662–3, about 80 days at Buckland (also in Hertfordshire) in 1783–4 and 102 days at another point in southern England in 1657–8. These may be compared with the general experience of between 50 and 65 days in 1962–3 and about 40 days in 1978–9. The great winter of 1683–4 was also remarkable for the recorded fact that the ground was frozen to a depth of nearly 4 ft (more than 1 m) where it was snow-free in southwest England (Somerset). In 1683–4 also belts of sea ice 5 km broad appeared along the Channel coasts of south-east England and France; at the North Sea coast of the Netherlands the ice belt is believed to have been 30–40 km broad (see also pp. 238–40). Shipping was halted, as in the Baltic. Similar conditions probably occurred in the winter of 1607–8.

The lowered summer temperatures in and around the 1690s were probably more important economically than the severity of the winters. We have reported the failures of the harvest in Scotland in those years and the similar difficulties in Norway and Switzerland. In England the growing season was presumably shortened on the long-term (30–50 years) average by about 5 weeks in comparison with the warmest decades of the twentieth century, and the yearly total accumulation of summer warmth for the crops correspondingly reduced. In the coldest individual years such as 1695, 1725, 1740 and 1816, when spring, summer and autumn temperatures were low and the summer months mostly about 2.0 °C (3.6 °F) or more below the modern normal, the growing season was probably shortened by two months or even rather more. The effects on crops in the lowland countries of Europe, particularly the continent's main 'breadbaskets' on the eastern part of the great plain in Poland and Russia, and in France, seem not to have been by any means as serious as in the uplands, but in 1695 the harvest failure was more general and from 1695 to 1697 there was famine in eastern Europe, e.g. Estonia.

There is an apparent anomaly in that the years between 1680 and 1720 saw the first great growth of merchant shipping in Norway, the first steps to that country's later possession of one of the biggest merchant fleets in the world. It seems from the local histories recorded around the coasts of southern Norway that the impetus came to an important extent from the decision of coastal farmers whose crops failed to turn instead to selling their timber and constructing vessels to transport it themselves.

ARTISTS' IMPRESSIONS

The impression that the onset of the severest phases of the Little Ice Age climate made upon the artistic and cultural life of the time is illustrated by the influence of the 1564–5 winter on the painter Pieter Brueghel the Elder, which started a whole new artistic tradition. It was in February in that winter, which exceeded in length and severity any winter since the 1430s, that he painted his famous picture 'Hunters in the Snow' (fig. 84). This may have been the first time that the landscape itself, albeit in this case an imaginary landscape, had been, at least in essence, the subject of the picture rather than the background to some other interest. And Brueghel went on to paint a series qf landscapes which dramatized each of the four seasons of the year. It is interesting to notice from the two pictures illustrated in fig. 85a and b how the bitter Flemish winter evidently

Fig. 84 'Hunters in the snow', an imaginary landscape picture painted by Pieter Brueghel the Elder in February 1565 during the first of the great winters of the next two hundred years. The picture set a fashion for landscape painting and of severe winter scenes in particular. (Original in the Kunsthistorisches Museum, Vienna.)

(a)

(b)

Fig. 85 Two paintings by Pieter Brueghel the Elder to depict the visit of the three kings to the infant Jesus, painted before and after the great winter of 1564–5: (a) 1563 version, in which the surrounding landscape and weather play no part; (b) 1567 version, in which the Flemish winter is used to emphasize the poverty and exposure of the accommodation.

suggested to the painter a recasting of a religious subject which he had previously painted in order to dramatize the poverty surrounding Christ's birth.

There is not space in a book of this length to examine equally thoroughly the evidence of the Little Ice Age period in other parts of the world. A few points must, however, be mentioned.

SOUTHERN EUROPE, NORTH AFRICA AND INDIA

In southern Europe we find records which indicate an enhanced variability from year to year and decade to decade, particularly as regards rainfall, and the difficulties it caused. In Spain there were some runs of drought years and others characterized by flooding of the rivers. Neither Spain nor the south of France escaped altogether the incidence of severe winters, which froze the rivers. And in southern France Ladurie[18] discovered that individual cold or wet years, when the harvests were disappointing, lay behind many short-term crises. Farther east, in Turkey, travellers and European consular representatives described large areas of the plateau of Anatolia in the late sixteenth and early seventeenth centuries as becoming desiccated, with empty villages and deserted agricultural land. This is an area where tree ring research in the coming years may usefully supplement our knowledge and verify the suggestion that environmental stresses of this kind underlay the riots and disturbances in Turkey at that time. Farther south, in Ethiopia a European, Manoel de Almeida, in 1628 reported snows, believed to be permanent, on the peaks at a level where it no longer occurs; and in Mauretania, in West Africa, reports of oak woods in the seventeenth century suggest a cooler and wetter regime than now, although the fall of the empire of Mali in 1591 suggests that the region was becoming drier than it had been.

The situation in this part of Africa is made clearer by the chronology of floods and famines in the region of the great bend in the course of the River Niger, which can be constructed from reports still extant. Timbuktu (16° 37′ N 2° 36′ W) lies at the northern limit of the zone affected by floods of the river. The records make it clear that the floods in the seventeenth and early eighteenth centuries habitually penetrated farther into the city than those of modern times, even reaching the citadel, the ancient palace of the kings of Mali and Songhay, and causing the population to flee.[19] The first of these great floods was on 16–17 December 1592, going beyond all previously known bounds. Others occurred in the winter of 1602–3, in all three winters between 1616 and 1619, and four or five more times in that century, between 1640 and 1672. There were three more cases between 1703 and 1738, although not quite so extensive; and none have been so extensive since. The floods result from unusually heavy rains in the previous summer over the upper basin of the River Niger far away in westernmost Africa, in Guinea, in latitudes 10–12 °N. Farther north around Timbuktu, near the bend where the river turns back towards the southeast, there was an abnormal incidence of famines due to droughts in about the same period, between 1617 and 1743, often in the very same years as the great floods of the river but also including 1695 and 1697. The chronology indicates that the severest phases of the Little Ice Age coincided with a time when the summer rains over west Africa were being held closer to the equator, rather than migrating seasonally to 15–20 °N or beyond as in this century before 1960; in those circumstances there were in the seventeenth century frequent severe droughts in the Sahel zone at Timbuktu. In the central longitudes of Africa, Lake Chad (13–14 °N 14–17 °E) was at a level possibly 4 m higher than today, though falling ultimately to today's level; and there seems to have been a migration of peoples towards the moister south going on during the seventeenth and early eighteenth centuries. But in east Africa at those times, unlike today, the situation evidently differed from that in west Africa at the same latitudes. The very high levels of the yearly floods of the River Nile at Cairo, especially in the late seventeenth century, indicate that the summer rains over Ethiopia were heavy,[20] while the low levels of the Nile then prevailing at other times of the year point to less rain at the equator over east Africa than in this century. It looks as if the Little Ice Age regime pressed especially far south in longitudes near the Greenwich meridian.

In India examination of the seventeenth century records indicates more frequent interruptions and failures of the monsoon than in our times, and according to Bryson the abandonment of the great city of Fatepur Sikri in 1588 only sixteen years after its construction can be attributed to failure of the water supply. A deficiency of the summer monsoon in the Indian subcontinent is consistent with the evidence of expansion of the polar cap and of the circumpolar vortex, although it is likely that great meridional (north–south) distortions of the upper atmospheric flow in that sector as elsewhere caused marked variations from year to year and sometimes within each year. Thus, it is not altogether surprising to learn that there is evidence, from travellers' observations of the landscape and lakes in Siberia, that northwest Siberia experienced great warmth (presumably due to southerly winds) in and around the 1690s when northern, western and central Europe had their coldest time, with frequent northerly and northwesterly winds from the Norwegian Sea.

THE FAR EAST

I am indebted to Dr J. L. Oosterhoff of the Department of History in the University of Leiden for the preliminary information from his studies of the Dutch East India Company's archives that the southwest monsoon in the seventeenth-century summers in Taiwan, off the east coast of Asia,

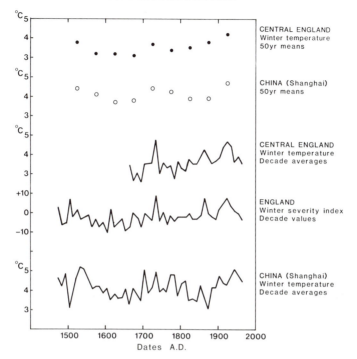

Fig. 86 The winter temperature records derived for central England and China (Shanghai) since 1500 compared. (The central England temperatures are from the series carefully homogenized by the late Professor Gordon Manley, and the Winter Severity index for England is the one described in chapter 5 (pp. 84–5): both items were used in the construction of fig. 30 (p. 84). The Shanghai temperatures are taken from a paper by Chang Chia-cheng, Wang Shao-wu and Cheng Szu-chung, given at the World Climate Conference, Geneva 1979.) Temperatures shown for Shanghai before the introduction of the thermometer were derived by methods similar to those used in England.

as in India, was also apparently frequently interrupted by northerly winds. At the same time China, like Europe, was experiencing a markedly colder climate than in the present century (fig. 86). A succession of winters with damaging frosts between 1654 and 1676 caused the cultivation of mandarins and oranges in the Kiangsi province, where they had been grown for centuries to be abandoned. Japan seems to have been less severely affected than China in the seventeenth century, although colder than now; the records of the dates of freezing of the small Lake Suwa in central Japan indicate that the greatest severity of the winters there occurred between 1500 and 1520 and again around 1700–10 and 1850–80.

Unlike the period around AD 1000, the course of the temperatures in China, as seen in fig. 86, shows a generally close parallelism with the temperatures in Europe.[21] Sometimes, however, there are differences between

237

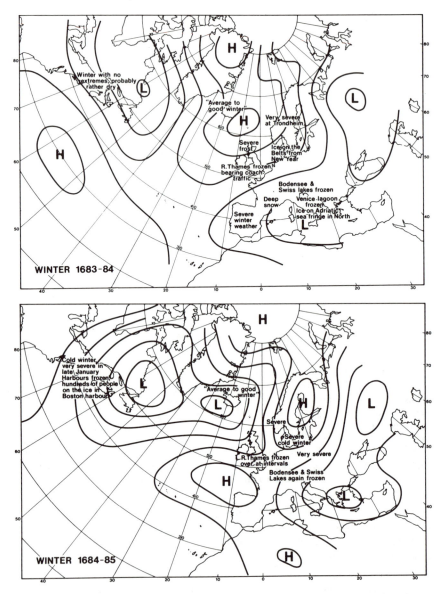

Fig. 87 Tentative analysis of the wind flow pattern and probable surface barometric pressure distribution in (a) the winter 1683–4; (b) the winter 1684–5;
with weather reports entered on the maps over the areas for which observation
reports have survived.

238

Europe and China of as much as ten to twenty years in the occurrence of the warm peaks or the coldest dips of the curve. This undoubtedly has to do with the prevalence of 'meridional' or 'blocking' type atmospheric circulation patterns with wide-ranging excursions of the flow in the circumpolar vortex towards lower and higher latitudes (see fig. 10, p. 37). Another thing which points to the frequency of such patterns in the Little Ice Age climate, and the fact that the southward bulges of cold air (or the northward thrusts of warmth which are their counterparts) occurred in different positions in different years, comes to light when the Swiss snow and English temperature records are compared. The severest winter in the English thermometer record was 1683–4; in Switzerland it was 1684–5 which produced outstanding length of snow cover. How this sort of thing happens may be seen by comparing the positions of the northerly wind streams in the maps of the two successive winters in figs. 87a and b. An apparently equally severe winter occurred in England in 1607–8 (before the availability of thermometers); in Switzerland it was 1613–14 which produced one of the three cases of longest snow cover. In England 1783–4 was the coldest winter in the 1780s, although 1784–5 was only a little less cold; in Switzerland 1784–5 was the other outstanding case of long snow cover. The proximity of outstanding cold and warmth in neighbouring sectors of the hemisphere with this type of wind circulation pattern has been illustrated in this book from modern years in figs. 20, 21 and 22 (pp. 57–9). Here we see it again in figs. 87a and b. It is, of course, this characteristic which makes the big variations between neighbouring years possible, which we noted in tables 2 and 3.

The 1683–4 winter in England was severe and long, the severest in Manley's assessment of the monthly mean temperature values from 1659 to our own times. At one place in southwest England, in eastern Somerset, the frost was found to have penetrated 1 metre (3 to 4 feet) into wet ground (although it only reached about half that depth where the ground was dry).

A note in a letter written by one Richard Freebody from Lydd in east Kent on 9 February 1683–4 (Old Style) – i.e., 19 February 1684 according to our modern calendar – which was discovered in 1861 and printed in the Royal Society's *Notes and Queries*, vol. 11 (2nd series), January–June 1861, lets us glimpse something of the dynamic nature of the winds and ocean current pattern that produced that winter. The following excerpts are taken from the letter as printed in Dr G. T. Meaden's *Journal of Meteorology*, vol. 11, no. 105, pp. 18–19, in January 1986:

> the tide for some dayes [*sic*] had not been seen . . . near folstone towne . . . by reason of the ice which lay there. . . . Islands of Ice, one to the West of the Light many miles long . . . the next day when I was at the light, I took a boat hook . . . and seeing the ice lying soe thick, I went on till I was about 2 rods upon the sea . . . the

flakes joined so close together and where I put my staff between them I felt [more] ice underneath. Old ____ observed some flakes . . . begin to come about 12 dayes before from the Eastward.

About 2 houres after I was upon it, I observed that when the wind and tide went together, then all the ice moved as fast as I could ride foot pace along by the side of it . . . I judg [*sic*] it must come from holland or some other eastern pts, w'ch by reason of a continued eastally [*sic*] wind was brought this way. . . .

Given what is now known about the increase of ice round Iceland in the seventeenth century, and especially after 1675, affecting all the coasts of the Norwegian Sea, but which would have been quite unknown to the writer of the letter in Kent in 1684, and noting the implied thickness of the ice, as well as the reported abundance of it that winter on both sides of the English Channel and southern North Sea, it seems far more likely that the ice in that year on these southern waters had come from the Arctic, passing between Iceland and Norway, driven on its way by general northerly winds.

The failure of the cod fisheries in the region all round the Norwegian Sea, and as far south as Shetland, in the last quarter of the seventeenth century, is believed to indicate that the sea surface temperatures north of about 60–61 °N had fallen below 2 °C.[22] This evidence from the fisheries and the ice – compiled in the last twenty years – indicates that prevailing sea surface temperatures in the region between the Faeroe Islands and Iceland in the last quarter of the seventeenth century were of the order of 5 °C lower than today. The development seems to have begun between about 1600 and 1615. The recovery was by no means complete before 1750 to 1840, and there were further setbacks in the 1880s, when the Arctic ice reappeared near the Faeroe Islands, and in the late 1960s. This is not just a bizarre discovery but demonstrates a latent instability in the ocean current climate that should not be overlooked. The very marked chilling of the waters in this northeast Atlantic area means that the warm North Atlantic Drift (Gulf Stream water), on which the familiar climates from the British Isles to Iceland and north Norway depend, must have been largely diverted elsewhere in the period(s) discussed – presumably a smaller scale shift of the same nature, and in the same region, as has been shown by the CLIMAP project work at the Lamont–Doherty Observatory in the United States to have characterized the last major glaciation twenty thousand years ago.

NORTH AMERICA IN THE SIXTEENTH AND SEVENTEENTH CENTURIES

For conditions in North America in the sixteenth and seventeenth centuries, we still have to rely mostly on various types of fossil or 'proxy' data, although some documentary reports begin to be available from the east and even a few from the centre of the continent. We know from the reports of early settlers and explorers from Europe that there were some notably severe winters. The severity of the winter of 1607–8 in eastern North America as in western Europe is attested by a number of reports. In Maine we are told of the persistent northerly winds and such severe frosts that many people died both among the Europeans and the Indian population. At Jamestown it was reported that 'the extraordinary frost in most of Europe . . . was as extreme in Virginia'. And Samuel Champlain, the founder of Quebec, found bearing ice on the edges of Lake Superior in June 1608. Reconstructions of the atmospheric circulation by Professor H. C. Fritts of the University of Arizona at Tucson and his collaborators, based on their surveys of tree rings at numerous sites over North America, further indicate enhanced prevalence of northerly surface winds over the eastern and central parts of the continent. There seems to have been a general reduction of the westerlies, which were shifted to a lower latitude over the Pacific. Most of the continent was colder than in the present century, but there are some hints of warmth associated with a weak circulation of southerly winds over the area of the southwestern United States. It is known also from air surveys and on-the-ground measurements of lichen growth on the rocks that a huge area of Baffin Island in the northeast Canadian Arctic was 70 per cent covered by permanent ice and snow, where today only 2 or 3 per cent of the area is under persistent snow-beds (although there has been some increase since 1960). Also the glaciers in northwestern North America registered advances to far beyond their present limits.

13

THE RECOVERY,
1700 TO AROUND 1950

SOME ASPECTS OF THE LITTLE ICE AGE WORLD

Around 1700 the prevailing temperatures in all parts of the world for which assessments are available were below twentieth-century levels. The difference registered by the thermometer record in England for the last forty years of the seventeenth century (see p. 211) may reasonably be taken as a first approximation to the world average. The cooling was certainly greater in parts of the Arctic fringe and less – it seems to have been almost zero – in parts of the subtropical oceans. Only in the Antarctic are there suggestions that at this time conditions were somewhat milder than a few centuries earlier or than they became in the nineteenth century.

In many parts of the world there were colder and wetter landscapes than at the present time owing to reduced evaporation and the accumulations of snow and ice built up over previous cold decades. We have mentioned the frequent avalanches, floods, landslides and rock-falls in Norway (p. 226). In the Alps the advancing glaciers in some places crossed a valley floor, damming up the streams to form a temporary lake. Such lakes repeatedly burst the ice dams causing disastrous floods down-valley. There were many such occurrences in Saasthal in Switzerland and in the Ötztal in Austria. Evidence of similar occurrences has been found in the Himalayas. In the British Isles in several places the peat-bogs became so sodden and swollen with moisture that they burst, as happened at Charleville in Co. Cork in 1697, in the Solway Moss (near the England–Scotland border) in 1771 and 1772 and at Haworth in the Yorkshire Pennines in 1824. All these cases occurred after great rains following a run of mostly cold summers. In the 'irruptions' of the Solway Moss in 1771–2 over 200 hectares (about 500 acres) of farmland were overrun by black peaty mud up to 4 m thick.

In the low ground of southern and central England, however, where rainfall seems to have been 5–10 per cent less than modern averages, the evaporation may have been more effective. In the 1740s the rainfall deficiency was certainly greater than this and may have amounted to 15–20 per cent over the entire decade. The choice of many house sites of the sixteenth and seventeenth centuries in these areas suggests that the valley bottoms tended to be drier than in later times. The same may apply to parts of central Europe. We also have a description by John Evelyn of the frequency in his day of dust storms on the heaths of the Breckland in East Anglia.[1]

WARMING SETS IN ERRATICALLY

All the temperature and proxy-temperature records at our disposal, however – in England, in central and northern Europe and in China, and from the ice in north Greenland, the trees on the heights in California and the stalactites in a New Zealand cave – indicate a sharp change (cf. figs. 28, 36, 52, 86) to much warmer conditions which lasted just a decade or two some time soon after 1700. In the case of north Greenland the warmth built up more gradually to an impressive peak in the late eighteenth century. In Europe it was the 1730s which produced a run of warmth equalling the warmest part of the present century. The warmth of that decade produced a significant improvement in the health and length of life statistics in Sweden and Iceland. In Scotland the annals of Dunfermline record that moves were made in 1733–4 to relax the strict puritanical rule which had long prevented the holding of an annual dance in the Town House, but after one year the ban was imposed again. The same annals record also that in 1733 'wheat was first grown in this district' (though it seems almost certain that it had been grown in the sixteenth century and earlier). Everywhere, however, one or more abrupt reversions to climatic conditions not unlike the coldest periods of the sixteenth and seventeenth centuries followed. And – apart from rather notable warmth of the European summers in and about the later 1740s and 1750s, and around 1780 and 1800–8 – it was not until the late nineteenth or early twentieth century that a more lasting warmth was established.

In the eighteenth and nineteenth centuries the characteristic wide variability of the Little Ice Age kept recurring. Thus, even in the warming period there were several more cold winters. The winter of 1708–9 was of historic severity in Europe, although Ireland largely escaped and even in Scotland it was hardly severe. People walked across the Baltic on the ice, and there was once more ice along the coast of Flanders. In England and Scotland it was very snowy, but in France it was dry and tremendous numbers of trees were killed there by the severity of the frost. Vine cultivation was permanently given up in the northernmost districts of France, and all the orange trees were lost in Provence. This is the pattern of a winter dominated by continental east winds. Seven years later, in 1716, the River Thames in London was frozen again, so firmly this time that a high spring tide in January lifted the ice 4 m without interrupting the frost fair. There was so much activity on the ice that London's theatres were almost deserted. Yet the winter of 1723–4 and no less than eight of the winters in the 1730s in England would rank with the mildest winters of the present century. Similarly, the summers of 1718 and 1719 produced great heat and drought over most of Europe and the summers of the late 1720s and 1730s in England would rank among the best of the present century, yet 1725 produced the coldest summer in the entire thermometer record with a mean temperature over June, July and August of 13.1 °C. July 1725 was described in London as 'more like winter than summer'. The mean temperature of that summer in central England was what is now regarded as normal for the north of Scotland. In Paris the remarkable feature was the continual rains. Nevertheless, the predominance of warm years about that time, with a month or more added to the growing season, with abundant sunshine and good harvests, must have made a very strong impression on the minds and spirits of those who had lived through the latter end of the previous century. Perhaps it was the sunshine of those years which determined the planning of the famous Georgian terraces of the city of Bath up the open hillside, a novel fashion indeed for any English city at that time.

The differences in the range of variability from year to year of temperature and rainfall, which we shall have to refer to at several points in the centuries covered by this chapter, included several runs of decades – around 1700 over half a century – of enhanced variability. These variations are confirmed both from thermometer and rain-gauge records and from tree ring series.

This improving tendency of the climate was rudely interrupted by the great winter of 1739–40, another east wind winter,[2] followed in England by an unbroken run of cold months all through 1740 and into 1741. This made 1740 the coldest calendar year in the English temperature record from 1659, with an overall mean temperature 6.8 °C (44.0 °F) in central England that would be about normal in Shetland. The winter, spring and autumn produced about what is expected for those seasons in the south of Sweden, the summer, though sunny, yielded temperatures which we regard as normal for central Scotland. Manley calculated that the fuel demand for heating in such a year would be just double that in the warmest years of this century. The next several winters were also cold (though no match for 1740) and it was in those years that the last wolves were seen, and shot, in Scotland and Ireland. The mostly warm summers and autumns which followed in the 1740s and 1750s, and the low rainfall of those years, made this an easier and pleasanter time in the experience of people then living in the countryside of England and elsewhere in Europe and in the eastern United States, where instrument records had also begun by that time.

DEVELOPMENTS IN AGRICULTURE

These variations, particularly the warmth of the 1730s and the prevailing summer warmth in many later parts of that century and a little after (see fig. 28, pp. 81–2), must have provided a helpful background to the agricultural improvements then being introduced by some Norfolk landowners and which made their names famous throughout Europe. It was from about 1730 onwards that Lord Townshend ('turnip Townshend') was diversifying the crops, particularly by the introduction of roots. And Thomas Coke in the last decades of the century further pioneered new crops and the improvement of the land by marling and shelter belts. Between the early years of the century and 1795 the average weight of sheep and cattle sold at Smithfield market in London more than doubled.

The potato, discovered in South America and grown in Ireland (where the climate was rather too wet for wheat to do well) on a considerable scale already in the later part of the seventeenth century, may have been largely responsible for sparing the Irish the famine which afflicted Scotland so direly in the 1690s. The potato blight (*Phytophthora infestans*) did not appear until much later. Adoption of the potato, spreading eastwards and northwards across Europe during the eighteenth century, did much to reduce the threat of famine in wet years when the wheat harvest failed or fell short. After a famine in 1772 the government of Hungary ordered the growing of potatoes. In Russia, too, it was taken up with government encouragement after failures of the grain crops in the 1760s and 1830s. And it came gradually to its ultimate importance in the northern districts of Scotland and Norway by the example and encouragement of the landowners. It was actually introduced to North America in 1718 by Scottish and Irish emigrants to New England who had already discovered its advantages in Europe. The potato had one other virtue for the rapidly growing and increasingly urban populations of European countries under the Industrial Revolution: it could produce several times as much food on any plot of land as any grain.

In the case of another crop from the New World, maize, its adoption in Europe may have been delayed by the colder climate which set in in the late sixteenth century and persisted in many aspects until around 1900. Maize was grown in parts of the south of Europe by the 1670s, but was not liked, being considered too hard to digest. In the south of France it was called 'Spanish corn', and John Locke was told that it served the poor people for bread. But by the 1780s it was grown widely and was a staple food in Spain, Portugal and Italy. Soon after that it had spread through all southeastern Europe.[3]

The average temperature of the summers in England and central Europe (e.g. south Germany and Austria) in the late eighteenth century was generally a little above twentieth-century levels. This may have been the reason why a house longhorn beetle (*Hylotrupes bajalus*), which is found only in areas where the warmest month averages more than 16.5 °C (62.0 °F), caused much damage to the woodwork in houses in southern England, particularly London, in late Georgian times. It later died out, but reappeared as a pest in the twentieth-century warmth between 1934 and 1953.

FURTHER CLIMATE DISTURBANCE IN THE LATE EIGHTEENTH AND EARLY NINETEENTH CENTURIES

As we have indicated, however, there were interruptions to the eighteenth-century summer warmth, and the other seasons of the year were generally colder than now. It was, in fact, alarm about the apparently increasingly erratic behaviour of the climate and increased incidence of extreme seasons of various kinds that prompted the organized collection of daily weather observations, with instrument readings, in France in 1775. Pfister[4] has noted from Swiss observations that from 1764 for about fourteen years the summers became generally cold and rainy in the lowlands and snowy in the Alps. At the height of this fluctuation, in the years 1769–71, the winters were long and snowy, and the summers were too short and cool to melt the snow from the upper Alpine pastures. The glaciers advanced markedly and food production dropped: wheat, potatoes and milk were all so adversely affected as to produce famine. In England the wettest series of summers in the rain-gauge record (effectively since 1697) was 1751–60, ten wet summers in a row averaging 127 per cent of the modern mean; the summers from 1763–72 were not far behind, with 117 per cent, and 1775–84 with 115 per cent. The last of these groups was generally warm as well as moist. 1763 and 1768 ranked respectively second and sixth for wetness in the whole gauge record, with 181 per cent and 164 per cent of the modern mean respectively. The latter part of the eighteenth century could be described as a time of mainly warm summers and cold winters, though a few of the summers in northern Europe were either so wet or cold enough (or both) to produce failures of the grain harvest and, in areas where the potato had not yet been adopted, famine. In Scotland the harvests of 1781 and 1782 were a faint reminder of the evil experiences of the 1690s, 1709 and around 1740–2, sufficient to give a spurt to emigration. In 1781 the summer was cold and too dry for the grass or corn to grow, and 1782 was such a cold backward season in Stirlingshire that the unripened corn was buried by snow on 31 October. 1783 was not much better because of the haze of dust which obscured the sun for three weeks and the sulphurous fog from the great volcanic eruptions in Iceland in May and June of that year, though the summer was hot in England. Some of the ash-fall damaged the crops in Caithness, in the north of Scotland. The sulphurous atmosphere smelt noxious, made the eyes smart and damaged plants in Holland. In the south of France the sun was not visible in June of that year until it was 17° above the horizon because of the density of the upper haze. Indeed, a good case can be made out for attributing many (or most) of the reversals of the climatic recovery in the eighteenth and nineteenth centuries to the extraordinary frequency of explosive volcanic eruptions, which maintained dust veils high up in the atmosphere, particularly between 1752 and the 1840s.

The persistent dust veils from the eruptions in the early years of the nineteenth century, after a lull in the northern hemisphere between 1783 and about 1802, produced optical effects which were described by many observers at the time, among them John Constable's friend Luke Howard[5] in London. It is reasonable to assume that it was the optical effects of this volcanic activity that started J. M. W. Turner rendering the sunset colours for which his pictures became famous. This aspect of Turner's work became noticeable about 1807 and was particularly prominent in the 1830s, coinciding with the most sustained period of stratospheric dust. The climatic effects were perhaps most remarkable in 1816, which became known in much of Europe and eastern North America as 'the year without a summer'. This was after the enormous eruption of the volcano Tamboro in the East Indies (8 °S 118 °E) in April of the previous year, a time lapse which was evidently right for the dust in the stratosphere to spread into a world-wide veil, reducing the penetration of the sun's rays, cooling the Earth and distorting the global wind circulation pattern. At least fifteen cubic kilometres of solid matter seem to have been blown up into the atmosphere by this eruption. There had also been major eruptions in 1812 on St Vincent Island in the West Indies and Awu in Celebes and in 1814 in the Philippines, from which the dust veils in the stratosphere cannot have cleared entirely. In June 1816 snow fell with a northeast wind over a wide area of eastern North America and covered the ground as far south as latitude 42 °N from the 6th to the 11th except near the coast. Quebec city had some days when the temperature remained near or below the freezing point all day. In Connecticut, where frosts after April are extremely rare, there were frosts in every month of 1816 (see pp. 298–9). In Europe generally the grain harvests and vintages were late. In parts of the west of the British Isles, owing to continual rains and low temperatures, the grain failed altogether, and families in central Wales took to the roads over long distances begging for food. But the wind pattern produced a hot summer in the Ukraine. And in the north of Scotland the weather was fair and in Shetland 'a beautiful summer after a rough spring'. This seems to tell of anticyclones in the sub-Arctic zone that summer after northerly winds which had troubled Shetland in spring and kept the Arctic sea ice at the coast of Iceland until June: in the summer it was European Russia that had the northerly winds. The Asian monsoon seems also to have had a distorted pattern in 1816, heavy rains in Korea and the Far East, but in India concentrated in the south.[6]

This pattern of anomalies in 1816 has been blamed for the severity of the typhus epidemic of 1816–19, the most extensive in the history of Europe, for the plague which raged through southeast Europe and the eastern Mediterranean in the same years and for the first great epidemic of cholera, which started in Bengal in 1816–17 and swept the world. Taking account also of the famines of 1816–17, this was one of the very great world disasters associated with climate,[7] almost comparable with the events of 1315–50. The harvest failures and famines of the 1580s–90s and the 1690s seem to have been more narrowly confined to Europe and, perhaps for that reason, their effects were not compounded by major spreads of disease. In a lighter vein, we may note that Mary Shelley is said to have been inspired to invent *Frankenstein* by the events of 1816.

It is necessary to return briefly to the weather of the 1780s, a decade which was also marked by an exceptional amount of volcanic dust in the high atmosphere after two eruptions of the first rank in 1783, in May–June in Iceland and in August in Japan. Whether or not this was the whole reason, it was certainly a decade of abnormal climate and wind circulation. Daily weather maps from 1781 to 1786, analysed by Mr J. A. Kington in the Climatic Research Unit at the University of East Anglia, Norwich, show a remarkably low frequency of days with general westerly winds over the British Isles. The average for those six years was 66 westerly days against 91.5 over the 118 years 1861–1978. In 1785 there were only 45 westerly days. Among the many extremes registered during the 1780s, central England experienced the warmest month (July 1783) in the 320-year thermometer record, with a mean temperature 18.8 °C (66.0 °F) that would be expected 100–200 km (say 100 miles) south of Paris, but also no less than four very cold winters with temperatures more than 2 °C below the modern average. Besides this the same decade included in 1788 possibly[8] the driest individual year in England and Wales in 250 years of record, with only 63 per cent of the modern average for January to December. It was in France, however, that the effects of the weather extremes in this decade seem to have been most serious. The coldest March in the record in 1785 across much of Europe extended what was already an outstandingly severe winter, and this was followed by a year of drought (67 per cent of the expected twelve-month rainfall in Paris). This resulted in a forage crisis on the French farms, and many cattle had to be slaughtered. The French peasants at that time ate bread made of rye or oats, only the upper classes being able to afford wheaten bread. Even so the dearth produced by this situation meant that about 55 per cent of the poorer classes' earnings went on bread alone. And after another year of drought in 1788, and another severe winter in 1788–9, the French workers were having to spend 88 per cent of their income on bread.[9] This is not to attribute the revolution which changed French society to the weather, but it can hardly be gainsaid that the weather intensified the pressures that released the explosion.

The summer of 1789 was not a bad one in that it was of about normal warmth. Indeed the storming of the Bastille in Paris in July took place in fair weather. But the rainfall places it about thirtieth in the list of wet summers of the last 250 years in England and Wales. In eastern Norway July 1789 produced an exceptional kind of disaster: after heavy rain the River Glomma rose to a flood of unheard-of proportions, apparently because the subsoil was still frozen and impervious after the length and severity of the previous winter.

CHARLES DICKENS AND THE ARTISTS AS CLIMATE REPORTERS

The 1790s and the first years of the new century produced a number of pleasant, warm summers in England and on the continent of Europe. And apart from the severity of 1794–5 the winters were less cold than in the 1780s. But 1809 brought the first of a long series of colder summers, and the decade of 1810–19 produced mostly cold seasons, for which the volcanic dust in the atmosphere has been blamed. Indeed, the descriptions of 'old-fashioned winters' for which Charles Dickens became famous in his books may owe something to the fact – exceptional for London – that of the first nine Christmases of his life, between 1812 and 1820, six were white with either frost or snow. That decade from 1810 to 1819 was the coldest in England since the 1690s. The colder seasons returning in the early nineteenth century led to the designing of certain articles of warm underwear for women, notably the 'bosom friend', and brought to an end the daring fashions begun in the post-revolutionary 1790s in France, which 'exposed the person' a good deal. It was remarked at the time that it was the north wind which enforced a return to modesty in women's dress.

It appears from statistical studies of the changes of fashion in landscape painting that John Constable's pictures, like those of others of his contemporaries and of the Netherlands School in the seventeenth century, tell us something about the characteristic summer weather of their times (figs. 88a and b). Surveys of the cloud cover in European representational style paintings of various periods have shown averages of nearly 80 per cent cover in pictures from the period 1550–1700, 50–75 per cent at various times in the eighteenth century, 70–75 per cent in Constable's and Turner's time (1790–1840) and 55–70 per cent in the twentieth century.[10] These variations recorded by artists working out of doors, generally in the summer half of the year, are in line with what might be supposed from the known variations of summer temperature, though it appears that the swings of fashion registered by the artists probably exaggerated the variations of mean

(a)

(b)

Fig. 88 Landscape pictures with typical examples of the cloud cover shown by painters in seventeenth-century and early nineteenth-century Holland and England: (a) 'A view of Deventer' by Jacob van Ruysdael (1628–82); (b) 'The cornfield' by John Constable (1776–1837)

cloudiness. In this they may well have been faithful to the subjective impressions of the people living at the respective times. Another aspect of the nineteenth-century scene which we find portrayed, apparently authentically, in the work of painters down the ages is the increasing smoke pollution of London's air and that of other European cities. Brimblecombe[11] has traced this in paintings from the fifteenth century to our own times. The prevailing sky colour gradually changes from blue to yellowish and then to pinkish greys. This and the other types of historical evidence indicate that the pollution of London's air rose sharply with the introduction of the burning of coal from Newcastle and Tyneside in Tudor times and from 1690 to 1900 maintained a nearly constant high level.

TOWARDS THE MID-NINETEENTH CENTURY AND THE BEGINNINGS OF THE GREAT RECESSION OF THE GLACIERS

The 1820s and 1830s introduced a return to greater warmth in Britain and Europe and were distinguished, particularly in the 1820s, by genial warm springs and autumns. And 1826 produced the warmest summer in the whole 300-year series of temperature observations in England, apparently slightly exceeding the record of 1976. But there was still a great variability from year to year.

In August 1829 in Scotland the weather turned cold at the beginning with northerly and northeasterly winds, and the forty hours of rainfall on the 3rd and 4th produced unheard-of flooding of the rivers all over the northeast of the country which washed away a huge number of bridges and river-side buildings and altered the course of some estuaries. Rain fell on twenty-eight days of that month in the usually dry lowlands of northeastern Scotland. The measured totals for the month, where available, seem to have been up to 2½ times the long-term average. The winters of 1821–2 and 1833–4, like that of 1845–6, were close to the mildest in the record. But those of 1819–20, 1822–3, 1829–30 and 1837–8, as well as several in the 1840s, were very cold. In 1829–30 the Bodensee (Lake Constance) in central Europe froze over completely for the first time since 1740 – it did not happen again until 1963, though it had happened five times in the seventeenth century and four times between 1563 and 1600. And 1837–8 was such an extreme winter in Scandinavia that there was ice all the way from Skagen (the north tip of Denmark) to the southernmost point of Norway and round along the southwest coast of Norway as far out to sea as the eye could see. (In March 1838 the ice on this Atlantic coast was drifting back towards the south again.)

Many of the springs and autumns in the 1820s, 1830s and 1840s in England were wet, and this combined with the experience of cooler and wetter summers after 1810 and in the 1840s (apart from 1846) seems to have led to general abandonment of the practice of irrigation on the farms which had begun in the drier periods of the eighteenth century, notably the 1740s. There were other, notably prolonged, wet seasons in individual years later in the nineteenth century in England, particularly in 1848, 1852, 1872, 1877 and 1882.

The continuing variability of the seasons in England in the 1840s brought several more unpleasantly cold summers but also one more very hot summer, in 1846. In that year the heat seems to have extended far across northern Europe and Asia to melt some of the permafrost in the tundras of northeast Siberia, where the commander of a small Russian survey ship in the River Lena described his difficulty of finding the river in the vast flooded landscape. The river was identified only by 'the rushing and roaring of the stream. The river rolled against us trees, moss and large masses of peat.' At one point 'an elephant's head' reared at times out of the water, and being ultimately washed against the side of the ship, and there secured for a time, the ship's company were able to examine the newly released mammoth before it sank once again into the mire. The winters of the 1840s in England were also a 'mixed bag', including at least three that were very cold but one (the winter of 1845–6) that was so mild that it led Sabine, then Foreign Secretary of the Royal Society and soon to be one of the founders of the Meteorological Society, to observe that the Gulf Stream extended far beyond its usual bounds.[12]

THE IRISH POTATO FAMINE

The human history of the 1840s has generally been written in terms of the ideas and rising pressure of movements towards democracy and universal suffrage that were occupying the nations of Europe, until they broke out in the year of revolutions in 1848. It might be worth investigation nevertheless to discover what part, if any, the weather and its effects upon agriculture and the urban poor played in all this. In one corner of Europe, at least, it had a critical effect, the turning point in Ireland's history brought by the great potato famine. The summer of 1846, which was warm in Europe generally, was humid, with moist southerly winds, and cyclonic at the Atlantic fringe and also at times in much of northern Europe. This provided ideal conditions for the potato blight fungus (*Phytophthora infestans*), which had made its first appearance in Europe (in a shipload from America which included diseased tubers) in 1845 and spread quickly. The organism multiplies rapidly in periods of some days in succession with temperatures continuously above 10 °C and relative humidity never below 90 per cent saturation, permitting the exposed plant surfaces to remain wet. We read in a farm diary from as far away as Jaeren in southwest Norway that in 1846 the alternations of rain and sun, always with warmth, ripened the corn quickly and it was safely got in by 29 August, but 'the potatoes rotted again'. In Ireland, where the potato was the staple crop on the multitudes of small farms, 80 per cent of them under 6 hectares (15 acres) and many only a fifth of that size, the effect was devastating. Despite relief measures, particularly large imports of maize from the United States, enormous numbers of the people died. Over six years of continuing outbreaks, aggravated by an epidemic of typhus which also was not confined to Ireland, it is estimated that there were a million deaths in Ireland, and the flow of emigration began. The population in 1851 had already dropped by nearly a quarter from its peak of 8½ million in 1845, and by the twentieth century it had fallen by a half and has never since approached the 1845 level.

MID-NINETEENTH CENTURY IN THE UNITED STATES

The climate of the period from the 1830s to the 1860s in the United States has been investigated by Eberhard Wahl[13] and associates at the University of Wisconsin in Madison, using a network of official weather station records which included the earliest of their kind from the middle of the continent. The results showed a climate that was colder than the 1931–60 averages over the eastern and central parts of the country by between 1 and 2 °C in the interior in each season of the year and by over 2.0 °C (3.6 °F) in the early autumn. In the 1850s and 1860s, for which data extend to the Pacific coast, it is seen that the mountain states were on the other hand up to 1 °C warmer than in 1931–60, up to 1.5 °C in spring, summer and early autumn. Precipitation was around 20 per cent greater than in recent times over the same area, but in winter the north–south belt of up to 40 per cent greater down-put of rain and snow lay over the Middle West. This distribution makes it clear that, as we also deduced for the sixteenth- and seventeenth-century climax of the Little Ice Age, the wind circulation was more meridional (with fewer west winds) than in the twentieth century. In particular, there must have been more northerly winds over the eastern and central parts of North America and more southerly winds over the west. But some changes of longitude of the main features of the pattern must have taken place; for the waggoners trekking out west to California in 1849 found the Middle West a virtual desert.

EUROPE AND THE ARCTIC FRINGE

In 1840 a shift in the ocean current pattern seems to have taken place, which gave Iceland fifteen years of near immunity from the ice of the Arctic seas and for a few years (1845–51) made a large cod fishery possible off west Greenland. At the same time, Europe experienced on balance a somewhat colder climate. The wind circulation features involved seem to have been frequent 'blocking' anticyclones over northern Europe, with easterly winds over much of the continent but giving southerly winds in the Iceland region: these warm winds occasionally extended east to Europe, however. In 1855 a further shift seems to have renewed the strength of the ice-bearing current off east Greenland and brought the ice back to Iceland. And soon the westerly and southwesterly winds over the North Atlantic Ocean were regenerated, bringing warmer seasons to Europe and starting a recession of the glaciers. This lasted through the 1860s. The summer of 1868, in particular, produced a remarkable number of hot days with temperatures over 30 °C in England, including the record value of 38.1 °C (100.6 °F) at Tonbridge, Kent on 22 July. The winter of 1868–9 in England was the warmest of the entire record with a mean temperature of 6.8 °C, which is more normal for the west coast of Ireland and warmer than some springs of the past in England. Also in the 1870s Europe enjoyed mostly warm seasons and mild winters, apart from some severe weather in the Februarys of 1870 and 1875 and a longer frost in December 1870 and January 1871. The year 1872 was very wet in England and Wales, and from 1875 onwards most of the summers were wet.

TOWARDS THE END OF THE CENTURY

A more serious reversion to colder climate came with the year 1879, a year well within the class of the 1690s. Through December 1878 and January 1879 the temperature in England stayed mainly below the freezing point, and it was very snowy; the spring was cold, with May colder than many an April; the summer was the wettest and one of the seven coldest in the long instrument records for England; it was followed by a notably cold autumn and another near freezing month in December. The cold wet weather delayed the ripening of the harvest, so that even in East Anglia in some places the corn had not been gathered in by Christmas. There were to be a number more skating winters in England and Holland, and amusements on the ice on the Swedish Baltic coast (fig. 89), before the stronger warming in the twentieth century came in. The decline of English agriculture, which lasted for fifty years, dated from this time. The harvests had been affected by difficult seasons from 1875, and the competition on Britain's free trade market of cheap North American wheat from the prairies was beginning to be felt. 1879 turned the decline into a collapse. Within a few years the cornlands of the northwestern half of England had been converted to grass, and soon that brought no profit as frozen meat began to come from Australia, New Zealand and South America. The farmworkers began to leave the land for the towns and to emigrate overseas in great numbers. Other European countries protected their peasantry against the American competition by import dues. But the effects of 1879 and the difficult years with cold winters and wet summers which followed were not

Fig. 89 Fun on the ice-covered Baltic off Malmö, Sweden in 1924. People could walk to Copenhagen. Elsewhere along the coast, off Lund, cars were driven on the ice. (Photographs reproduced by permission from *Sydsvenska Dagbladet* and Scania Photopress AB. Malmö.)

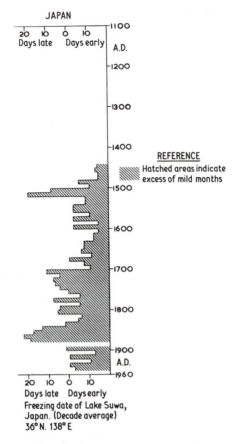

Fig. 90 Average freezing date of Lake Suwa in central Japan decade by decade since 1440. The mean for each decade is shown as the number of days departure (– early, + late) from the overall average date 15 January.

confined to England. The peak emigration of people from the countries of northern, central and western Europe was in the 1880s.

The years 1876–9 also brought droughts, monsoon failures and famine in China and India. The old stories of medieval Europe's famine situations of outbreaks of cannibalism and children sold into slavery repeated themselves in these years in the Far East. The temperature records in China (fig. 86) and indicators such as the freezing dates of Lake Suwa in central Japan (fig. 90) show that this was one of the severest phases of the Little Ice Age in the Far East. The deaths due to famine in the late 1870s in India and China have been estimated at 14–18 million.

The historical documentary information which begins to be available from the southern hemisphere in the centuries described in this and the previous chapter seems to confirm that there too a colder climate developed during the last millennium. Glaciers advanced in South America and New Zealand, and there were appropriate changes in the New Zealand forests. But the timing of the severest phases was different, it seems almost opposite, to that in the northern hemisphere. We have referred to evidence of this in chapter 3 (p. 39). Captain Cook's voyages in the 1770s and others on to the 1830s confirm that the Antarctic sea ice was more restricted and open sea extended farther south, although those were times when the northern polar ice was well forward and troubling Iceland. Later in the nineteenth century, in the 1850s and around 1900, the southern sea ice extended farther north and there were many accounts from the sailing ships of those days of sightings of the great tabular icebergs calved from the Antarctic inland ice drifting to much lower latitudes, off the River Plate and approaching the other southern continents.

After 1894–5, when there was a good deal of ice on the Thames in London, there was a long respite from severe winters in England and in Europe generally. Not again was there a month with mean temperature below the freezing point in England until January 1940. Only the winters of 1916–17 and 1928–9 during that interval of forty-five years could be considered in any way severe, the February in both cases coming near to being a freezing month in England and causing some ice to appear on the Thames. The much more severe winter of 1962–3 (3-month mean temperature in central England −0.3°C, January −2.1 °C) never brought the water temperature in London's river below about 10 °C (50 °F), owing to all the industrial and urban effluents now passed into the river. (That winter was colder than some in which frost fairs were held on the river in London in the past. The progress of urbanization suggests rather that the pastimes in future cold winters will be to skate on the Thames at Hampton Court – at the western limit of the metropolis – and then swim in it from Westminster pier!)

It should plainly be desirable to update our portrayal in fig. 91a of the course of world-average temperature at the surface of the Earth, as indeed has been attempted in various quarters. The most authoritative version is due to the (WMO/UNEP) Intergovernmental Panel on Climate Change[14] (IPCC for short). The curve here shown as fig. 91b represents the IPCC figures when looked at as the successive five-year means from 1860 to 1989. The three-year mean for the remarkably warm years 1990–2 is the last point at the right-hand end of the graph. The overall shape of this historical curve is the product of successive revisions adjusting the values for urban and industrial warming and any other possibly distorting influences at the observation sites – not least the changes that have taken place in the observing practices at sea with ever bigger ships, changes in the height of their decks above the water, and measurements in recent decades being made within the vessel in intake pipes instead of in open buckets. This

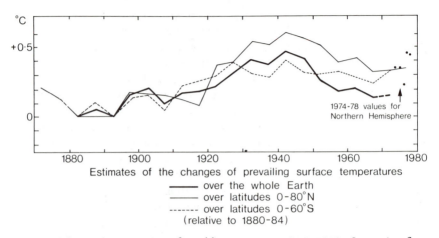

Estimates of the changes of prevailing surface temperatures
——— over the whole Earth
——— over latitudes 0-80°N
------- over latitudes 0-60°S
(relative to 1880-84)

Fig. 91a The apparent course of world temperature since 1870. Successive five-year averages assessed for the whole world, for 0–80 °N and for 0–60 °S, and estimates for the complete northern hemisphere for each year from 1974 to 1978. (The five-year values are considered to be more reliable than the estimates for individual years.)

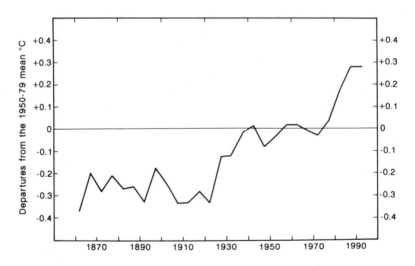

Fig. 91b Changes of the world average temperature from 1861 to 1992 as indicated by the figures given by the *Intergovernmental Panel on Climate Change (Supplementary Report)*.

writer is inclined to the belief that, however careful the observing pro-
cedures and however carefully studied the adjustments applied afterwards
to the observations, to declare a value for the world average or an area
average to within some hundredths of a degree centigrade is an unattain-
able ideal. All responsible researchers agree that the temperatures measured
must be adjusted for peculiarities of the site and the changes of these peculiarities with time. All these things have led to revisions of the data. Urban
and industrial influences also change with time, as does the pollution of
various kinds which may influence the data. Also, it is now appreciated
that the climates of islands, big and small, differ from those over the open
sea as well as from the climate of the nearest extensive land-masses. Even
the inhabited camps in polar wastes create their own climates through the
artificially generated heat, smoke and pollution, all of which tend to be
trapped locally and held beneath the temperature inversions. The light wind
speeds below the inversion also lead to a strong local concentration of the
effects. Hence, adjustments must be attempted even though they introduce
an arbitrary element into the results.

Our fig. 91 was derived from the average surface temperatures for ten-
degree latitude zones around the whole Earth between latitudes 80 °N
and 60 °S, presented in 1963 by the late J. Murray Mitchell Jr to the
WMO/UNESCO Rome symposium on changes of climate, for successive
five-year periods from 1870 (in the northern hemisphere from 1840) to
1959 and later extended to the 1970s. The revised version put before us
in fig. 91b, using for the first time the now available results of the fine
collection and survey of the world's ocean surface temperatures since the
1850s by C. K. Folland and his associates in the British Meteorological
Office and the work on land stations all over the world by P. D. Jones and
others in the Climatic Research Unit, Norwich, gives a disconcertingly
different picture of the course of world temperature history that should
not pass without notice. The differences between the course of world
temperature in the twentieth century as displayed in figs. 91a and 91b
must be partly explained by the fact that the survey represented in fig. 91
omitted the Antarctic and all latitudes south of 60 °S, where all our evidence
makes clear that very substantially higher temperatures have been observed
since about 1950, as is also true for New Zealand. The other main contribution to the discrepancy is the significant warming of the tropics, though
by only some tenths of a degree but applying to the great area of the
tropical zone.

There are, nevertheless, points of agreement between the two versions.
It is agreed that the 1880s and early 1890s were a cold time, though not
everywhere in the northern hemisphere oceans, and that the twentieth
century has been generally warmer. Warming was rapid from about 1920
to 1940. The cooling which set in in the 1940s had a wobbly course, but
the climatic record continued generally colder in the northern hemisphere

until some time after 1970. In the southern hemisphere, particularly the Antarctic and the sub-Antarctic ocean zone, there was a rapid warming going on from about 1950 onwards. Despite the rapid rise of world temperature after 1975 indicated by our fig. 91b, there has been a noteworthy occurrence – seen, for example, in the Danish temperature record here reproduced in fig. 28a (p. 80) and in other records in North America and Europe[15] of further cold events or some continued colder conditions until 1985 to 1987. The state of affairs at the time of writing (1994) seems to be that, after truly exceptional warmth in the years 1989–91, there has been some fall of temperature world-wide, which has been attributed by many to the effects of the great volcanic eruption of Mount Pinatubo in the Philippines in June 1991.

THE TWENTIETH-CENTURY WARMTH

It was during the second and third decades of the new century that the climatic warming became noticeable to everybody. The phenomenon is well seen on figs. 28, 30, 36 and 86. In England, and probably in many other places, the temperature jump from one decade to the next was not as great as the change from the 1690s to the first decade of the eighteenth century, but it was to be much longer sustained. The apparent changes of world temperature over about the last hundred years are shown in fig. 91. Of the various curves shown, that for the northern hemisphere is doubtless the most reliable. (The difficulties of estimation over parts of the world where regular observation sites are few and unevenly scattered, because of extensive oceans or ice, restrict the accuracy and have therefore restricted the areas of the Earth which these curves are designed to cover.) The change of long-term prevailing temperature in England over this period is evidently similar to that averaged over the Earth as a whole. Places near the Arctic fringe – such as Iceland, Spitsbergen and even Toronto – experienced warming that was from twice to five times as great. This was accompanied by, and its magnitude in part explained by, withdrawal northwards of the boundary of persistent ice and snow. In Spitsbergen the open season for shipping at the coal port lengthened from three months in the years before 1920 to over seven months of the year by the late 1930s. The average total area of the Arctic sea ice seems to have declined by between 10 and 20 per cent over that time. Even the equatorial oceans and small islands near the equator warmed up by about the same amount as the computed world average. In a few areas in low latitudes, however, where there were changes in the location or extent of upwelling colder water from the deeper ocean, because of the upwelling there was some cooling. When account is also taken of the changes in the atmospheric circulation, and hence in the distribution of rainfall and its variability as well, it is hardly too much to say that the twentieth-century climatic regime from 1920 to 1960 changed the world.

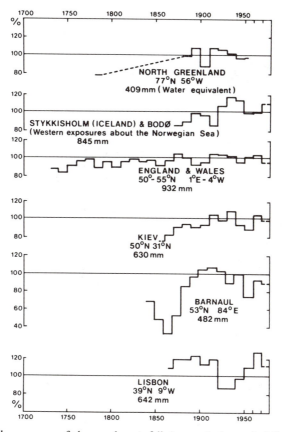

Fig. 92 Decade averages of the yearly rainfall (or equivalent rainfall deposited as rain and snow) at selected places in the northern hemisphere since 1740.

The prevailing westerly winds of middle latitudes increased their dominance (fig. 17, p. 53, illustrates this in the case of the British Isles), as the cyclones of the subpolar zone became bigger and spread their wind circulation farther into the Arctic. In consequence of the more frequent westerly winds, all those places which derive their rainfall from moisture transported from the west experienced increased and more reliable rainfall. This is illustrated in fig. 92 for places near 50 °N across the plains of Europe into central Asia. The same happened at places with western exposures in corresponding latitudes in the southern hemisphere. And more moisture penetrated into high latitudes to be deposited as snow on the ice-sheets in north Greenland and Antarctica. In middle latitudes in the Americas the situation was different, because in the rain-shadow of the Rockies and the Andes increased westerly winds brought dryness, culminating in the disastrous droughts of the Dust-Bowl years in the 1930s in the United

261

GL D'OTEMMA

GL D'EPICOUNE

GL DE CRETE SECHE

GL DE FENÊTRE

GL DU MONT DURAND

GL DE LA TSESSETTE

(a)

(b)

Fig. 93 Two views of the same panorama in the Alps in southwest Switzerland (Val de Bagnes), (a) from a watercolour, from the Alpine series by the careful draughtsman H. C. Escher, in 1820, (b) from a colour photograph by Dr W. Schneebeli of Zürich in 1974. (Reproduced by kind permission of Dr Schneebeli and the publishers of *Die Alpen*, the Swiss Alpine Club's journal.)

Fig. 94 A comparison of favourite house sites: (a) (top) typical hilltop sites chosen for luxury housing at Guildford, Surrey in the 1930s; (b) sixteenth-century houses in the street of Shere, Surrey in a valley bottom only about 5 km from (a).

States Middle West. In fig. 92 it may also be noticed that at Lisbon, as in other places in subtropical latitudes, rainfall also decreased. This was due to the increased size, and some northward displacement, of the anticyclone belt. Correspondingly, south of the Sahara the monsoons of west Africa penetrated farther north at that time. And similarly the monsoon in India was at its most reliable, with only two partial failures in thirty-six years between 1925 and 1960.

In temperate latitudes the growing season increased in length – in England, and perhaps typically, the average length increased by about two weeks.[16] The frequency of snow and frost decreased generally. The dates of last frost in spring became earlier and of first frost in autumn later, commonly by two weeks or more. And the retreat of the glaciers after about 1925 became rapid. It was almost entirely during the twentieth century warming that the Alpine glaciers disappeared from the valley floors up into the mountains. Similarly great retreats occurred in Scandinavia, Iceland, Greenland, in the Americas and on the high mountains near the equator. The upper limit of trees on the mountains of Europe and the northern forest limit in Lapland were affected by the warming. Fig. 93 illustrates the overall change of scene from the early nineteenth-century landscape in the Alps. Similar changes are recorded in pictures from many other valleys and in other parts of the world. The ranges of birds and the northern and southern limits of various fish species in the oceans moved poleward during the twentieth-century warming, but this movement has been gradually reversed since about 1960. Fig. 94 contrasts the open hilltop sites preferred for the luxury housing of the 1930s in England, following the trend of fashion set in Bath in the 1730s, with the sheltered valley-bottom sites which were favoured in the sixteenth century.

Part III

CLIMATE IN THE MODERN WORLD AND QUESTIONS OVER THE FUTURE

14

CLIMATE SINCE 1950

ANOTHER TURNING POINT

Over the years since the 1940s it has become apparent that many of the tendencies in world climate which marked the previous fifty to eighty years or more have either ceased or changed. It is undoubtedly this that has stimulated interest in climate and increased effort in climatic research in recent years. It was only after the end of the Second World War that the benign trend of the climate towards general warming over those previous decades really came in for much scientific discussion and began to attract public notice. Attention at that time was focused on where continuation of the trend might lead: on the possible disappearance of the Arctic sea ice by the end of the century, and what effect that might have on agriculture, and the possibilities which might open up farther north to grow food and settle a growing population. Of course, apart from the question of trend, there have been since 1950, as before, the usual swings of climate from one year to the next and from one group of a few years to those immediately following. These make it difficult to discern the direction of any trend until it has already been established for some time, perhaps even for some decades. And, indeed, there have been many suggestions that the range of these short-term variations has widened since the middle of the century.

VARIABILITY INCREASES

Such world-wide surveys as have been attempted seem to confirm the increase of variability of temperature[1] and rainfall, as illustrated in figs. 95a and b. Much the same applies to the atmospheric pressure distribution except that variability was low in the 1950s. In Europe, as has been noted elsewhere by Professor Flohn and by Dr C. J. E. Schuurmans of the Netherlands meteorological service, there is a curious change in the pattern of variability: from some time between 1940 and 1960 onwards the occurrence of extreme seasons – both as regards temperature and rainfall

(a)

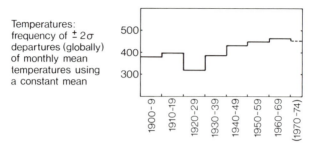

OBSERVATIONS OF EXTREME MONTHLY
MEAN TEMPERATURE VALUES

Temperatures:
frequency of ± 2σ
departures (globally)
of monthly mean
temperatures using
a constant mean

(b)   OBSERVATIONS OF MONTHLY RAINFALLS IN THE EXTREME
(TOP AND BOTTOM) QUINTILES

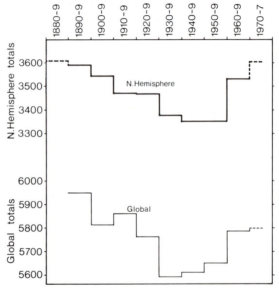

Fig. 95 Changes in the frequency of occurrence of extreme months: (a) temperature – a global survey by decades from 1900–9 to the 1970s; (b) rainfall – northern
hemisphere and global networks surveyed for the frequency of reports of extreme
months, wet or dry, from 1880–9 to the 1970s.

– has notably increased, while the overall variability as determined largely
by the remaining years has been rather lower than in the preceding decades.
Summer is an exception to this and even in Europe shows the more general
experience of increased variability.

A world-wide list of the extreme seasons reported since 1960 makes
impressive reading. Among the items included are:

1960–9	Driest decade in central Chile since the 1770s and 1790s.
1962–3	Coldest winter in England since 1740.
1962–5	Driest four-year period in the eastern United States since records began in 1738.
1963–4	Driest winter in England and Wales since 1743; coldest winter over an area from the lower Volga basin and Caspian Sea to the Persian Gulf since 1745.
1965–6	Baltic Sea completely ice-covered.
1968	Arctic sea ice half surrounded Iceland for the first time since 1888.
1968–73	Severest phase thus far of the prolonged drought in the Sahel zone of Africa, surpassing all recorded (twentieth-century) experience.
1971–2	Coldest winter in more than two hundred years of record in parts of eastern European Russia and Turkey: River Tigris frozen over.
1972	Greatest heat wave (in July) in the long records for north Finland and northern Russia.
1973–4	Floods beyond all previous recorded experience stretching across the central Australian desert.
1974–5	Mildest winter in England since 1834. Virtually no ice on the Baltic.
1975–6	Great European drought produced the most severe soil moisture deficit that can be established in the London (Kew) records since 1698.
1975 and 1976	Greatest heat waves in the records for Denmark and the Netherlands and, in some particulars, for England also.
1976–7	Severest winter in the temperature records (which begin in 1738) for the eastern United States.
1978–9	Severest winter and lowest temperatures recorded in two hundred years in parts of northern Europe and perhaps also in the Moscow region. Snowfalls also extreme in some parts of northern Europe.

This shortened list omits most of the notable events reported in these years in the southern hemisphere and other parts of the world where instrument records do not extend so far back. Cases affecting the intermediate seasons, the springs and autumns, have also been omitted.

These variations, perhaps more than any underlying trend towards a warmer or colder climate, create difficulties for the planning age in which we live.[2] They may be associated with the increased 'meridionality' of the general wind circulation, the greater frequency of 'blocking', of stationary high and low pressure systems giving prolonged northerly winds in one longitude and southerly winds in another longitude sector in middle latitudes. The corresponding decline of the westerlies since about 1950 has

been illustrated in fig. 17 (p. 53). Other factors believed to be of importance can be identified in studies which have monitored the global patterns
of the wind circulation and the pressure of the atmosphere. Thus, both
the Chilean drought of the 1960s and the strong warming (by about I °C
in the overall average) of New Zealand since about 1950 can be linked to
a southward shift[3] of the southern hemisphere zone of strongest upper westerly winds and the sub-Antarctic storms and cyclonic activity. Over both
hemispheres there has been more blocking in these years. Average atmospheric pressure has been higher than in the first half of the century in all
latitudes from 67 °N to 22–25 °N. It has been lower than before nearer
the equator and also in the Arctic except over the areas commonly occupied by blocking anticyclones (near or over northeast Canada and
Greenland, northern Europe and northeast Siberia). The most remarkable
feature seems to be an intensification of the cyclonic activity in high latitudes near 70–90 °N, all around the northern polar region. And this
presumably has to do with the almost equally remarkable cooling of the
Arctic since the 1950s (fig. 96), which has meant an increase in the thermal
gradient between high and middle latitudes.

The more complex patterns of the wind circulation which occur when
there are stationary anticyclones in parts of the middle latitudes and
subpolar zone have been able to produce an increased frequency of the
occurrence of cyclonic centres in places in the middle and subtropical latitudes, despite some rise in the overall average pressure level, and increased
rainfall in parts of the Mediterranean (notice also the Lisbon record in
fig. 92) and much of the southern United States. The long record of wind
and weather patterns over the British Isles (latitudes 50–60 °N) shows the
twenty-year period 1960–79 as having the greatest frequency of both anticyclonic and cyclonic centres over the country since the nineteenth century.

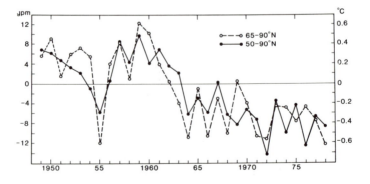

Fig. 96 Changes of prevailing temperature over the higher latitudes of the northern
hemisphere, 50 °N to the pole and the Arctic and sub-Arctic regions north of
65 °N, yearly from 1949 to 1978. (Adapted from a diagram published by Professor
H. Flohn.)

It is presumably a consequence of the enhanced frequency of depression centres passing in these latitudes that counts (from the long series of daily weather maps available here) of the occurrences of gale and storm force wind[4] situations between latitudes 50 and 60 °N, over the British Isles and eastern North Atlantic and over the North Sea, show a variation very similar to the global rainfall variability curves in fig. 95b. The frequencies in the 1960s and 1970s – averaging about thirty days with gale situations a year over the North Sea and 50–55 over the easternmost Atlantic – represent a return to the level of the 1880s and 1890s and the earliest years of this century after 20–25 per cent lower frequencies in most of the decades in between. The 1960s and 1970s have also seen an increased frequency of northwesterly and northerly winds in this region (part of the compensation for the fall off of the westerlies, seen in fig. 17), and these have often been strongly developed in the rear of the depressions here discussed. This seems to account for the increased roughness of the North Sea observed by the German navy. Indeed, river-gauge observations in the Elbe at Cuxhaven and Hamburg show that the North Sea storm flood frequency in the winter 1972–3 was the greatest since 1792–3, although the flood levels were less severe.[5]

COOLING IN THE ARCTIC

The cooling of the Arctic since 1950–60 has been most marked in the very same regions which experienced the strongest warming in the earlier decades of the present century, namely the central Arctic and northernmost parts of the two great continents remote from the world's oceans but also in the Norwegian-East Greenland Sea. (In some places, e.g. the Franz Josef Land archipelago near 80 °N 50–60 °E, the long-term average temperature fell by 3–4 °C and the ten-year average winter temperatures became 6–10 °C colder in the 1960s as compared with the preceding decades.) It is clear from Icelandic oceanographic surveys that changes in the ocean currents have been involved. Indeed a greatly (in the extreme case, ten times) increased flow of the cold East Greenland Current, bringing polar water southwards, has in several years (especially 1968 and 1969, but also 1965, 1975 and 1979) brought more Arctic sea ice to the coasts of Iceland than for fifty years (fig. 97): in April–May 1968 and 1969 the island was half surrounded by the ice, as had not occurred since 1888.

Such ice years have always been dreaded in Iceland's history because of the depression of summer temperatures and the effects on farm production. In the 1950s the mean temperature of the summer half year in Iceland had been 7.7 °C and the average hay yields were 4.3 tonnes/hectare (with the use of 2.8 kg of nitrogen fertilizer); in the late 1960s with mean temperature 6.8 °C the average hay yield was only 3.0 tonnes/hectare (despite the use of 4.8 kg of fertilizer). The temperature level was dangerously close to

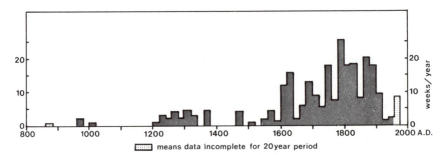

Fig. 97 Variations in the occurrence of Arctic sea ice at the coasts of Iceland. (Twenty-year averages from L. Koch's *The East Greenland Ice*, Copenhagen, 1945, updated to 1975 with information kindly supplied by the Icelandic Weather Bureau.)

the point at which the grass virtually ceases to grow. The country's crop of potatoes was similarly reduced. The 1960s also saw the abandonment of attempts at grain growing in Iceland which had been resumed in the warmer decades of this century after a lapse of some hundreds of years. At the same time the changes in the ocean have produced changes in the spawning grounds and seasonal range of migration of fish stocks – a not much publicized aspect of the international wrangles and 'cod wars' of recent times. With the fall by over 1 °C in the mean sea surface temperatures off west Greenland from the peak years in the 1920s and 1950s, the cod fishery there declined by the early 1970s to a tiny fraction of what it had been in those times.[6] The Greenland cod migrated to Iceland waters, and for a few years (1967–71) offset the declining stocks there; but since 1974 the spawning stocks in Iceland waters have been only a tenth of what they were in the late 1950s and the total stocks have fallen by almost a half, the decline being probably attributable to combined effects of the change in water climate and over-fishing. Similarly, herring stocks have moved from Iceland waters to the wider reaches of the Norwegian Sea farther east, south and north and to the North Sea, while a southward shift of the southern limit of cod seems to have led to increased catches in the North Sea since about 1963.

An interruption of the colder regime introduced by the 1960s affected Europe and Iceland, part of east Asia and the eastern United States in the early–mid 1970s and was perhaps too hurriedly hailed as a reversal of the trend. Most of Europe and parts of the other regions named experienced between 1971 and 1977 four to seven mild winters in a row, largely thanks to repetitive occurrences of anticyclones in positions which gave them southerly or southwesterly winds. One or two of these winters produced extreme phenomena such as the roses still blooming in the parks in Copenhagen in late January. But much of the remaining areas of the northern hemisphere, in Asia and Africa and including the polar region and the two great oceans as well as eastern Canada, had a straight run of colder than usual winters in the same years. As the pattern depended so largely on the positions of stationary ('blocking') features in the wind circulation in middle latitudes, no great surprise should have been caused when conditions were reversed again in many of these regions in the immediately following years later in the decade. By the end of the decade in Iceland, as in other regions of the Arctic fringe, it had to be concluded that the colder regime which set in in the 1960s seems to be continuing; and after notably cold years in 1979 and 1980 the widely debated expectation of global warming setting in as a result of the impact of the man-made increase of carbon dioxide on the world climate is being called in question in these countries.

WORLD TEMPERATURE

An assessment of the changes in overall world temperature over a hundred years past has been shown in fig. 91. Whatever reservations one must have about our ability to gauge the global temperature average to the degree of accuracy necessary to identify the changes, there is no doubt that the main features of fig. 91 are essentially right as regards the land areas, especially the five- or ten-year means indicated by the northern hemisphere curve. Other details known show that the recent cooling of the Arctic has been broadly matched by a warming in the Antarctic and in the sub-Antarctic ocean zone, which in one sector extends far enough north to embrace New Zealand. The changes in high latitudes north and south are several times greater than elsewhere and so we can be surer of them and detect them more easily. But it is clear that over the rest of the world there has been some net cooling since the warmest decades of the century. In Europe the peak years, and the most summer warmth and sunshine, were between 1933 and 1952. The change back to a cooler climate generally took place between about 1950–3 and the late 1960s; up to the end of the 1970s or later, there was no significant change.

Some studies[7] have suggested that the increased warmth of our cities; with their well-drained, paved surfaces and extravagant heating of buildings, which has made urban observation sites commonly 1.0-2.0 °C (up to 3.5–4.0 °F) warmer (in the averages for the whole year) than the surrounding countryside, enters into enough of the temperature statistics to account for most of the difference between the apparent overall world averages for the 1960s and 1970s and the level of the late nineteenth century.[8] Even some towns with only 50,000–100,000 inhabitants now show this 'urbanization effect' and are liable to be overall 0.5–0.7 °C warmer than the open country. On still, cold winter nights and hot summer afternoons, when the sky is clear, the temperature drop between inner city sites and the country outside is usually much greater and may exceed

5 °C. The twentieth-century warming was not entirely a fiction due to urbanization of the observation sites, however, for it affected Valentia Obervatory in southwest Ireland on the edge of the Atlantic and is impressively shown by the world's glaciers.

The temperature changes since 1950, small as they look in terms of averages, have affected the length of the growing season. In England many farmers and gardeners are familiar with the turn to colder springs. Between 1938 and 1953 all but two of the sixteen years had warm springs, warmer than the 1920–60 average. Since 1962 up to 1980 only one spring has been up to that average level, and in the sixteen years 1965–80 none at all. The warmest spring (March to May) in the earlier group was 1943 with an average temperature in central England of 10.5 °C (51.0 °F), the coldest in the later group 1978 with 6.3 °C (43.0 °F). At the same time there have been some runs of notably warm autumns; only two were below the 1920–60 average in the decade 1945–54 and there were three or four warm autumns in a row in the later parts of each decade between 1950 and 1979, while 1969 produced the warmest October (average 13 °C, or about 55 °F) in the 320-year record for central England. But warm autumns in England in the past as in Switzerland (see figs. 28 and 76) seem rather to have presaged some of the more notable drops in the annual temperature curve. The net effect of these changes in England has been that the growing seasons (defined by the duration of temperatures above 6 °C) have been on average about nine to ten days shorter since the mid-1950s than in the previous warmer decades, and the mean date of onset of spring (similarly defined) at Oxford has changed from 4 March between 1920 and 1950 to about 20 March between 1963 and 1980.[9]

Other effects of the temperature changes since 1950 which have been reported include a notable delay (and an increased year-to-year variability) in the arrival of the first summer day with maximum temperature 25 °C (77 °F) or above in the Netherlands: in each decade from 1910 to 1949 the average was between 9 and 17 May, in the 1950s and 1960s 22 May and in the 1970s 3 June. On the other hand, the wheat growers on the Canadian prairies have been troubled in the 1970s by earlier arrival of the first frosts of autumn in September. And despite all the variations of autumn and winter temperature in Europe, the first snowfalls in autumn have on average come earlier and the last snowfalls in spring have come later in the 1960s and 1970s.

Temperature changes in and near the tropics are harder to establish than in northern latitudes because of the great extent of ocean and because they are often smaller (though not where changes in the upwelling of cold water in the oceans are involved). The 1970s seem to have been cooler on overall average than the previous thirty to fifty years by about 0.3 °C between latitudes 20 and 40 °N and by up to 0.5 °C (about 1 ° between the equator and 15 °S, presumably indicating greater cloudiness in those zones. A smaller increase of temperature seems to be indicated around 15 °N and 30–40 °S, suggesting reduced cloud amounts there.

EFFECTS ON RAINFALL

It is only in recent years that attempts have been made at mapping the world distribution of rainfall – or, strictly, the total downput of rain and snow expressed as equivalent rainfall – in different periods for comparison. Fig. 98 shows the results of comparing the distribution by latitude in various runs of years since 1950 (from provisional surveys of a few hundred observation points) in terms of the change from the 1931–60 average. What is most apparent here is that the equatorial rains have been concentrated nearer the equator than in the earlier decades of the century. This is the counterpart of the rainfall deficiency in latitudes near 15 °N , which has been seen in its most serious form in the prolonged drought in the Sahel-Ethiopia zone of Africa, and between 20 and 30 °S where Rhodesia/Zimbabwe and parts of South Africa have at times been seriously affected.

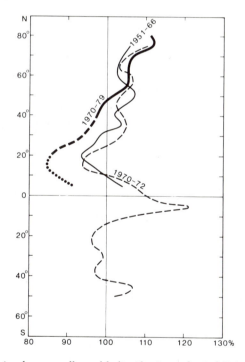

Fig. 98 Changes in the over-all world distribution of rainfall (total down-put of rain and snow as equivalent rainfall) by latitude. The experience of different periods between 1950 and 1979 is expressed as percentage of the average experienced in the period 1931–60. (Note the increases in high latitudes and near the equator and the deficits near latitude 20 °.)

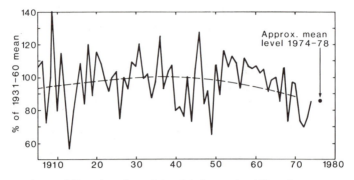

Fig. 99 Yearly rainfall at the edge of the Sahel zone in Africa from 1905 to 1974 and the five-year average for 1974-8. Rainfall averaged for five places between about 12 and 14 °N: Zinder, Niamey, Sokoto, Kano and Maiduguri. The smooth curve is a best fitting sine curve applied to the figures from 1905 to 1974. (Adapted from a diagram published by A. H. Bunting, M. D. Dennet, J. Elston and J. R. Milford in the *Quarterly Journal of the Royal Meteorological Society* in 1976.)

Bryson has shown how the northernmost limit of the monsoon rains in Africa near the southern fringe of the Sahara progressively retreated from near 22 °N in 1952–8 to about 19 °N by 1972. Fig. 99 illustrates the history of the rainfall at five places in west Africa between 12 and 14 °N on the fringe of the Sahel with records from the beginning of the century, showing the apparently continuing decline from the years of maximum rainfall between about 1915 and 1960. There is some suggestion from the best-fitting smooth (sine) curve shown that the variation may be an aspect of a two hundred year oscillation. In the late 1970s it has come to appear that the equatorial rain belt over Africa has become further restricted so that even its yield near the equator has declined; but this may be part of another (possibly much shorter-term) process, since in just the same years rainfall near the equator on the opposite side of the globe (Indonesia, western and central Pacific island groups) seems rather to have become more abundant.

Along with the more limited seasonal movement of the equatorial rain system and the increased variability associated with meridional wind flow patterns in middle latitudes, the average rainfall over northern India has declined and the southwest monsoon has become less reliable in recent years, as may be seen in fig. 100. The same applies to some extent also to the monsoon in East Asia: M. Tanaka[10] has found that the large-scale rainfall patterns over the whole region show a linkage with the position and westward extent of the North Pacific anticyclone, so that the rice yields over the whole monsoon region of Asia have some tendency to suffer simultaneous variations. It was Bangladesh, India, Burma and Korea that were found to experience the greatest variations of rice yield due to climatic

276

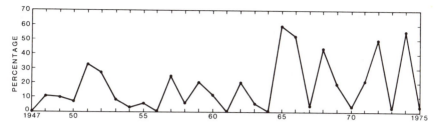

Fig. 100 Variations in the performance of the Indian southwest monsoon from 1947 to 1975. Percentage of the area of India with 'scanty' rainfall in the summer months each year. (Notice the increased variability after the mid-1960s.) (From a diagram published by K. R. Saha and D. A. Mooley in K. Takahashi and M. M. Yoshino (eds), *Climatic Change and Food Production*, University of Tokyo Press, 1978. Supporting detail given by K. R. Saha elsewhere in the same publication shows that the summer monsoon precipitation over India from 1965 to 1976 has been less than at any time since about the beginning of the century and possibly since the period 1840–60, which also showed a comparable year-to-year variability.)

fluctuations. The total yield of the summer monsoon over India in 1965 and again in 1972 was reported to be less than in any year since 1918, ranking with that year and the worst years of the later nineteenth century (1848, 1855, 1877 and 1899). Other recent years, such as 1975, have produced very high rainfall and disastrous flooding on the river plains of northern India.

In the same sector of the globe the variability in middle latitudes has produced some great droughts that have interfered with grain production in Soviet central Asia. The overall average rainfall also seems to have declined there, though not to the levels which prevailed for some decades in the middle of the last century (fig. 101). In 1972, from May to September, a huge area of the Soviet Union covering most of European

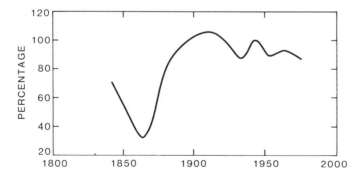

Fig. 101 Rainfall at Barnaul (53 °N 84 °E) in the region of the Soviet grainlands in central Asia. Variations of the yearly totals, smoothed and expressed as a percentage of the average (482 mm) for 1900–39.

Russia and stretching into central Asia had under half the usual rainfall, and in most of that area the totals were under a quarter of the expected amount. With the average temperature of the summer up to 3.7 °C above the long-term averages for this century, the great drought ruined the crops, caused extensive forest fires and even set the dried-up peat-bogs on fire. This was a year when the Soviet Union was apparently obliged to make massive wheat purchases in the West. But, as it also coincided with monsoon failures in India and west Africa, the food shortage was more widespread and had serious effects on world trade.

The same summer of 1972 in the neighbouring sector embracing central and western Europe produced temperatures 1 °C or more below the usual level and, though dry in the British Isles, it gave over twice the usual rainfall in parts of Italy and the Balkans. The whole eastern half of North America was also cooler and generally wetter than usual. Western Europe and North America, both east and west of the Rockies, experienced in their turn great droughts and flooding in different years in the same decade. In 1973 the Great Lakes of North America and the Mississippi River were at their highest levels since 1844.

EFFECTS ON GLACIERS, ICE-SHEETS AND SEA LEVEL

The cooling of the general run of European summers since 1953, albeit with the year-to-year variations stressed in this chapter, has had another consequence which marks out the change of climatic tendency since the middle of the century. The long retreat of the glaciers in the Alps first slowed down; then in 1965 some, mostly small, glaciers which were evidently nearly in equilibrium showed advances; and since 1972 in some regions, 1975 in others, in most years the majority of the glaciers, including the big ones, in Italy, Austria and Switzerland have been advancing. Also in west and north Norway these years have produced the first general advance of the glaciers for many decades past. Similarly, in North America the earlier twentieth century predominance of glacier retreat has been followed by advances in some areas, in the Cascades Range in the northwest of the United States from as early as the 1950s. And on (the large) Baffin Island in northeast Canada, in the central part of which 70 per cent of the highland region seems to have been covered by 'permanent' ice and snow between two hundred and four hundred years ago and where this had been reduced to 2 per cent by 1960, the 'permanent' snow beds have been increasing again since and incipient new glaciers have been found. This has been brought about by a lowering of the summer freezing level by nearly 300 m (1000 ft) in the later years.

With the melting back of glaciers all over the world from their maximum Little Ice Age positions, most of the retreat taking place in the earlier part of the present century, one should expect that the general world sea level was rising (although the biggest element in deciding this question could be the state of balance between nourishment and wastage of the great mass of the land-based ice-sheet covering Antarctica). In fact, sea level does seem to have been rising, although unfortunately our longest tide-gauge records are from the North Sea basin where a great part of the changes measured may have to be attributed to the warping of the Earth's crust in that region. The earliest of these records, a gauge installed at Amsterdam in 1682, points to a rise of the sea level in that locality by about 18 cm up to 1930. The rise in the region of the mouth of the Elbe seems to have been as much as 37 cm between 1825 and the 1970s. From 1830 onwards gauge measurements were made at an increasing number of places, becoming ultimately a more or less world-wide network. It is clear that the main rise, amounting in the region of the British Isles to 15–20 cm, took place between about 1895 and 1960. (Some places, e.g. on the continental shore of the North Sea, reported a slowing of the rise towards 1960.) These observations show such good conformity with the time of general warming and most rapid retreat of the glaciers in temperate and higher northern latitudes that it suggests this as the main cause. Since about 1960 this trend also seems to have ceased. At least for the time being a levelling off of the sea level curves is reported – though there are still, and seem to be always, some irregularities around the southern North Sea – and it has been suggested that the beginnings of a reversal may be detected in the 1970s. At the time of writing (1981), it is too early to be sure of any such favourable trend, far less to predict its continuance, which demands a climatic forecast. And while the general sea level remains 20–40 cm higher than it was at various times in the last century, this increases the danger of storm surges leading to sea floods.

MONITORING THE DEVELOPMENT OF WORLD CLIMATE

As we find ourselves driven towards the question of prediction of future climate, it is important to consider how we can most effectively and economically keep a watch on the basic tendency of the climate. In this chapter, as throughout the book, we have treated any change in the overall average temperature level as the most fundamental index of the state of world climate. Through any resulting change in the gradient of temperature between high and low latitudes, and the location of the main part of this gradient, this affects the strength and patterns of the wind circulation – and thereby the transport of heat poleward and the more detailed distribution of prevailing warm and cold, wet and dry, calm and stormy weather. We have seen that the changes produce much bigger changes of temperature level in high latitudes than elsewhere, not necessarily of the same sign in the Arctic and Antarctic (although the main ice age to interglacial variations seem to have been roughly contemporaneous in both). It is in the Arctic that the variations, at least in historical time, judged by observations in its Atlantic and European fringe, seem to be in the same direction as those prevailing over the rest of the Earth. There have been some significant temperature variations in other latitudes, including the tropics, although in more complex regional or localized patterns related to the wind and ocean circulations.

Changes in raininess can be both strong and localized where changes in the frequency of winds blowing towards one or the other side of a mountain range (and even smaller hills) are involved. Thus in the 1960s and 1970s, with the decline of the North Atlantic westerlies in latitudes between 45 and 60 °N, rainfall has decreased by a few per cent in western Europe and has decreased more on the western than on the eastern side of the British Isles. In the northeastern segment of North America from New England to Labrador there have been net increases. In the corresponding sector of northeast Asia, in the Soviet Far East and around the Okhotsk Sea and Kamchatka, the increases have in some areas exceeded 20 per cent, presumably registering an increase of the rain and snow-bearing easterly winds from the Pacific and less frequent winds from the west. Around the fringe of the Arctic Ocean the coasts exposed to rain and snowfall with onshore winds associated with the increased cyclonic activity in the highest northern latitudes have had increases of up to 20–50 per cent in the total down-put in the 1960s and 1970s compared with the previous three decades. We have also noted significant changes of rainfall at the fringes of the desert zone in low latitudes.

Thus, our simplest indicators of the state of world climate seem likely to be found in

1 the temperatures prevailing in high latitudes, particularly in the Arctic (the strong variations with time offer a strong 'signal', but the big variations from year to year in any individual area in the Arctic, associated with shifts of position of the coldest conditions which develop wherever little disturbance happens to be felt from other latitudes, may however cause confusion);
2 some indicator of the overall character of the wind circulation in middle latitudes, such as the frequency of the westerlies over the British Isles or the frequency of blocking;
3 the temperatures in low latitudes (if we can sense the rather small net variations with sufficient accuracy).

It is possible that the overall rainfall in the inner tropical zone or the yield (and latitude range in Africa) of the monsoons, as well as the down-put of snow and rain in the highest latitudes, may also serve as indicators of the energy of the global wind circulation.

We have noticed in this chapter the fairly quick response of the Arctic sea ice to the fall of temperature from the 1950 level, just as it had retreated quickly in the earlier twentieth-century warming (see figs. 96 and 97). It has, in fact, long been supposed by investigators that the variations of the Arctic ice could be used as an index of world climate (even though there may be danger in the fact that the only long data series are for the Iceland and Greenland sectors, which may not at all times be representative of the whole Arctic). We must also notice the generally parallel course of the frequency of the middle latitudes westerlies as indicated by that of westerly wind situations in the British Isles (fig. 17, p. 53) with the course of world temperature since 1870–80 (fig. 91a). If we may safely use the data from the limited period of history for which instrument observations and numerical assessments of the ice exist, then we do find some warrant for using this and a few other items as a world index, as table 4 shows.

In discussing our list of possibly simple indicators of the state of world climate in these pages, and the associations between them, we have touched on items which are clearly among the things which it would also be most desirable on directly economic and social grounds to be able to forecast.

Table 4 Correlation coefficients indicating associations in world weather

Items related and period covered by the data	Time units (i.e. the successive non-overlapping blocks compared)	Correlation coefficient	Statistical significance level (%)
Arctic sea ice at the coasts of Iceland and world temperature 1880–1974	5-year	–0.64	99.0
Arctic sea ice at Iceland and temperature in central England 1870–1974	5-year	–0.53	almost 99.0
Arctic sea ice at Iceland and number of westerly days in the year over the British Isles 1870–1974	5-year	–0.54	99.0
Temperature in central England and northern hemisphere temperature 1870–1974	5-year	+ 0.71	99.9
Number of westerly days in the British Isles and rainfall in the Sahel zone (10–20 °N) of Africa 1900–73	1-year	+ 0.56	99.0
Number of days with south-westerly surface wind in London and amount of snow deposited at the south pole 1760–1957	10-year	+ 0.75	99.9

In the remaining chapters of this book we must review the range of impacts of climatic developments on human affairs both now and in past history. And we must survey the possibilities of forecasting the future tendencies of the climate, including any side-effects of man's activities which might affect the trend. We must also consider the application of climatic knowledge to how else we may best plan our affairs to allow for the impact of climate and its future development.

15

THE IMPACT OF CLIMATIC DEVELOPMENTS ON HUMAN AFFAIRS AND HUMAN HISTORY

GENERAL INTRODUCTION

In our survey of human history against the background of what is so far known of the past record of climate, we have remarked only on some of the most obvious or interesting hints of relationships between the two. If we wish to assess the impact of climatic shifts and changes upon human history, or on human affairs today, we must first recognize the many different ways in which an impact may occur. Second, we must be prepared to treat as separate issues the case of peoples directly hit by a climatic event and the more difficult problem of tracing the influence of a climatic event upon societies which were, or are, only indirectly affected or affected much less severely by it. This is also a matter of recognizing that there are situations in which some development of the climate may completely bar certain previously accustomed human activities; in many other situations there is no compulsion, the influence of the climatic event or trend is only a matter of degree, of increasing pressure or difficulty of some operation, leaving the human societies concerned a wide choice. In these cases, their reactions will be decided in large part by other pressures or opportunities. And the weaker, or the more remote the origin of, the climatic stress, the more difficult it must be to trace the working of its effects upon society and the economy. It may be useful to think of the cases of direct impact as 'first order effects' of climatic fluctuation or change and the more indirect impacts as second, third or fourth order effects according to the number of links in the chain by which the impact is transmitted. Third, we must distinguish between the effects of short-term climatic or weather stresses and those of long-term changes, whether gradual or abrupt.

Since there is no difference in general nature between the climatic (or meteorological) events which impose stresses on settled ways of life today and those which did so in the past, the impact on history and the difficulties for human society now or in the future can usefully be considered in the same chapter. Such differences as there are depend on the shield provided by modern technology and our increasing knowledge and ability to adapt our ways or take suitable precautions. But there is reason to ask whether, and in what ways, our vulnerability to climatic events may now be increasing again.

If we analyse the climatic stresses, in origin they are of course physical – a matter of the freezing up or evaporation of waters, of wind stresses, of rainfall and flood levels, or the energy and power of storm waves and tide, and so on. Both physical effects and direct biological consequences come in with consideration of changes in the accumulated warmth of the growing season or the occurrence of prohibitively high or low temperatures inimical to the life of human beings, animals or disease organisms, as also with some of the effects of rainfall, flood levels, drought, and so on. But many significant results may come about through economic and even psychological effects on societies. In some cases it may not be so much the climatic event itself as how it is interpreted, and what it is thought at the time to portend, that influences human action.

Some combinations of the physical variables produce effects on the environment and on man-made structures which have to be considered. For instance, as Professor Flohn has pointed out, variations of rainfall commonly result in much bigger percentage variations in soil moisture, run-off and river flow because of the reduced evaporation in heavily clouded, rainy weather. The effects of rain and snow driven by strong winds are also different – in respect of penetrating walls and loading roofs – from those of similar falls in calmer weather. With temperatures below the freezing point of sea water (about −2 °C), any wind strong enough to produce spray may have a lethal effect on ships by the accumulation of frozen spray on the upper-works and rigging, causing the vessel to capsize: there have been many disasters of this kind on the Arctic fishing grounds. And as is well known to dwellers on the plains and prairies of North America and from the experience of polar expeditions, strong winds greatly increase the physiological effects of low temperatures. Put simply: in the Antarctic – as on mountain heights and in frozen landscapes in winter elsewhere – it is the wind (or really the combined effect of wind and low temperature) that is the killer. A 30-knot wind with a temperature of −5 °C has about the same cooling power as temperatures approaching −30 °C with little wind. The range of ambient temperatures within which the human body is comfortable is also much affected by the humidity of the air, because the body's cooling mechanism depends on sweating. Studies on twentieth-century white European populations have indicated that in still air and out of the sun the average upper limit of 'comfortable' temperatures is about 22 °C (72 °F) for 100 per cent relative humidity, rising to 27 °C when the relative humidity is 66 per cent and 38–39 °C (102 °F) with very dry air. The upper limit of what is 'just bearable' appears to go from 38 °C (100 °F) with 100 per cent relative humidity to about 56 °C (over 130 °F) in very dry air. Since humidities over 90 per cent are found to induce feelings of lethargy whatever the temperature, and cause temperatures below about 7 °C (45 °F) to be felt as 'raw', the range of temperatures which are comfortable is narrower the higher the humidity. A full consideration of the influence of climatic conditions on comfort and the ability to work would have to include the effects of exposure to solar radiation as well.

Considerations such as these have been thought to have to do with differences in the energy of nations.[1] Similar studies might throw light on differences in the energy, ability to acclimatize, and tolerances of different human racial types, if such exist. The comfort and well-being of animals is determined by the same elements of the climate, though the thresholds differ for different species. There is a like need for knowledge and understanding of the climatic tolerance ranges for different food crops as well as the optimum conditions for each. No less important are the climatic conditions which favour, and the limiting conditions for, various pests, diseases and disease-carriers.

We may list the main ways in and through which climatic fluctuations and changes impinge on human affairs, as follows:

1 Water supply, particularly affecting ground-water levels and soil moisture, well levels, river levels, lake levels, also glaciers and, of course, the availability of water for water-power (from mills to hydroelectricity).
2 Temperatures prevailing and their direct effects on human and animal comfort,[2] and hence on fuel demand, and on crop growth.
3 Sunshine, humidity and cloudiness and their effects on health and growth, also the potential of solar power.
4 Windiness and its effects in either damaging structures or the availability of wind and wave power. The effects on evaporation, and hence on vegetation and crops, and on the breeding conditions for insects and bacteria may also be important.

The specific fields in which the impacts are felt can be summarized as:

1 Agriculture and horticulture, including fruit and vine cultivation.
2 Forestry.
3 Insects (e.g. locusts) and other pests, blights, mildews, and their control.
4 Plant, animal and human health and diseases.
5 Weather-sensitive manufacturing and construction industries (textiles, civil engineering, etc.).
6 Trade (national and international trade and planning, quotas and their fulfilment, planning of the locations of agricultural cropping and industrial concentrations and emergency back-up measures) and effects on prices.
7 Travel and communications (opening and closing of mountain passes, of ways across deserts and marshlands, and of routes across seas

threatened by storms or ice), costs of highway clearing and maintenance, of telegraph and cable lines, electricity lines, oil and gas pipe-
lines, etc., and in some regions of ice-breaking.

8 Tourism (summer and winter sports, arrangements for travel and
cruises, health resorts, and the costs of investment in equipment and
maintenance).

9 Disasters and difficulties caused by avalanches, glacier surges, mud-
flows, landslides, rock-falls, floods or parched ground, subsidence and
frost-heave, exceptional snowfalls, violent windstorms, etc. Disasters
to wildlife (fauna and flora) may also be of economic importance
(as when the walnut trees of Europe were so devastated by the great
winter of 1708–9 that the crisis in the furniture industry caused
France to ban the export of walnut for twelve years, and importation
was begun in Europe first of black walnut from America and later of
mahogany in the ships of the East India Company).

10 Coastal flooding and erosion, sand and gravel movements, silting of
estuaries and harbours, associated with either sea storm surges or longer-
term changes of sea level.

11 Arrangements for, and costs of, insurance and safeguards (insurance
industry, storage and stockpiles of food and water, irrigation, building
of coastal defences, climate monitoring and research).

12 Arrangements for, and costs of, relief measures and resettlement of
refugees, settlement of disputes and containment of riots, and threats
of revolutions and wars.

IMPACTS OF THE FIRST ORDER

In our survey of the past we can notice a number of cases where climate
exercised a compulsion on human affairs. The great rise of world sea level
progressing over thousands of years, which followed the ending of the last
ice age and submerged formerly inhabited lowlands and coastal plains, is
one example. The later drying up of the north African, Arabian, northwest Indian and central Asian deserts ended the human activities and
cultures there and must have caused at first famines and ultimately a like
shift of populations. It has even been suggested that the refugees may have
provided the source of slave labour that made the highly organized river-
valley civilizations possible. Other examples of climatic compulsion are
provided by the loss of access to the high-level mines in the Alps about
800 BC and again in the later Middle Ages; the flooding of the prehistoric
lake villages in central Europe at various earlier times and also around
800 BC; and probably the variations of moisture and forest growth in the
valley of Mexico and Yucatan, as well as in Cambodia and elsewhere in
southeast Asia. We have noticed the climatic developments which seem
likely to have caused the abandonment of the old caravan route of trade

between China and the Roman empire, the Great Silk Road through central Asia, and the cities along it; and those which cut off and doomed the Old Norse colony in Greenland and caused the late medieval depopulation of the uplands in central and northern Norway, the abandonment of tillage in many other parts of Europe, and the retreat of the northern limit of vine cultivation.

In these last-named cases it may be held that we have entered a 'grey area', where other causes for the change can be alleged. It is often said that the demise of the medieval vineyards in England and northern Germany and elsewhere was due to economic causes and most particularly that good wine could at last be transported from Bordeaux and southern France; and this is continually repeated as the 'obvious explanation', a matter which can be readily understood by the ordinary man. But from analysis of the data it seems undeniable that climate tipped the economic balance. The purely economic explanation does not square with the fact that in the twelfth and thirteenth centuries, when these French wine districts were under English rule, diplomatic pressure was exerted to try to get the king to suppress the English vineyards. It seems more likely that the Bordeaux trade gained and the terms of trade changed when the English vineyards increasingly failed to produce an acceptable harvest in the fourteenth and fifteenth centuries. Indeed the failure of at least one of the English vineyards, at Ely, seems to be fairly well documented to the bitter end in 1469. Long before that, it is recorded in 1341 that there was still much land uncultivated all over England which had been cultivated before the disastrous summers, and the severe death roll from starvation, in the decade around 1315.

Suspicions of a more far-reaching influence of climate on history have inevitably been aroused by the apparent correspondence between the high points of cultural achievement in northern Europe in the Bronze Age (particularly in the development of sea-going trade), in late Roman times, and the high Middle Ages, and the crests of the temperature curve. There is suggestive further detail too in the coincidences of decline and unrest with a number of the known climatic shocks, particularly around 800 BC in central Europe and more widely in the fifth, sixth, fourteenth, fifteenth and seventeenth centuries AD. In one decade, the 1430s, characterized by a majority of severe winters in much of Europe (and evidently remarkable frequency of 'blocking' in the atmospheric circulation), the summers setting examples of both extremes of temperature and rainfall, we find the Scottish Highlands and Bohemia in civil turmoil, the capital of Scotland moved south for greater security to Edinburgh, a particularly savage phase of the Hundred Years War between England and France, and the collapse of a period of Chinese expansion on land and sea under the Ming dynasty because of internal troubles. And, as we have seen in the last chapter, over the time between the late thirteenth century and the fifteenth, the cultural (and in some senses the political) capital of northern Europe moved south in successive stages from Trondheim to Bergen to Oslo and thence to Copenhagen. Finally, in 1536 Norway ceased to exist as a separate country. Iceland was also subjected to more and more absolute rule from Copenhagen. In 1707 a like move ended the independence of Scotland, which was absorbed in the United Kingdom and ruled from London. At each stage in these developments other, non-climatic causes can usually be alleged, and climatic stress was seldom mentioned as the reason for decisions taken by the people at the time, except in relation to events in Iceland and Greenland or near the glaciers of Europe and in cases of harvest failure.

And if we look at the history of the Far East, the time of drought around AD 300 in central Asia coincided with conflict there leading to the destruction of the Tsin dynasty in northern China by invading nomads. Refugees poured into south China and contributed to the cultural development there, while others fleeing to Korea and western Japan figure prominently in the peopling of those countries. Something like this history was repeated with the Manchu invasion of China, ending the Ming dynasty in 1662 in the midst of one of the severest parts of the Little Ice Age period.

If there is any reality in the web of climatic influence appearing to show this much control over history, it is certainly not simple in its working and comes to light only as the net outcome, the statistical result of an enormous diversity of movements, choices and activities.

MORE COMPLEX CONSEQUENCES

There is not much difficulty in finding cases of contrary effects and opposite movements. Clearly, Denmark and England gained from the decline of their northern neighbours under the difficulties and disasters with which they were beset in the advancing Little Ice Age between 1300 and 1700. And thanks to the mortality experienced in those countries as well as almost all of Europe in the famines and epidemics of the second decade of the fourteenth century, and in the Black Death and subsequent plague epidemics which followed, the surviving population of Europe seems possibly to have been better nourished in the later fourteenth century than in the seventeenth and eighteenth centuries. Thus Slicher van Bath[3] cites the quantities of meat, fat, bread, etc, and estimated total protein served in a fourteenth-century hospital in Nuremberg, totalling about 3400 calories per diem, compared with that in a hospital in Munich four hundred years later, totalling an estimated 1900 calories per diem. Shifts of the northern Atlantic codfish stocks in the fifteenth century are thought to have encouraged the exploration by English, French and Portuguese fishermen of new areas of the ocean, until at some apparently unrecorded date in the middle or late 1400s they began fishing on the Newfoundland Banks.[4] And in the early part of the next century the abandonment of the Baltic by the herring caused the North Sea fisheries to spring into importance, giving a great boost to English and Dutch seafaring activities. Holland became very prosperous by the early seventeenth century, though the later part of that century saw some decline also there owing to troubles from the great storms and sea floods which broke the dykes and with disruption of the fisheries and on the farms. Already long before, in the decline of the Old Norse colony in Greenland, the fading out of the hunting along the northern reaches of the west coast near Disko Bay seems to have led to an outburst of renewed foraging farther afield, at first in 1267 north and west into Baffin Bay and as late as 1347 west to Markland (Labrador). And it seems that the furthest explorations achieved in the history of the colony were made at that late stage.

Other examples of curious and complex phasing of population movements during the development of the Little Ice Age can be found in Scandinavia itself. While farms were being abandoned in the fifteenth and sixteenth centuries in north Norway, the fisheries along the coast were being increasingly developed and population was increasing there; in part this is thought to have represented an influx of different people spreading northward from the coastal fishing settlements farther south. At the same time, while even in England and central (and parts of southern) Europe farms and villages on the uplands and elsewhere were being abandoned, settlement was still advancing in places in northern Sweden and Finland. This may have been, in part, because population had always been so sparse there that even many favourable localities had never been occupied. But the evidence of tree rings in Lapland, in fact, indicates a predominance of good years for growth there right on until 1580 or almost 1600. There is similar evidence from Alaska and the Yukon. Taken together, this is meteorologically suggestive of blocking anticyclones commonly giving warm sunshine over those areas, while the northerly and easterly winds on their eastern and southern sides carried Arctic air into Russia and central Europe and similarly into central and eastern North America. In the late sixteenth century, however, there was some migration of Finns into central Sweden and Norway and towards the Atlantic coast farther north. And in the more widespread and well-documented cold regime in the seventeenth century, particularly towards the end of that century, the tide of settlement went into retreat over the whole of northern Scandinavia and Finland.

Thus, while we reasonably look for the most direct effects of climate on human history – and on human affairs in any age – among peoples living at the poleward or hot desert margin, there is no lack of complex and contrary movements and activities in the regions from which we have drawn examples. One of the most remarkable responses to the climatic stress of the climax of the Little Ice Age is reported in southern Norway all around the coast between Trondheim and Oslofjord. In the late seventeenth century, when the harvests were poor and the grain sometimes failed to ripen on the farms even in the most favoured areas along the southeast coast, as mentioned briefly in chapter 12, the farmers took to trading abroad the timber on their land, notably to England, and those near enough to the coast built their own ships to carry it in. Those, particularly in the south, who had oak were in the best position. This seems to have been the beginning of what became two of Norway's greatest industries, the timber trade and her merchant fleet which by our own century was one of the biggest in the world. And so it came about that the years 1680–1709, which seem clearly to embrace the bitterest period of the climate in northern Europe, are described as 'the first great period of Norwegian shipping'.[5] Of course, those were also times when the great powers farther south in Europe were at war: the war boosted the trade in timber, and together with the activities of pirates it encouraged those not involved to protect themselves in an armed neutrality. A report on Stavanger briefly indicates the situation at an early stage in this development: 'despite the town's miserable condition in 1685, it managed to keep one defence ship with 25 pieces'. In some years in the 1690s the death rate greatly exceeded the birth rate, and there was a net fall of the population of the town from 1685 to 1701. The numbers and sizes of the ships kept at nearly every port along the coast increased greatly in the next twenty-five years. The numbers were swollen by incoming Dutchmen, who took Danish-Norwegian nationality in order to sail under the 'flag of convenience' of a neutral country and who then stayed on to become big shipowners in Norway.[6] (The Dutchmen came especially to Bergen, where there was some involvement in the whaling up to Spitsbergen though not on the scale of the operations from the Dutch and German ports.) In Sweden it is recorded that Dutch shipwrights in the seventeenth century took a leading part in shipbuilding and contributed significantly to the strength of the Swedish navy.

At the same time as these developments were going on in Scandinavia, in Iceland the reaction to the climate stresses of the seventeenth century was very different. According to the Icelandic historian Gisli Gunnarsson, the strength of the landowners' position in what was very much a feudal society enabled them to oppose in several effective ways the drift of labour from the farming areas, where difficulties were increasing, to take up fishing on the coast. This opposition is documented in the records of the courts. All sections of Iceland society – in its depressed and fossilized state at that time – seem to have been against technical innovation. Only open-decked boats were used. And this was compounded by crippling restrictions, permitting no more than one hook on a line and forbidding the use of worms as bait. It is clear that the difficulties were always greatest in the north and east of Iceland, where the polar water from the East Greenland Current is liable even now to come in along the coast; but in the worst phase, between 1685 and 1704, not only the hay harvests but also the cod fishery were poor, or failed completely, even in the southwest of the island.

In the late eighteenth century, when the government in Copenhagen was trying to stimulate recovery in Iceland by encouraging fishing and seamanship with the introduction of decked sailing vessels and more hooks on the lines, it had a long struggle against general opposition to any change.[7]

Over the whole Little Ice Age period the population of Iceland was falling. At its peak in the eleventh to thirteenth centuries it can be estimated from tax records at between seventy and eighty thousand. At the first census in 1703 it was 50,358, but was reduced four years later by about a third in a smallpox epidemic. It rose in the warm years in and around the 1730s to about 48,000 by 1755 and again in the 1770s to 49,863, after a dip in the severe years in between. But the severe seasons which followed, and the volcanic effluents which poisoned the pastures and the cattle, reduced the population again to its lowest, about 38,000, in 1784–6. Once again we see a historical development which runs fairly closely parallel to the temperature curve, even though the apparent link operates in various ways, through undernourishment and starvation, through illness and emigration.

Resistance to change is, of course, familiar enough in other parts of the world. As if to parallel and explain the seventeenth-century Iceland situation by a current anxiety in the modern world, the 1980 Report of the Brandt Commission[8] stresses that human energy and ability to innovate depend on adequate nourishment and good health; yet most people in today's poverty belts suffer from long-standing malnutrition and parasitic diseases, and for that reason cannot help themselves unaided to set up a new economy that might better withstand the pressures of overpopulation and the harsh climates of Africa and south Asia. We have noted (p. 245) the slow progress in Europe in adoption of the potato. In the moist climate of Ireland it was so much more reliable than either wheat or even oats that it was soon taken on – already in the seventeenth century – as the 'bulwark against famine' and gradually eliminated grain growing over wide areas. But in France it became regarded as suspect because of its botanical family relationship to the native belladonna (deadly nightshade).[9] Some other crops from the New World, such as string beans, were taken on readily enough in southern Europe; and John Locke after travelling in France in the 1670s recommended putting leaves of kidney beans under your pillow, or in other convenient places about your bed, to concentrate the bed bugs and save yourself from being bitten. But maize was not liked and its progress seems to have been delayed partly for that reason and partly because of the cool summers of the Little Ice Age.[10]

We have taken our examples of the more complex involvement of climate in human decisions in the past mostly from those parts of Europe which were particularly vulnerable to climatic change and for which we have good information. If we are to attain a fuller understanding of the lessons for our own day and for the future, we must proceed to more specific detail of how the climate works upon food production and health. Among the indirect and subtler influences are, of course, many that can hardly be measured, such as the effects upon art (p. 233) and architecture. Was it, for instance, just a coincidence that the widespread introduction of glass into windows in the houses of Europe coincided with the late sixteenth and seventeenth-century privations of the Little Ice Age? Instances are also not hard to find of influence upon fashions in dress, particularly as regards warmth – often adjustments to an event that has already taken place and which therefore may or may not be repeated. Among the subtler influences, one may perhaps detect the optimism engendered in Europe by the glorious summers of 1718 and 1719, the warmth of the 1730s, and more good summers in 1759 and around 1778–80, in the writings and perhaps in the music of the time. The psychological effect must have been particularly strong on those who had lived through the 1690s.

EFFECTS ON GRAIN HARVESTS

We have seen how in the late Middle Ages wheat cultivation was given up in Norway and in much of Scotland. In Iceland, and on difficult land in many other areas of Europe, grain cultivation was given up altogether for a long time. Elsewhere oats or barley were kept on (in one or two places in Iceland until the sixteenth century and in Scotland and Norway throughout), and rye was brought in or increased. Sooner or later in the progress of the climatic recovery after 1700 these changes were reversed, except that in many areas rye had won a permanent place. What do we know about these crops' requirements that might explain the linkages which we must suppose to exist in these parallel histories?

M. L. Parry[11] has shown how the matter may be investigated. Grain crops, like any other plant, have certain requirements as regards overall warmth, moisture, sunshine and not too much wind in the growing season, if they are to come to fruition. Parry considered in some detail the case of oats, which were the major grain crop at the upland limit of cultivation in Scotland over many centuries past and were important also in Iceland and north Norway. The varieties grown were changed in the nineteenth century, the older varieties being shorter in the stalk and less liable to shaking, but there is no evidence to suggest that their warmth requirement was any less, or their tolerance of wetness any greater, than the modern varieties. The differences were probably slight but such as to make the modern oats somewhat hardier on the high farms except in regard to wind speed. Comparison with climatic atlas data for the Lammermuir Hills in southeast Scotland showed that the upper limit of oats cultivation in 1860, at about 320 m above sea level, corresponded closely with the 4.4 m/sec mean wind speed line. Mean wind speed increased about 1 m/sec for every 80 m increase of height above the 200 m level. Changes of solar radiation

with height appeared too slight to be a limiting factor: the increased intensity of the radiation at the higher elevations is slightly more than offset by
reduced duration owing to hill fog, the net reduction being only of the
order of 5 per cent on the upper levels of these hills. Soil moisture increases
rapidly with height and for this reason oats, which are intolerant of waterlogging, have an absolute limit at 425 m above sea level on the hills of
southeast Scotland in the present climate. The temperature requirement
generally limits the possibility of growing the crop at well below that level;
moisture may, however, also contribute to failure of the crop at still lower
levels in some years.

Parry proceeded to examine the moisture and temperature involvement
more closely. There is liable to be a spell in early summer, even on the
heights and near the northern limit of cultivation, when the potential
transpiration of moisture through the stems and leaves of the plants exceeds
the rainfall and leads to drying of the soil. What is liable to damage crops
on the heights is the water surplus produced in the later part of the summer,
when the crop is ripening and the rainfall exceeds the potential for
evaporating transpiration (a quantity known as the 'potential evapo-
transpiration'). The wetness of a summer on the heights can usefully be
measured, according to Parry, by the difference between the water surplus
at the end of August and the greatest potential water deficit which occurred
earlier in the summer. Study of present-day oat cultivation in the hill
country of southern Scotland established that the limiting conditions corresponded to an average value of 60 mm of water for this difference, and a
mean wind speed of 6.2 m/sec, while the minimum accumulated warmth
requirement for the growing season was about 1050 day-degrees C above
the 4.4 °C threshold of growth. For commercially viable cropping the
critical figures could be taken as 20 mm accumulated water surplus, 5 m/sec
mean wind speed and 1200 day-degrees C. The zone with conditions
normally between these two sets of criteria can properly be described as
marginal land.

Wind speed and humidity records are too short to provide the relevant
figures on greatest potential water deficit and later water surplus for summers
in past centuries to compare with the harvest records; but the accumulated
summer warmth can be calculated for summers in Scotland in the late
eighteenth and nineteenth centuries and can be estimated for earlier times.
This is not such a serious restriction as it might seem because there is a strong
correlation between warmth and the dryness of the summers. Parry therefore went on to consider the probable frequency of failures of the oat harvest
in earlier centuries by reference to the temperature changes known or derived
for central England (fig. 30 in this book, p. 84).

Total failures of the oats may have been rare where the crop was grown
for consumption by the family on the farm. There are cases recorded where
the crop was reaped in December and even January, though in a poor state,

liable to be mildewed or sprouted in the ear and much of it lost by wind-shaking. Such were the dreaded 'green years' when the crops failed to ripen. And in such cases recourse would be had to eating some of the seed reserves from the previous year. A sequence of such harvests, as in several reported runs of two or three bad years in Scotland (e.g. in 1740–2, and 1781 and 1782, let alone the seven years out of eight between 1693 and 1700, when in the upland districts overall perhaps a third of the population died), would soon produce famine and tend to put some farms out of business.

After establishing the accumulated warmth figures for some historic summers, such as 1782, 1799, 1816, when the harvest was not got in until the end of November or later on the hill farms in southeast Scotland, Parry was able to calculate the probable frequency of such summers at various times in the past by assuming the changes of mean temperature level to have been the same as those affecting central England and taking the variability (standard deviation) as constant. The calculations produced the curves seen in fig. 102. The curves show the most probable intervals between harvest failures in a single year or between failures two years in succession on these assumptions, when and wherever the longer-term average summer warmth gives the accumulations of day degrees above the 4.4 °C datum specified. With the temperatures prevailing in recent times the average accumulation can be taken as about 1150 day-degrees at 300 m above sea level in the area investigated, for which the graph indicates an average expectation of a harvest failure about one year in seven. With the temperatures derived for the thirteenth century, giving an average of 1200 or more day-degrees at the same height, this expectation might be reduced to one year in about twenty. But with the climate as it was in the second half of the seventeenth century, the average would be about 975 day-degrees and harvests likely to fail two years running once in about four years. Clearly agriculture could not then be sustained at the 300 m level. Crude as the assumptions are on which these calculations are based, they give a firm enough glimpse of the compulsion to abandon the upper areas of former cultivation under such circumstances.

An attempted similar use of the available rainfall estimates indicated no significant changes in the historical course of the apparent upper limit of possible oat cultivation. A temperature curve resembling fig. 30 in this book, but with the values reduced by the difference between central England and the southern uplands of Scotland and converted to average yearly accumulations of day-degrees above the 4.4° datum, which we take to represent zero growth, should give us a history of the frequency of harvest failures. This is the message of fig. 103, which Dr Parry has kindly allowed me to reproduce from his book. It indicates the dates at which a farm which is now near the limit of cultivation at about 320 m was (a) no longer marginal around AD 1200, when the limit was more than 400 m above sea level, and (b) when it encountered increasing difficulty and presumably became

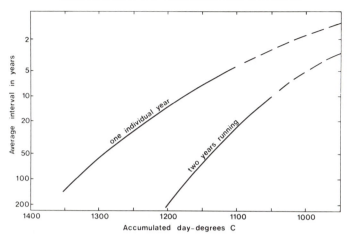

Fig. 102 The frequency of failures of the oat harvest in the hill country of south-east Scotland according to the prevailing average level of accumulated warmth of the growing season. (Adapted from a diagram by M. L. Parry.)

no longer viable from around AD 1400 till the nineteenth century. Parry was able on this basis to map the probable cultivation limit at various stages of the climatic cooling between AD 1300 and about 1700 and of the subsequent warming, mainly since 1900. As it was also possible to map the former settlements abandoned at various dates – fifteen of them before 1600 and twelve more between 1600 and 1750 – in the Lammermuir Hills study area, and compare these maps with Parry's theoretical limits, the thesis can be regarded as having been vindicated by test.[12]

Barley also is one of the most important crops in most parts of northern Europe. Studies in the 1970s of its responses to weather have indicated that in eastern England it does best in cool years. High yield was found to be favoured by lower than average temperatures and dryness in spring, items tending to produce slow progress at that stage. In the next stages high rainfall was beneficial, but the strongest relationships were between high yield and Julys that were cooler, duller and more humid than average. In Scotland, where barley is now grown mainly in the broad eastern lowlands, over about the same years it was found that the average yield in tonnes/hectare was about 20 per cent higher than in England and was less variable from year to year. Evidently this is the climate that suits the crop; significant sensitivity to year-to-year weather differences was indicated only by the positive response to the sunniest years in the growing season, particularly in June and July, in Scotland.[13]

In the grain growing areas on the extensive low ground of eastern England, anxiety is nowadays more often caused by drought in the growing season, although waterlogging in wet autumns can hinder, or perhaps even

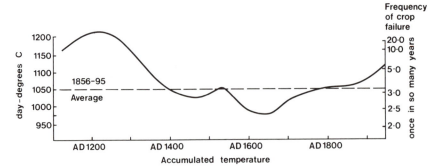

Fig. 103 The estimated average accumulated warmth of the growing season prevailing at sites near the upper limit of cereal cultivation in the hill country of southeast Scotland in the period 1856–95 (horizontal line) and its variations over the last thousand years. The probable frequency of failures of the oat harvest at such sites in different periods can be read off the scale at the right of the diagram. (From a diagram by M. L. Parry.)

prevent, the autumn sowing of grains just as it does the lifting (especially with modern heavy machinery) of root crops. A history of the soil moisture deficit built up over the four months of the growing season from May to August in southeast England has been reconstructed from the long series of temperature and rainfall data for London (Kew).[14] The result, illustrated in fig. 104, reveals the periods of greater moisture – enough to produce troubles with the grain harvest in parts of the region – in the 1760s–early 1770s and before 1740, as well as in the period 1810 to the 1830s and the 1870s and 1880s. But the most notable feature is the increased frequency of droughts in the warmer years of the present century, to some extent paralleled two hundred years earlier in the 1740s. Indeed, it is reasonable to consider whether the difference apparent between recent times and the eighteenth-century incidence of drought could largely be explained by the growth of London and the now pronounced urban effect on the temperatures measured at Kew (now on average about 2 °C above those observed in the surrounding country).

DETAILS FROM SWITZERLAND IN THE EIGHTEENTH CENTURY

The history of the renewed deterioration of the climate in the latter half of the eighteenth century in Switzerland and its effects on agriculture has been closely studied by the Swiss historian, Christian Pfister,[15] with the aid of daily meteorological observations and the agricultural and economic reports collected at the time by the then newly established Economic Society of Bern (Ökonomische und Gemeinnützige Gesellschaft des Kantons Bern). There were short runs of warm years between 1759 and 1763 and between

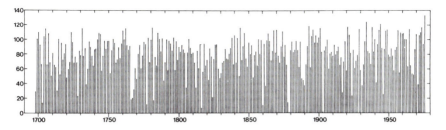

Fig. 104 Soil moisture deficits over the months May to August each year from 1698 to 1976 at Kew (in the London area), as calculated by Drs T. M. L. Wigley and T. C. Atkinson (1977). (Reproduced here by their kind permission.)

1778 and 1784, but in the colder periods outside those dates some of the severest conditions of the Little Ice Age occurred. After 1764 the summers were generally cold and rainy in the Swiss lowlands and there were up to twenty-four days of new snowfall between mid-May and mid-September on the heights between 1500 and 2300 m. Those summers were too short to clear the snows that had accumulated on the upper Alpine pastures, an experience to some extent repeated after a long respite in 1978. In 1770 the Stockhornkette chain at 2000 m remained snow-covered all summer, and 1771 was little better. Similar years are known to have occurred around 1713, 1740, 1792–5 and especially from 1812 to 1817. Several of the winters were longer and delayed the coming of spring even on the lowlands until later than has occurred in the present century. In the worst years the harvest was not brought in until after the long, snowy winters and the seed was found to have rotted under the snow cover, probably due to the parasite *Fusarium nivale* (see p. 216). The total harvest of bread grains and the yield of the tithes in the cantons of Bern, Vaud and Emmental fell by a quarter to a third in 1769–70, and the price of bread grains more than doubled in 1771. The prices of hay and animal products – beef, butter and cheese – were similarly affected. A 40–60 per cent drop in the yield of the tithes in 1785 betokens another very bad year after the exceptionally long winter of 1784–5 and a spring and summer that were wet except in the extreme west of the country. Grain prices reached another sharp peak in 1789 and again doubled in 1795, when animal products were also affected after another severe winter followed by great wetness, but the situation was eased by the potato harvest. Prices continued abnormally high in 1796.

These were the worst years in the forty-year series of data for Switzerland tabulated by Pfister. Although the long-term average levels of the tithes and the harvest yields showed a minimum centred about the 1770s, and the glaciers were advancing, the other years were by no means so bad. Thus the yield of the tithes in the sampled areas was about 10 per cent above the smoothed average in 1786 and 1787 before another sharp drop (10–30 per cent deficiency) in 1788 and 1789.

It can, of course, be argued that neither Iceland nor Norway nor the uplands of central Europe, let alone Scotland, provides a meaningful test of the effects of climatic vagaries on the wider community of Europe. But the effects of the severe winters of 1784–5 and 1788–9 seem to have been harsher in France than in Switzerland, perhaps because of the summer droughts in 1785 and 1788, and we have seen in chapter 13 how the consequential rise in the price of bread may have played a part in the French Revolution.

THE TIME AROUND 1816

The anomalous weather of the years 1812–17, which accompanied the exceptional outburst of volcanic activity in those years with tremendous injections of matter into the stratosphere, reached its climax in 1816, as described in chapter 13. In the summer of that year the usual sub-Arctic cyclonic activity, with its rainfall and storms, was concentrated in a belt from near Newfoundland crossing England into the southern Baltic. In central England the average temperature of the summer three months (June–August) 1816 was 13.4 °C (56.0 °F), almost as low as the coldest years in the Little Ice Age period (13.2 °C in 1695, 13.1 °C in 1725) and a figure bettered by many a September and one or two Mays. The overall climate, but especially the summers, averaged almost 1 °C colder in England in that decade and again between about 1835 and the late 1840s (also associated with volcanic dust loading of the stratosphere) than in the preceding and following decades.

Severe cold and harvest difficulties were reported from many other regions, especially in Europe and the northern United States in 1816 and Japan in 1836. The monsoons were disturbed in India (see p. 248). Can it really be unconnected that 'the years 1812–17 introduced three decades of economic pause punctuated by recurring crises, distress, social upheaval, international migration, political rebellion and pandemic disease'? The writer of those words[16] does not think so. He goes on:

> Those who account for this period by citing the nettlesome decades of early industrialization should recall that these phenomena were not limited to western Europe. Although the numerous crises, popular disturbances, and rebellions between 1812 and 1848 are well known, the epidemiology of these decades is not. . . . the meteorological patterns of 1816 induced the first modern pandemic of cholera which began in Bengal in 1816–17. The most extensive typhus epidemic in European history struck in two waves, an earlier one in 1813–15, and a more severe contagion in 1816–19 . . . an epidemic of plague

raged in the Balkans, along the Adriatic coast, and in the lands of the southern Mediterranean, during the last half of this decade.

He adds that the connection between typhus or cholera and cold, wet vegetative seasons is now well understood, but the ecological conditions which favour plague are not and the plague outbreak may well not have been so directly attributable to the weather. Any direct connection with the Napoleonic war which ended in 1815 is at least equally unlikely, since, for instance, in the Swiss records studied by Pfister in the parish of Alpenzell there was a 50 per cent reduction in the birth rate which culminated sharply in 1819. And it was from about 1818 to 1855 that the Alpine glaciers showed perhaps their most continuous advance.

The effect of these years on the price of rye in Germany is marked by the sharp peak in 1816 and 1817 seen in fig. 33b (p. 88) and is probably the main contributor to the great peak of the wheat prices in the parts of Europe covered by fig. 33a, even though many of the war years were included in the same twenty-five-year mean. J. D. Post has described these years as the last great subsistence crisis in the western world. The effects were already mitigated, however, in all those areas, notably Ireland, where potatoes were already grown.

In England the practice of irrigation, which had been begun on the farms in the pursuit of agricultural improvements in the eighteenth century and was no doubt given an impetus by the dry years in the 1740s and 1750s, seems to have been given up in the cooler, wetter summers of the nineteenth century, particularly from the second decade onwards. Spring was also more frequently wet than before. Similarly in the Far East the double rice cropping regime which had been adopted in the lower Yangtze valley in the eighteenth century – giving its greatest yield of 7.6 tonnes/hectare in 1718, which was a warm year there as in Europe, and an average of 6.2 tonnes/hectare – failed in the early nineteenth century owing to the climate turning cold.[17] In Japan this was perhaps the coldest part of the Little Ice Age with great famines caused by harvest failures and shortfalls in the cold summers of 1782–7, 1833–9 and 1866–9 produced by cold northeast winds and excessive rains. (At least the first two of these groups of years were characterized by exceptional loading of the stratosphere with volcanic matter after very great eruptions.) Much of the rice crop never ripened, and the poor were driven to gather nuts and roots for food and to eat dogs and cats. And as in Europe in the famines of the late Middle Ages there were some reports of cannibalism. The population of districts in northern Japan fell by about 10 per cent partly due to deaths and partly through vagrancy. Also, as in Iceland in the seventeenth century, it appears that feudalism and the imposed isolation from contacts with the outside world aggravated the disaster and told against any adjustment and innovation which might have improved the situation either in the short or longer term.[18]

1879 AND THE DECLINE OF BRITISH AGRICULTURE

Later in the nineteenth century, the disastrous harvest of 1879, another cold year, with a mean summer temperature of 13.7 °C in central England and nearly twice the normal rainfall – all the other seasons of that year were also cold – cut the wheat harvest by half. With similar effects in other European countries it precipitated a change that would otherwise have come gradually in any case, the large-scale importation of cheap wheat from the North American prairies where the beginnings of mechanization had already appeared. There had been wet summers in three of the preceding four years and the run continued unbroken to 1882. With this England's agriculture went into crisis and a decline that continued for fifty years. Other European countries protected their small peasant farmers by tariffs on imports, but Britain went on with the policy of free trade which had helped build up her manufacturing industries. The result was a drift of population from the land to the industrial areas and to the British colonies overseas. The area on which corn was grown shrank by a quarter in the last thirty years of the century, and the rural population declined by some hundreds of thousands.

FOOD SHORTAGES IN MODERN TIMES

The near-immunity to food shortages which came about in the present century in North America, Europe and other advanced countries, and has come to be taken for granted as a benefit of modern scientific know-how, can rightly be attributed to the advances of science and technology. However, as McQuigg and others have pointed out,[19] the doubling of wheat and corn yields in the states of the United States Middle West over the period 1955–73 not only was achieved by technological innovations but owed a good deal to the long run of benign, drought-free years (see fig. 105). Similarly agriculture and husbandry in western Europe gained from the warmth maximum of the period from 1933 to the 1950s. The growing season in Ireland, which had averaged eight months of temperatures above 6 °C around 1900, increased to almost nine months, resulting in a 20 per cent reduction in the season of winter cattle feeding: the growing season has since shortened again in Ireland, as in England, by about two weeks.[20] With the development of new high-yielding strains of rice and other crops and the ability to ship food in bulk, and in emergency quickly, around the world, the benefit has increasingly been spread to much of the Third World also. But the increases of crop yield won by improved scientific knowledge cannot for ever be followed up by further increases, and already there are some poor countries in Africa and south Asia where population growth has been outstripping the increase in production of food.

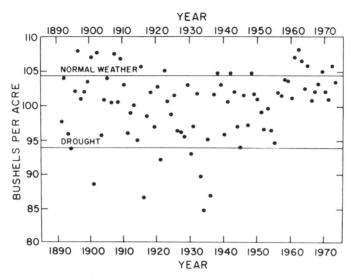

Fig. 105 Calculated yields of corn (maize) each year since 1890 in the five main corn growing states of the United States (Ohio, Indiana, Illinois, Iowa and Missouri), based on the hypothetical use throughout of the methods and technology of the 1970s. (The calculation used a model due to Professor L. M. Thompson of Iowa State University, Ames, Iowa. The diagram was first published by Dr J. D. McQuigg of Columbia, Missouri and others in 'The influence of weather and climate on United States grain yields: bumper crops or droughts', (Washington, DC, NOAA Report, 1973) and is reproduced here by permission.) The drought years of the 1930s – the so-called 'Dust-Bowl' disaster – and in two earlier decades stand out. One also sees how the unprecedented run of drought-free years from 1955 to 1973 contributed to improved yields.

The whole favourable development is, in fact, threatened on a world scale by the growth of population. And since the best land for agriculture has already been taken into use, one must expect lower returns from any further increases of the acreage sown. In this, one finds that man's vulnerability to climatic fluctuations is bound up with, and intensified by, the population explosion. My colleagues Jean Palutikof and Graham Farmer in the Climatic Research Unit at the University of East Anglia have recently pointed out that this is now seen in a particularly stark form in the drought-prone areas of East Africa.[21] Traditionally the population of these areas guarded against disaster by planting a wide variety of crops, spread out over many weeks, so that at least some were likely to survive any drought periods and come to fruition. For the same reason they kept great numbers of cattle, and they could roam over an extensive area for whatever nourishment was to be had. The increase of population and modern political and organizational developments since 1950 have made these safeguards largely impossible. Governments prefer to organize cash cropping, taking

301

in larger areas which were formerly used for grazing, and concentrating on few varieties if not only a single crop. And national and other boundaries now restrict migration.

Some aspects of the threatening situation are well illustrated by Kenya, which has been one of the most stable countries in post-imperial Africa. Kenya, with a high birth rate and falling death rate, now has the most rapidly rising population in the world, the first nation in recorded history to achieve an annual increase of 4 per cent. The total population was 10,943,000 in 1969 and over 15 million in 1979. In some rural areas the population density already doubled in seven years during the 1960s, and at this rate Kenya may be trying to feed a population of 50–60 million by the year 2000. With a run of three or four good rainfall years up to the time of writing, the rural population must have begun to make use of, and perhaps settle in, areas where the same rainfall cannot be expected to continue.

If a good deal of the food security of earlier decades of the present century was due to the combination of some already improved scientific knowledge in agriculture with a still diversified husbandry in the advanced countries, and a lower density of population in the Third World, the more recent development of rationalization on a world scale – with concentration on just one or two crops in each extensive region where they are supposed to grow best – constitutes a threat to this security. Monoculture was at the root of most of the great famines of the past. And it should be noted that the selection of areas where each crop grows best implies a forecast of no climatic change, indeed of no fluctuation beyond a certain expected range.

This increase of vulnerability is real and, together with the increased pressure on grain supplies arising from the growth of world population and the demand for a rising standard of living everywhere, it must give rise to serious anxieties about the future – indeed about the not distant future. The remedy seems to lie in deliberate choices of more diversification (rather than planning maximization of a single crop) and a curb on the feeding of grain to animals for meat production for the wealthier countries, for in that way seven times as much grain is needed for the same amount of protein production for human food as when the grain is consumed for itself. Moreover, the risk is not limited to the case of a greater than expected climatic deviation alone: monoculture and specialization of crops over great areas must increase the probable scale of the disaster if any new crop disease or mutant of a known disease should be spread by winds or other weather conditions which are themselves within the expected range. This seems to have been an important aspect of the Irish potato famine of the 1840s. The success of the potato in the moist climate of the Atlantic fringe of Europe, and the growth of population in Ireland in the eighteenth and early nineteenth centuries, had meant that this was the one crop which

Table 5 World wheat and coarse grains[a] trade (millions of tonnes)

	Production averages for 1974–5 to 1979–80	Exports 1973[c]	Exports 1934–8[c]
USA	251.3		
USSR	184.8		
Western Europe	138.1		
(European Community)	(104.8)		
China	116.8[b]		
World total	1086.0		
North America		88	5
USSR and Eastern Europe		–27	5
Western Europe		–21	–24
Asia outside the USSR		–39	5
Africa		–4	1
Latin America		–4	9
Australia and New Zealand		7	3

Sources: US Department of Agriculture and Overseas Development Council

Notes:
a. Rice not included
b. The total Chinese grain crop including rice in 1979 was 332 million tonnes
c. The minus signs in these columns represent net importation

could produce enough food for a family on the very small farms, many of which were of only one hectare. So the potato had become the only crop grown in much of the country. And when the previously unknown blight appeared, and was quickly spread by the winds of the autumn of 1845 and the moist summer of 1846, all was lost.

It may be useful to review more recent developments in the light shed from this background. In 1913 Imperial Russia was still the main producer and exporter of surplus grain, especially wheat.[22] And up to the 1930s many other countries, including some in Europe, produced surpluses for export. But despite very great increases in production world-wide, population and consumption have increased so much that now only North America produces substantial surpluses of grain for export.[23] But if the United States should follow Brazil's lead in growing crops on a large scale to produce liquid fuel and lubricants (oils, alcohol or methanol), there might soon be no surplus even there. The North American surplus may disappear in any case within about a decade because of its own growing population.

Since 1960, and even since 1970 despite the development of high-yielding crops in the Green Revolution, the world's total production of grains has barely, and certainly not consistently, kept pace with the growth of world population. In 1960–1 the end of season world stocks of grain were estimated at 222 million tonnes, representing 26 per cent of a year's requirement. Ten years later, in 1970–1, the figures were 166 million tonnes or 15 per cent and in 1974–5 and 1975–6 the end of season reserves, amounting to only 131 and 138 million tonnes respectively, were reckoned as 11 per cent of a year's requirement. Of course, changing policy decisions in some countries had affected the issue, and in 1979–80 the figures had recovered to 195 million tonnes end of season stocks or 14 per cent of a year's requirement. In the 1970s the Soviet Union, although still the world's second biggest grain producer , had become a net importer of grain. Table 5 gives a brief survey of the changed world situation. The relative bulk of the different chief crops is given in table 6.

Weather has come into the situation chiefly through the year-to-year variability, which as we have seen seems to have increased in some regions, although the Canadian wheat crop has been affected by recurrent earlier autumn frosts and crops in Africa south of the Sahara by the increased incidence of drought in the 1970s. Fig. 106 shows the variations of the Soviet grain harvest from 1960 to 1980. In all the years with major short-falls substantial grain purchases were made from the West. In 1980 another shortfall of well over 20 per cent – the second year running – has been reported; and this time political actions make it unlikely that purchases can be made on the same scale. In some ways the most serious case was in 1972, when there were also other severe weather-induced shortages in food production elsewhere around the world.

The Soviet Union seems to be afflicted by the nature of climate in ways that cause the year-to-year variability of the total harvest, surveyed over

Table 6 World's twenty-five chief food crops (millions of tonnes grown in 1976)

Wheat	417	Potato	288
Rice	345	Sweet potato	136
Maize	334	Cassava	105
Barley	190	Grapes	59
Soybean	62	Cane sugar	52
Sorghum	52	Tomato	41
Millets	51	Banana	39
Oats	50	Beet sugar	34
Rye	28	Oranges	34
Cotton-seed oil	25	Coconut	33
		Watermelon	23
		Apples	22
		Cabbage	21
		Yam	20
		Peanuts	18

Source: FAO Production Yearbook 1976

many years, to be about twice as great as for North America. Looked at in another way, the percentage probabilities of a single poor wheat harvest or of two or three poor harvest years in succession – poor harvests being defined by production more than 10 per cent below expectation – over the longest spans of years in the present century examined were as shown in table 7. It will be observed that the overall variability of the total Soviet crop is somewhat less than for either the spring or winter-sown crops singly, as there is some apparent compensation in the variations in the different seasons. The vast west-to-east extent of the plains of Eurasia and of the mountain massifs on their southern border means that, when blocking situations occur, the stationary anticyclones may be elongated in such a way as to bring drought (or severe cold weather in winter) over a very wide sector and yet cover a somewhat different span of longitudes in another year. Indeed the anticyclones are generally differently located in winter, spring and summer of the same year. The frequencies of two or three years in a row of poor harvests indicated in table 7, being relatively rare (though very serious) events, are undoubtedly affected by sampling problems – i.e. the erratic results usual in small samples. The figures must have been raised in the United States by the historic drought in the 'Dust-Bowl' years in

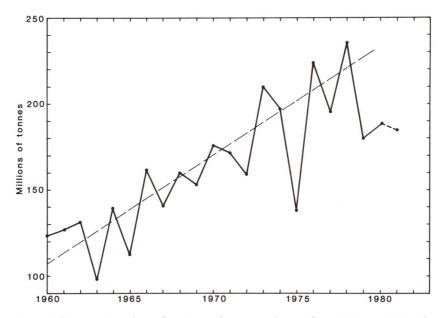

Fig. 106 Soviet Union's total grain production each year from 1960 to 1980. The broken line indicates the rising production expected from increasing acreage sown and increasing technological input. A third successive bad harvest in 1981 is expected to total about 170 million tonnes. (Data kindly supplied by the United States Department of Agriculture.)

Table 7 Percentage probability of unfavourable wheat harvests (more than a tenth below expectation) over the years 1900 or 1910 to 1974 or 1975

	One single bad harvest	Successions of	
		2 years in a row	3 years
USA	17	9	5
USSR (spring wheat)	44	13	4
USSR (winter wheat)	32	9	3
USSR total wheat crop	27	6	2

Source: Data as given by C. Sakamoto, S. Leduc, N. Strommen and L. Steyaert in an article entitled 'Climate and global grain yield variability', *Climatic Change*, vol. 2, no. 4, pp. 349–61, 1980.

the 1930s. In order to eliminate the political effects on the magnitude of the Soviet harvests in the revolutionary years, the figures used for those years have been derived by simulation using a meteorological model, i.e. a meteorological fiction rather than the reality of those harvests.

THE EXPERIENCE OF 1972

The events of 1972, briefly referred to above, caused a great deal of concern about tendencies of the climate that had escaped notice, or not been much thought about, until then. In that same year, with its extraordinary heat and drought in Russia (see pp. 277–8), when the Soviet grain harvest was about 13 per cent short of expectation, the drought belt continued eastwards in such a way that the Chinese harvest was also described as disastrously short and in northern India there was a deficient monsoon with a similar result. The drought, already then prolonged over several years, in another belt along the southern fringe of the desert zone also reached a climax in 1972 and 1973, with the result that an estimated 100,000 to 200,000 people and perhaps four million cattle died in the zone that stretches across Africa from the Sahel to Ethiopia (fig. 107). There was also a mass migration of people leaving their homes and accustomed land southwards, in some cases crossing the frontiers which are an awkward legacy of the former European imperial administrations of the region. The coffee harvest in Ethiopia, Kenya and the Ivory Coast and the ground nuts, sorghum and rice in Nigeria were also sharply reduced. And, to complete the picture, the Australian wheat crop in 1972–3, owing to drought there, was also more than 25 per cent below the previous five-year average; and an irregular fluctuation (known as El Niño) of the ocean currents off Peru and Ecuador ruined the usually abundant anchovy fishery there.[24] The net effect was that the world's total food production in 1972, although the second greatest ever achieved, fell nearly 2 per cent below the 1971

Fig. 107 In the parched landscape of Niger in the Sahel zone in the great drought of the 1970s. (Photograph kindly supplied by Oxfam.)

achievement. This was the first drop that had occurred in any year in the period of technological advance since 1945. (Over many of those years up to 1972 world production had been increasing by about 3 per cent a year.) And there was a scramble among the countries most directly hit to purchase food from the American reserves, a scramble in which Russia was able to buy up a quarter of the United States wheat crop of that year as well as buying elsewhere in the West, with the result that the world price of wheat doubled within a few months and the difficulties increased for the poorest countries suffering shortage.

The 1972 case had other repercussions. Most directly, the stresses arising from the famine seem to have triggered the revolution which toppled the age-old imperial regime in Ethiopia. Those developing countries which have oil but a climate unsuited to agriculture developed a new appreciation of the climatic and other threats to the world's food supply and of the necessity of using their dwindling resources to diversify and strengthen their economies against the time when their oil is exhausted. And in the leading scientific, technical and administrative institutions in the advanced countries, there was some confusion about how to interpret the climatic event and revise attitudes to climate, even before the anxieties aroused by the unprecedented international economic crisis, which began to develop with the first (fourfold) oil price increase in 1973–4. Most immediately,

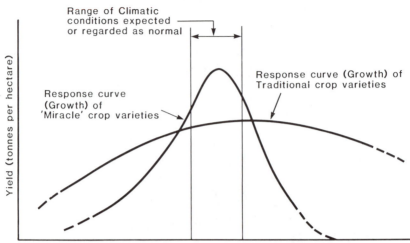

Fig. 108 Characteristic responses of traditional and new varieties of crops (e.g. rice) to climatic conditions within and beyond the expected range.

the hopes that had been raised by the Green Revolution of being able to meet indefinitely the food demands of the world's rising population were seen to have been unduly optimistic, particularly since the high-yielding new varieties of rice and other 'wonder crops' were often more sensitive than the traditional varieties to deviations from the expected climatic conditions (see fig. 108). There was also no ready means of being sure which aspects of the 1972 world-wide climatic anomaly pattern would prove to be a short-term fluctuation and which should be recognized as part of a probably much longer-term trend. Voices among the 'experts' in the scientific world ranged from alarm – exaggerated by distortions in some sections of the sensationalist press – to extreme complacency. To some extent this is still the case nine years afterwards – an unhappy position for science, which only emphasizes the need for more firmly based knowledge and understanding of climate, so long taken for granted and ignored.

TECHNOLOGY, CLIMATE AND FOOD SUPPLY

The basic lesson was, of course, learnt that despite the enormous technological advances in agriculture and average yields per acre that are in many areas over double what they were before 1945, the production of sufficient food for the present population cannot be guaranteed in every year. And it can be foreseen that even without climatic fluctuations or change, but with the further increase of the world's population that is inevitable, at least over the next few decades, some reduction in the feeding of grain to animals will be needed in order to reserve more for human mouths. Moreover, even the biggest countries are not immune to the direct effects of weather-induced variations of the harvest. Mistakes have been made, and have continued to be made, through over-confidence in the amount of security that has been won at any given time by advancing technology and larger-scale organization. Indeed, these very factors probably induce a tendency to press the rationalization of food production to extremes, both in respect of monoculture and by taking in marginal land, and relying on technology to secure the expected yield whatever the weather does.

This policy is reminiscent of that ruling in the Soviet Union in the enthusiasm of the first two or three decades after the revolution. The work of the Russian statisticians V. M. Obukhov and N. S. Chetverikov, which had shown in 1927–8 from pre-revolutionary years data that the variations of the country's overall grain production depended largely on the weather and the violent fluctuations in a few key provinces in the south of the Union, was effectively ignored. And when the first five-year plan was inaugurated in 1929, the variant which had allowed for possible difficulties due to the weather was dropped. In an official publication the editor wrote 'The question of the yield can be resolved only by Marxist dialectics . . . and . . . is closely related to the industrialization of our country. . . . The yield will become the object of the planned action of the productive forces of the Socialist state.'[25] A slow change of attitude began with the results of the 1936 harvest, officially described as 'satisfactory . . . if you take into account the complicated climatic conditions' and 'very similar to 1891', which was the year of a great famine!

Similarly, the successes achieved over many years since the late nineteenth century with the increasing scale of the mechanized farming operations in the United States Middle West, taking in more and more of the great grasslands, gradually caused the earlier droughts to be forgotten, until disaster struck in the 1930s. Drought had occurred about every twenty to twenty-two years[26] since the first European penetration of the region, but this one was more serious than any drought until then known. Whatever causes the apparent roughly twenty-year cycle in mid-western droughts, this time the climax nearly coincided with the climax of that longer-term variation, which we have noticed, that brought the mean position of the subtropical anticyclone belt a degree or two farther north and also showed in a more regular performance of the westerly winds in middle latitudes. Successive summers between 1932 and 1937 brought repetitions of hot, dry winds from over the Rockies which parched both vegetation and soil in the Middle West. Previously the native grasses of the region, when so parched, had produced a tough dried-up mat that protected the soil. Now the crops were killed, and the soil that had been disturbed by the plough

Fig. 109 A scene from the great drought years in the 1930s in the United States Middle West: an advancing dust storm of wind-blown soil.

just blew away. In 1933 and 1934 the wind-blown dust was readily traced to the east coast. On 12 May 1934 the New York Times reported that the cloud of dust coming from the 'drought-ridden states as far west as Montana, 1500 miles away, filtered the rays of the sun for five hours yesterday'. New York was in a half-light like conditions in an eclipse of the sun, and the dust-cloud was thousands of feet high.[27]

Thousands of farmers were ruined in those infamous years when the Middle West became a 'Dust Bowl' (fig. 109), many families migrated to seek a new living near the west coast, and farms inland 'that never should have known the plough' were abandoned. Soil rehabilitation programmes had to be instituted by the federal government, involving returning much of the land to pasture and planting trees as windbreaks.

A partly similar mistake, or misjudgement, had been made in the Sahel in the time of more abundant rains in the 1950s and early 1960s. International aid for the developing countries in the zone was used to drill deep wells in order to use (and ultimately use up) the great reserve of subterranean water – sometimes known as 'fossil water' – which accumulated in different climatic regimes thousands of years ago. This introduced a kind of short-lived prosperity to the region with greatly increased cattle herds and growth of the human population (the latter thanks also to the beginnings of satisfactory health services). The sparse vegetation was soon over-grazed, resulting in a spread of the desert.

And this, it seems, certainly introduced meteorological self-reinforcing mechanisms, which help to maintain the drought.[28] Through the greater reflection of the solar radiation by bare soil, the total energy absorbed in the ground and lower atmosphere is reduced and an anticyclonic tendency with dry air subsiding from aloft is introduced. At the same time there would be even less moisture than before stored in any vegetation and available for recycling. In these ways the whole region became more vulnerable than before to the next down-turn of the natural rainfall, which duly came from the mid-1960s onwards. And now we learn that, because the later years have shown the meagre recovery of rainfall in the Sahel from its 1971–3 minimum seen in fig. 99 (p. 276) and because some meteorological advice takes the complacent view that the recent extreme stress was oniy a random short-term variation, resettlement of the displaced population and rehabilitation of their cattle stocks is under way.

Underlying the events reviewed in this chapter there seems to be a sort of historical cycle, whereby human populations expand in periods of benign climate and occupy with increasing density lands which sooner or later fail to support the numbers by then dwelling in them. Similar expansions of population are seen to be introduced by advances in technology. When the bad years come, the population has always in the past been reduced or disappeared, partly through migration and partly through undernourishment, disease and death. The situation is doubtless characteristically compounded by the inflexible attitudes developed by the sufferers. Confronted by similar threats to the greatly inflated, and still fast growing, population of many regions in the world today,[29] some take the view that humanity has coped with all the climatic changes of the past, including the ice age to post-glacial change, and will doubtless do so whatever may befall in the future. This view overlooks the enormous human sufferings involved, ranging from the difficult lot of the immigrant and slave labourer to mass epidemics of disease and death. It is surely our duty, and a wise precaution of the advanced and more fortunately situated countries too, to do whatever can be devised to minimize such troubles by preparedness based on a more realistic understanding of climate.

CLIMATE AND DISEASE OUTBREAKS

Having dwelt so long in this chapter on the impacts of climatic variation on food supply, as the most basic aspect, there is no space here to go into many of the other types of impact on human society mentioned at the beginning of the chapter. The most serious among these in terms of death toll are those that occur through epidemics of disease and through episodes of flooding either by rivers or the sea of extensive, heavily populated lowlands. The latter have commonly been followed by disease epidemics, though modern advances in protective medicine can now be expected to contain the situation – so long as there is no breakdown of organization and the scale of the disaster is manageable. We have noticed in earlier chapters of this book how in some degree climatic fluctuations seem to have been involved in the disastrous plague that swept the Roman world in Justinian's time and in the Black Death in the Middle Ages, as well as in the great cholera epidemic that started in Bengal in 1816–17. It was well known, too, centuries ago in Europe that the recurrent outbreaks of plague seemed to be affected by the weather, flaring up in hot, dry summers and tending to die out in long, severe winters. It is not the intention of this book to present a theory of climate and history, nor to pretend that the linkages are simpler than they are. But in the case of links between climatic fluctuations and major outbreaks of disease, listing a few broad categories may help understanding. The circumstances conducive to such situations may be classed as:

1 events, such as some of the greatest droughts and floods, which cause a breakdown of sanitation and hygiene;
2 weather conditions exceptionally conducive to the breeding of certain insects and other disease organisms and vectors, of the hosts of various sickness organisms, and/or conditions which extend their geographical range;
3 weather conditions, and any weather-induced failures of the food and water supply, which lower the resistance of human populations to sickness and disease.

We may note in passing that the cases classed under 2 also apply to the diseases of animals and must include the breeding of insects, such as locusts, and parasites, blights, etc. which damage crops and other elements of the vegetation on which the human economy depends. Cases to be classed under 3 seem often to involve wet winter conditions (in almost any latitude) and the common infections which are among the first results of them.

In a great number of the phenomena included under 2 in the above list certain combinations of warmth, though not excessive heat, and a moist environment, or enough humidity to ensure at least some locally moist micro-environments, seem to be necessary. As examples of the weather-dependence of the abundance of insects associated with (a) human illness, (b) a devastating sickness of both animals and men, and (c) the large-scale destruction of crops, we may briefly consider the following:

1 *The flea that transmits bubonic plague* (*Xenopsylla cheops*) undergoes a speeding up of its life-cycle as temperatures rise in the range 20–32 °C (68–90 °F). Breeding is speeded up, but the death of each generation of the insect also comes sooner, the higher the temperature. At relative humidities below 30 per cent of saturation the life of the flea is reduced to a quarter of what it is in near-saturated air.

The *malaria-bearing mosquito* (*Anopheles*) does not breed at temperatures below 16 °C (61 °F) or with relative humidity below 63 per cent, and like all mosquitoes thrives in moist environments with stagnant water bodies. From time to time this sickness has been introduced by home-coming travellers (the mercenary soldiers of former centuries seem to have provided examples) to the mosquitoes of the northern Europe – Oliver Cromwell died of the ague, as it was called, caught from an English mosquito in the Fenland, and cases are known to have occurred as far north as Sweden – but it has always died out within a few years in conditions that were presumably too cold or too dry for it at some critical stage of the mosquito's life-cycle.

Rather similarly observation has shown that the average life of the *yellow fever mosquito* (*Aëdes aegypti*) is reduced from 7.0 days in near-saturated air to 4.5 days in dry air at 20 °C with humidity below 48 per cent, and to about 2.0 days if the temperature is 26 °C, thereby reducing the opportunity for reproduction. Different varieties of mosquito seem to be capable of transmitting yellow fever infection in rural conditions in Africa and South America, but their climatic preferences are evidently similar.

2 The *tse-tse fly* (*Glossina*), whose bite spreads the deadly *sleeping sickness*, similarly requires enough, but not excessive, warmth and humidity. There are several varieties of this fly involved in transmitting the sickness, all of which require some shade from the sun, though in differing degrees. Most of them therefore thrive best in areas with plenty of trees or shrub vegetation, often near rivers and lakes. Their biting and bloodsucking is done in bright conditions in the day time, but activity ceases at temperatures below about 15.5 °C (60.0 °F). Drought is very damaging to the fly populations in the larval andipupal stages. Hence their range is confined by the desert zone to low latitudes, and some areas have been improved for habitation by clearing of the vegetation.

The particularly deadly time in West Africa in the 1860s and 1870s, which earned the region the name of 'the White Man's Grave', when the average expectation of life of a European going there was six months, seems to have been a period when the equatorial rains were peculiarly active over Africa and the lakes were rising strongly. (Despite this evil health record at that time the Jesuit and Methodist missionary societies never had fewer than twelve volunteers waiting in London to go out to the mission field to replace those who died.) A little earlier, in the days of the old Danish colony in Ghana in the 1820s and 1830s, when lake levels in equatorial Africa were lower, and other evidence suggests that climates in this part of Africa were drier, the incidence of the sickness seems to have been by no means so bad. It also eased off after the 1870s–80s despite the fact that there had been no real advance of medical technique, but perhaps connected with the decline of the rains which became sharp in or around the 1890s.[30]

3 The *desert locust* (*Schistocerca gregaria*) also needs moist periods, after rains in the desert or desert fringe, to multiply. It is most active in temperatures between 25 and 35 °C (77–95 °F). At temperatures below 15 °C it is lethargic, and temperatures above 50 °C (122 °F) seem to be lethal for it.

In all these cases, therefore, the insects multiply when they find themselves in the optimal weather conditions mentioned, and are then liable to be spread by whatever winds blow. Locusts have from time to time turned up in many parts of Europe but very seldom in enough numbers to do significant damage. But there are many records of crop disasters and dearth caused by locust swarms in hotter countries in earlier times. The total area that has experienced invasions of desert locust swarms in recent times amounts to about thirty million square kilometres, and over an important part of the area the breeding of locust swarms is observed on average every second year. In all the cases mentioned modern control methods take advantage of the known environmental and weather dependence of the species concerned. In the case of locusts, the international anti-locust organization monitors the situation by continual mapping, including weather mapping, of the whole zone where the insects breed. Control spraying of insecticides can be guided to the actual breeding areas and to where the locust swarms on the wing are concentrated into a narrow zone[31] by the meeting of air currents from both hemispheres at the Intertropical Front or along lines of convergence within the Intertropical Convergence Zone. There has been clear evidence of success of the control campaigns launched in the 1960s and after, though it is acknowledged that the overall decline of the locust menace in these decades may be in large part attributable to weather less favourable to the insects.

Similarly, specific weather conditions promoting the development of potato blight (periods of forty-eight hours or more with temperatures continuously above 10 °C and humidity above 90 per cent saturation), or favouring cattle diseases in temperate countries such as liver-fluke and gastro-enteritis (and their hosts or vectors at some vital stage of the development cycle), can be defined and are used to issue warnings and initiate preventive measures.

THE IMPACTS OF FLOODING AND BITTER WINTERS

The only weather conditions which in the worst cases have directly caused within a single year or less disasters to humanity on a scale comparable with those occurring through starvation and disease have been, as stated earlier, vast flooding:

1 River floods, such as those of the Yangtse-kiang and Yellow River (Huang-ho) in Honan, China in June 1931, when more than one million drowned, and the similar disaster in the Yellow River valley in September–October 1887 when 900,000 were reported drowned. The river floods and subsequent disease in China in 1332–3 are said to have taken seven million human lives, with long-lasting devastation in parts of the country and destruction of many settlements; and it has been suggested that this may have been the starting point of the plague which swept the world as the 'Black Death'.

2 Coastal flooding by the sea in storm surges, propelled either by tropical cyclones and typhoons or by the cyclonic storms of middle latitudes. On 12–13 November 1970 Bangladesh was visited by a flood of this kind, due to a cyclone in the Bay of Bengal, which submerged a large fraction of the country. The death toll, originally estimated at 300,000, was finally put at about three-quarters of a million by the authorities. Many similar disasters were recorded on the low-lying coastlands around the North Sea in the Middle Ages and after, particularly on the continental side, with estimated death tolls from 100,000 to 400,000. Their non-occurrence in recent times is a tribute to the effectiveness of the sea defences that have been built over the last three hundred years.

The number of human lives lost in the worst phase of the drought in 1972 and 1973 in the Sahel-Ethiopian zone of Africa certainly came within the latter range.

The bitterest winters in Europe and North America seem never to have produced deaths on any such scale. Despite much misery and privation to the poor and the old, and numbers of people reported frozen to death on the roads, buried by snow in the countryside, and dying in the streets of the cities, the severest impact was usually quite localized or even just on scattered individuals; the scale was probably never worse than in the famous disasters to Napoleon's and Hitler's armies exposed on the plains of Russia in 1812 and 1942. There was, however, sometimes a more widespread indirect impact through food shortages and bread prices, etc., which we have noticed in the case of the French Revolution after the winter of 1788–9 and in some much earlier cases where both trouble with wolves and the eating of small children were reported.[32] And in our own day, of course, important economic losses can arise through severe winter weather, particularly in regions where this is somewhat exceptional; and there is much room for economic (and military) gain in careful planning to optimize capability in such weather and in the economics of provision for it. The cost of providing powerful snow-clearing equipment for highways (snow-blowers), of heating certain stretches of road surface and railway junctions against frost, of holding fuel reserves and standby services (helicopters, etc.)

for emergency food-drops and rescue work – much of which may only be needed in a small minority of winters – has to be weighed against the losses incurred in such winters. Even in countries where cold winters are common, severe losses can arise. The city of Buffalo in New York State was brought to a standstill for many days in the winter of 1976–7 when snow 3–4 m deep blocked the streets. And in the winters of the late 1970s in south Norway flat roofs on modern buildings collapsed under the weight of depths of snow which had not been experienced in recent times. In the winter of 1978–9 in England, when snow lay for about forty days over much of the low-lying parts of the country with depths ranging up to 50 cm, where the recent long-term averages have been seven to fourteen days, the cost of road clearing in one county in the eastern Midlands (Nottinghamshire) was 3.3 times the average; and over the whole United Kingdom the extra costs were estimated to total £500 million (or about £10 for every man, woman and child), most of which represented the 9 per cent greater than usual fuel consumption. Evidently the extra costs of a year like 1740 in England which was cold in every season and reasonably judged to lead to 50–70 per cent additional fuel demand could run to a much bigger figure.

Possibly it is because of the excessive concentration of much writing about climatic fluctuations upon exceptional winter snow and ice and the advances of glaciers, and the heart-rending accounts of individual suffering in such circumstances, which – unlike harvest failures – are nonetheless marginal to the wider community, that some historians have proclaimed that the history of Europe over the last thousand years would not have been much different if the climate had remained constant. What view of history is that at bottom? A political or constitutional historian's view? Certainly one biased towards the more secure and sheltered parts of Europe in the west and south: for in the north even the political and constitutional divisions and alignments may be held to have been affected. And certainly it is a view which ignores much that concerned the health, lives and happiness of Europe's people.

OTHER ASPECTS

For each development mentioned in this book, for instance, in the social history of Europe, and even in the outlying colonies in Iceland and Greenland during the late medieval decline, it is possible to suggest other, non-climatic causes. Certainly, all the other stresses involved need to be established. But, although we cannot, or cannot yet, establish in all necessary detail the reason (or the chain of events) whereby pestilence or social unrest broke out just when and where it did, there can be little doubt that the climatic shift that was going on in the late Middle Ages occupies a central place in the simplest explanation of the whole complex of events. And in some details, such as the epidemics of ergotism (see pp. 199–200) and the cessation of communication with Greenland when the sea approaches were usually blocked by ice, there can be no doubt at all that the development of the climate was crucial. In other parts of the world, too, on the arid fringe, the abruptness of some of the decreases of ancient populations may prompt reasonable questions about the possible involvement of climatic disasters or more indirect climatic pressures. A possible example of the latter affecting, at least to some extent, an invader and his victims in different ways may be the great massacre of its population by Mongol invaders which ended the medieval greatness of Baghdad in AD 1258 after Iraq's agriculture had long been in decline. About the same time the Mongol homelands in central Asia seem likely to have been thriving – and become overpopulated – under a moister than usual climatic regime.

Some developments were apparently caused by the mistakes made by man in what were probably deliberately attempted adaptations to changed climatic circumstances or to the prevailing impression of them. Thus, the adoption of sheep rearing in parts of Denmark and in the Breckland of East Anglia during the colder centuries of the past millennium on lands that had been tilled in the high Middle Ages did not turn out well. In the always rather dry, windy climates of East Anglia and Jutland the vegetation cover did not stand up to the grazing, and the land deteriorated to a sandy waste only reclaimed by planting trees to provide shelter belts, and later afforestation, in more recent times.

Even today, when our perception of, and ability to cope with, short-term disasters by mustering relief supplies and first aid from all over the world is so impressively improved, it is doubtful whether our ability to absorb long-term changes is significantly better than it ever was. We are clearly hindered by too much rigidity of planning and, for instance, allocating quotas of essential agricultural products to be met by monoculture to supply the world economy, and by the rigidity of national frontiers when the need for human migration arises. Even the effects of the rather noteworthy tendency to clustering of two, three or four years of similarly anomalous weather could impose (almost?) unmanageable stresses. One of the lessons from our summary in this chapter of the impacts of climate on human society must surely be that in the modern world the effects of climatic difficulties and disasters, and particularly the stringency imposed by harvest shortfalls, in any one region reverberate around the world and are liable at least to affect prices ruling in the whole world's economy.

Some of the side-issues can be quite interesting and may acquire an important cumulative effect with time. Thus, it was suggested by Dr E. J. Moynahan at the International Climate and History Conference held at the University of East Anglia, Norwich, in 1979 that through the famines of the Middle Ages and after there may have been a natural selection operating in favour of fat people, who would be better able to survive than their leaner fellows.[33] Indeed, the selective advantage of fatness may have operated on human populations from the earliest times. It is only with the much longer expectation of life in European and other populations in recent times that the advantage has gone to the leaner physical constitutions which place less burden on the heart.

Some have objected to the term Little Ice Age to describe the colder parts of the last millennium on the ground that different writers have placed the limits of it at different times, e.g. 1300–1900, 1430–1850, 1550–1700 or about 1800, and so on. But these differences are only a matter of which of the more sudden developments in the onset and recovery stages passed particular significant thresholds for the subject or region of interest. It has also sometimes been objected that the greatest growth of population, improvement of general health, and advances in industrial technology and agriculture, and in the extension of civilization to the whole globe, took place 'precisely' during the Little Ice Age, between 1700 and 1900. But this is to ignore the fact that all these developments took place during the long drawn-out, and erratic, recovery from the depths of the Little Ice Age regime. The 'parallelism of the climatic and cultural curves' is, in fact, remarkable and calls for some consideration.

Possible future climatic changes in marginal areas may also easily come to affect the whole world's economy. A conference on the World Food Supply in Changing Climate, held at the Sterling Forest Center in New York in 1974, estimated that the grain growing area on the Canadian prairies would be reduced by about 1 per cent by a 1 °C fall in the long-term average temperatures and production would fall by a similar amount; but the effect on production would become much greater with any further cooling. A 10 per cent decrease of the rainfall would lower production by several per cent, a 10 per cent increase of rainfall would increase the wheat production by a few per cent but have little effect on the oats and barley.

In sum, the impact of climatic fluctuations and change on history, and on human affairs today and in the times with which our future planning must be concerned, can best be seen as a destabilizing influence and catalyst of change. At the worst, we see reactions by human society which have amounted to shifting or concentrating the burdens of suffering on to the weakest members of the national and international community. This may be appreciated perhaps best when we consider the ugliness of the extreme case, the reported developments of cannibalism.[34]

16

THE CAUSES OF
CLIMATE'S FLUCTUATIONS
AND CHANGES

GENERAL

If we are to develop any sound scientific and reliable system of forecasting or even just advice on future climate, we must first understand
the causes of climatic fluctuations and change. Without knowledge of the
processes involved in the development of climate's variations, their normal
time-scales and the range of their effects, as well as some ability to monitor
the key elements in the progress of each, any forecast must be mere
guesswork. This criticism must apply even to apparently sophisticated
mathematical models of climate development, unless and until their results
can be demonstrated as realistic when compared with an assortment of
epochs in the known past record of climate. It is important also to gauge
the present and probable future limits of predictability. And if the interested layman, especially anyone involved in decisions affected by future
climate, is to be able to judge what may be possible in the way of
forecasting, or what else should be done to allow for the future behaviour
of the climate, the current state of knowledge of the causes and processes
of climatic change – and the prospects of advance – must be properly
understood.

We have had a first look at some of these questions in chapters 3 and
4 of this book. In this chapter it is time to summarize briefly the causes
and symptoms of change and how our scientific ability to handle them
has developed and is developing. Here too we must begin to consider not
only the natural causes of climatic change but also man's impacts on the
climate.

We have seen how the level of temperatures prevailing and their distribution over the globe can conveniently be treated as the most fundamental
things, since they explain so much else – the development of the general
wind circulation and, through it, the redistribution of heat and moisture
and the development and steering of weather systems. Even the yield of
the Indian summer monsoon seems to go up and down with the global
temperature level, particularly as represented by the shrinkage or expansion

of the region of Arctic cold to the north. The things that can change the prevailing temperature level may be summarized as follows:

1 Variations in the energy output of the sun (and possibly in the transparency of interplanetary or interstellar space).
2 Astronomical variations affecting the distance of the Earth from the sun at different seasons and the angle at which the sun's beam falls on the Earth at different latitudes and seasons.
3 Variations in the transparency of the atmosphere to either the incoming solar radiation or the outgoing Earth radiation.
4 Changes coming about in the internal heat economy of the oceans and atmosphere as a result of their circulation (in three dimensions) and whatever influences bring about changes in the circulation.
5 Changes in the absorption and re-radiation of incoming energy at and near the Earth's surface through
   (a) variations of cloudiness, and
   (b) changes in the nature of the surface itself – extent of snow and ice, of different kinds of vegetation and bare soil, desert or marshland and lakes and, on long time-scales, through changes in the distribution of land and sea, of mountains, plateaux and ice-sheets.

Let us consider these items one by one.

VARIATIONS OF THE SUN

The idea that climatic variations might have their origin in variations of the prime source of energy, the output of the sun itself, was obviously one of the first thoughts of people with a scientific concern about the subject. Riccioli suggested in 1651 that the temperature of the Earth should fall the more spots there were on the sun. The variable occurrence of dark spots on the sun had been a cause of fear and prognostications of doom to the peoples of Europe a generation or two after the Black Death, when some sunspots were so large in the 1360s–80s as to be obvious to the naked eye looking at the sun in foggy weather. One report tells of 'dark spots on the sun's face as big as the nails in the church door'. Galileo observed sunspots with his telescope in 1611, and an increasingly continuous record of their variations can be pieced together from that date. As mentioned in chapter 4, there seems to have been a prolonged period of almost no sunspot activity between about 1645 and 1715, and this phase coincided with a time of generally low temperatures prevailing over most of the world. It is now well known that sunspots are only one of a number of different types of solar disturbance, and when they occur the reduced energy output from the darkened areas of the sun is often more than compensated by intensified radiation from brightened areas round about, known as faculae. These have been systematically measured only since 1874.

Thus, unfortunately, our fine long record of sunspots is a very inadequate indicator of solar output variations, though the dates of maxima of the roughly eleven-year sunspot cycles have been tentatively established (partly by using reported observations of great displays of the polar lights or 'aurorae') back to 649 BC.

A better index might be one which measured the difference between the areas of faculae and sunspots. In fact, very few series of weather or climate data have shown any appearance of significant associations with the somewhat variable, but approximately eleven-year, sunspot cycles. The more prominent occurrence of more or less cyclic recurrences of weather patterns at about 20–23 year intervals – e.g. of droughts in the United States Middle West and of some features in the long temperature record in England – may be connected with the double sunspot cycle: but this is complicated by the fact that the sunspot cycle, although averaging about 11.1 years, actually varies in length. Extreme cases observed in the last two centuries have ranged from just under nine to fourteen years. The activity of the cycles also varies, the shortest cycles generally producing the greatest numbers of spots. And it appears that the sun's output may actually be greatest at middling sunspot numbers, about eighty on the internationally recognized scale of Zurich relative sunspot numbers, as compared with over two hundred at the greatest maxima (when, presumably, the effect of the darkened areas outweighs that of the faculae). There is some evidence that the longer-term variations of sunspot activity may be more simply associated with variations of the global temperature level. Thus, fairly short cycles and apparently rather high sunspot activity prevailed not only during the warmest period of the present century (average cycle length 1915 to 1964 was 10.2 years), but also during several other warm climatic periods in the past, in late Roman times and in the Middle Ages. And the so-called Spörer minimum of solar disturbance (with mean cycle length, between the successive weak sunspot maxima, of about twelve years) between AD 1400 and 1510, like the Maunder minimum in the seventeenth century, seems to have coincided with a notably cool period of global climate.

A better measure of the sun's luminosity, or strength of the solar beam, seems now to be available in the form of the ratio of the darkened area (umbra) to that of the grey area (penumbra) in sunspots, this being assumed to measure the rate of convective flux of energy from deeper in the sun.[1] The annual values of this solar luminosity index, plotted in fig. 110, seem to parallel rather well (or slightly precede) the global temperature rise and fall within the period from 1880 to the 1970s which we have presented in fig. 91a (p. 258). (Although this same index cannot be produced for the eighteenth century, the pronounced maximum of warmth that affected most of the known world in or about the 1730s bears a similar relationship to the sunspot record preceding by some 20–40 years a series of extremely

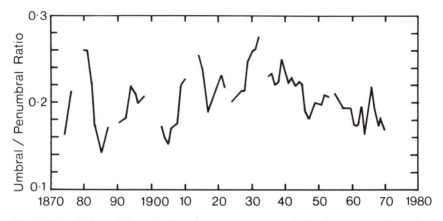

Fig. 110 An index of the variations from year to year of solar luminosity (i.e. solar output), based on the structure and gradations of luminosity within the sunspots, from 1880 to 1975. (After D. V. Hoyt – by permission.)

active sunspot cycles as in the twentieth-century case, just two hundred years later.)

VARIATIONS IN THE EARTH'S ORBIT

The nature of the astronomical variations concerning the Earth's orbit and axial tilt, which affect the strength of the solar beam in the ways mentioned above, has been explained in chapter 4. The agreement between the time-scales of these variations and the variations of the Earth's thermal regime and land-based ice-cover, indicated by isotope examination in recent years of the longest records available in ocean-bed sediment cores,[2] seems to put the thesis that these orbital variations control the timing of ice ages and interglacial periods beyond reasonable doubt. One may, however, assume that the step-by-step increase of the Earth's reflectivity (albedo) as the area of snow and ice extends, under declining radiation receipt, provides a necessary amplification of the effects of the radiation variations. There has been some interesting debate in recent years as to just where and at which season the response to these radiation variations should be most sensitive and significant to the climatic development over the whole globe.[3] As a result, the older notion that the most significant variations of snow and ice accumulation should be responses to the varying strength of the summer sun on the almost completely (and partly mountainous) land zone between latitudes 60 and 70 °N seems to need some modification. The most im-mediate response in growth of snow cover to the regular seasonal change of radiation from the sun is found in the autumn between latitudes 40 and 70 °N and particularly in the heart region of the Eurasian continent between longitudes 50 and 70 °E. In the melting season the response to radiation change is quickest in the southern hemisphere oceans – so also around October–November. This may be the time of year therefore when any long-term changes in the radiation available have most effect. It is logical to suppose that in full ice age conditions, with permanent ice covering North America north of 50 °N, the responses to radiation changes there would be less sensitive and that the summer peak would become more important there than what happened in autumn.

What concerns us here is that these astronomical variations provide the one entirely predictable element of the future. To this we shall return in the next chapter. But it is important to note that they do not explain the suddenness of some of the climatic shifts that are indicated by the geological record in the course of these 10,000–100,000 year cycles. For these we must look to other causes, such as possible variations in the sun's output or in the amount of volcanic dust in the atmosphere, happening to reinforce (or oppose) powerfully the slow trend induced by the orbital variations.

VOLCANIC DUST IN THE ATMOSPHERE

There is no doubt that the most important changes which nature produces in the transparency of the atmosphere from time to time, over durations that directly concern us, are those due to variations in the amount of volcanic material present. We shall return later in this chapter to effects on the atmosphere's transparency produced by man's activities.

Massive volcanic explosions such as that of Mount St Helens in May 1980, pictured in fig. 111, put myriads of submicroscopic-sized rock particles and aerosol derived from sulphur dioxide into the stratosphere, where they are beyond the reach of the rain which washes such impurities out of the lower atmosphere. The volcanic matter typically passes round the Earth in ten days to a few weeks, taking a different length of time at different heights owing to differences in the strength and sometimes also differences in the direction of the wind; these differences and diffusion processes, including convection and turbulence which transfer some of the material to somewhat higher and lower layers, gradually spread the material into an increasingly uniform veil which may cover the hemisphere concerned (or even the whole Earth) within about half a year. The greater the height to which the exploded material is thrown by the eruption, the longer the veil will last. The fall speeds of the minute particles are so small that they may take from twenty days to a year to fall one kilometre and are liable to stay for one to seven years, or more, in the stratosphere. The effect of their partial interception of the solar (mainly shortwave) radiation, while the Earth's outgoing (mainly long-wave) radiation passes nearly unhindered, is to warm the dust layer while at the Earth's surface and in the lower atmosphere temperatures fall somewhat below what they otherwise would be.

Fig. 111 The explosive eruption of the volcano Mount St Helens in Washington state on 18 May 1980. (Photo kindly supplied by, and reproduced by permission of, the United States Geological Survey.)

The cooling, at its maximum in the first year, after various great eruptions reported in the past has been assessed (averaged over middle latitudes) at from 0.1 to around 1.0 °C. In 1783, when there were two very great eruptions – in Iceland and in Japan – in the same year, the combined effect may have been a cooling of the northern hemisphere by 1.3 °C, gradually tailing away to zero over the following four or five years.

The Mount St Helens eruption of 1980 will probably not rank among the biggest eruptions in terms of stratospheric dust veils, in spite of the fact that one or two cubic kilometres of rock were blown into the air, because (unusually) a large proportion went in a nearly horizontal blast. But it may be regarded as part of a significant global trend towards increased volcanic activity since about 1960 after a marked lull which in the northern hemisphere had lasted nearly fifty years. On a scale[4] which ranks volcanic dust veils in terms of the mass of material initially ejected and the duration and maximum spread of the veil, the great eruption of Krakatau in the East Indies in 1883 is ranked as 1000, and the total veil from various eruptions in the 1880s reached about 1500. In 1902 a group of eruptions in the West Indies produced a new veil ranked as 1000, and this was renewed at least over the northern polar regions in 1912 by the great eruption of Katmai in Alaska. After that there was no big injection of dust that seems to have affected the northern hemisphere until the eruption of Mount Agung in Bali in 1963 (dust veil index 800), which with eruptions of other volcanoes in the following years once more produced a veil rated at over 1000 by the late 1960s. Two of the biggest eruptions in the early part of the nineteenth century had produced veils rated globally at 3000–4000 on the same scale,[5] and it seems clear that any bunching in time of such great eruptions must produce significant coolings and related effects on the climate lasting for periods from some years to a decade or two.

Some prehistoric and early historic eruptions, such as that of Santorin in the Aegean in the time of Minoan Crete and Vesuvius in AD 79, can be assessed on the basis of surveys of the dust (and larger ejecta) deposited and still indentifiable. Dust veil ratings from around 3000 to 10,000 seem probable for the Santorin eruption and between 1000 and 2000 for Vesuvius in AD 79. The dust veil from the eruption of Öraefajökull in south Iceland in AD 1362 may be tentatively put at about 500, the lower rating arising partly because of the smaller extent of the globe affected by the spread of dust from high latitude eruptions. One or two eruptions of Hekla in Iceland probably had a similar magnitude, e.g. in AD 1104 and about 750 BC.

A chronology of volcanic material, identified in the form of sulphuric acid in the year-layers of the Greenland ice-sheet by Professor W. Dansgaard of Copenhagen and his co-workers, C. U. Hammer and H. B. Clausen, shows a gratifying degree of agreement with the volcanic (global) dust veil chronology from AD 1500 referred to in these paragraphs. (A correlation coefficient of 0.46, statistically significant at the 99.9 per cent confidence level, was obtained for the whole span of the dust veil index chronology, despite the fact that the dust veils from some parts of the world could not be expected to be fully represented in a deposit on the ice-sheet at latitude 71 °N. Over the period 1770–1972, for which the dust veil chronology is presumably more reliable, being based on more nearly complete reporting, the correlation coefficent was 0.65.) The measurements of the acid deposit in the Greenland ice have been carried as far as the ice-layer laid down in the year AD 553, and the comparison shown in fig. 112 between the successive half-century values of the acidity (little acidity upwards, much acidity downwards, in the diagram) and an index of northern hemisphere temperature shows an impressive parallelism. It must surely be accepted that the variations of the amount of volcanic material carried in the atmosphere, and deposited by it, seem to have something to do with the climatic variations in the fourteen hundred years covered, even appearing as perhaps an important part of the causation of the Little Ice Age. This thesis is supported by other approaches used by Bryson and Goodman of the University of Wisconsin,[6] which indicate also that the cooling of the

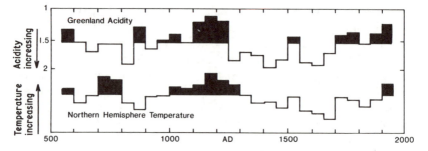

Fig. 112 The record of the amount of acid deposited in the year-layers in the Greenland ice-sheet (at the crest near 71 °N 37 °W) from AD 553 to 1950 compared with an index of northern hemisphere temperature according to Professor W. Dansgaard and his colleagues C. V. Hammer and H. B. Clausen. Both items here as fifty-year averages. The ice acidity measurements are thought to be more or less representative of the amount of aerosol present in the northern hemisphere atmosphere in general. The northern hemisphere temperature index is derived from the temperatures in central England (as in fig. 30), the tree rings at the upper tree limit in the White Mountains in California, and oxygen isotope measurements in the Greenland ice. (From a diagram kindly supplied by Professor Dansgaard.)

northern hemisphere since 1950 may be attributable to a doubling of the volcanic material in the northern hemisphere atmosphere over the same period.

Another volcanic chronology[7] has studied the variations of eruptive activity over the hundred years since 1880 in different latitude zones and in the northern and southern hemispheres separately in terms of a simple classification of eruptions as great, moderate or small ash producers, and gives some weight to numerous moderate eruptions which were largely excluded by the dust veil index whose chronology we have discussed above. There is a large measure of agreement between the two chronologies, but the new one reveals another feature which reinforces the apparent significance of volcanic dust for climate. The fifty-year lull in volcanic injections into the northern hemisphere stratosphere between 1912 and about 1960 was not matched in the southern hemisphere: in fact, the greatest peak of ash-producing volcanism in the southern hemisphere during the entire hundred-year span was between about 1925 and 1945. And whereas there was a great warming of northern hemisphere climates during the fifty-year quiescence of the volcanoes after 1912, and particularly the great warming of the Arctic which took place between 1920 and 1940, southern hemisphere temperatures showed a dip during the 1930s (see fig. 91a).

There are some details which show quite clearly that volcanic variations were not the only cause of the climatic variations in these years. For instance, the northern hemisphere cooling which set in about 1950 preceded any

326

significant increase of volcanic activity; and the rising trend of tempera-tures which affected the southern hemisphere, as well as the northern hemisphere, for forty years after 1890 was not accompanied by a decrease of southern hemisphere volcanic activity (nor of the northern hemisphere volcanism before 1912). Nevertheless, the evidence is strong that volcanic veils have played an important part in recent climatic history.

There have indeed been some studies – for example, a much longer, but less precisely dated, chronology of volcanic dust in the Antarctic ice-sheet – which seem to indicate greatly enhanced volcanic output during various main stages of the last major ice age. But here the cause and effect relation-ship is by no means clear; and it is possible that the changes of stress on the Earth's crust, when enormous masses of water from the oceans were converted to ice on land, produced waves of volcanic activity. Even if this be true, however, there may well have been a reaction – a 'feed-back effect' – of the dust veils in the atmosphere leading to a sharper cooling of the climate at the Earth's surface.

VARIATIONS OF THE CIRCULATION AND HEAT DISTRIBUTION IN THE ATMOSPHERE AND OCEANS

Much less can be said about the internal variations on time-scales from weeks to years in the heat economy, and the evolutions in the circulations, of atmosphere and oceans. Most meteorologists believe it necessary at present to treat these in relation to longer-term forecasting as random in their occurrence. There may nevertheless be some natural oscillation periods, such as one of thirty days (or very close to one month) which is prominent in the weather variations during the winter half of the year in middle latitudes of the northern hemisphere. Hints have been found of associations with (a) various shorter-term cycles of solar activity, (b) vari-ations in the tidal pull of the planets on the sun as their alignments change and which may have some effect on disturbances of the sun, and (c) cyclic variations of the combined tidal force of sun and moon acting upon the Earth and its atmosphere as well as on the oceans. The varying activity of solar disturbance may itself be partly associated with the (predictable) tidal pull on the sun of the planets as their positions vary. The likely period lengths are in many of these cases known but the correlations appear to be weak and unlikely to serve as a practicable basis for forecasting. There may be an exception to this in the case of the complex of small wander-ings of the Earth's rotation axis (and hence of the poles) that are known collectively as the Chandler wobble. This wobble is presumably related to readjustments of angular momentum (the momentum of spin) and of inertia between the solid Earth, and the fluid elements of its interior, and the atmosphere and oceans, at least partly under tidal forces. The compo-nents of the wobble include an annual cycle of displacement of the poles by a few metres and oscillations of other period lengths ranging from about thirteen to fifteen months. Several scientists in the United States and in Russia, notably I. V. Maksimov of the Main Geophysical Observatory, Leningrad, have been interested in the possible usefulness of the wobble in weather forecasting, since even such small displacements of the pole may produce enormously bigger displacements in the atmospheric circulation. This is because of the effect of any momentum exchanges between the massive Earth and its thin atmospheric 'skin'. Lately Bryson and Starr[8] in the United States have succeeded in resolving the wobble into discrete components, which facilitate prediction of it and seem to show useful associations with global weather development over some years ahead. The hope is that this may open the way to some, at least partial, success in forecasting the weather season by season over periods from one to ten years ahead – a time-span of much practical importance for which it has hitherto seemed impossible to cater.

The naturally occurring changes in the surface of the Earth which affect the absorption of radiation and the flow of the winds and circulation of the oceans, and hence must alter the development of climate, are chiefly the very long-term changes associated with the drifting of continents and mountain-building over tens and hundreds of millions of years. These do not concern us here. Some changes, however, produced by the weather itself or the circulation of the oceans, or by accidents such as screening of the sun's radiation by dense volcanic dust veils or blockage of certain channels by drifting polar sea ice, may have effects on the climate over a few months or a few years.

The greatest deviations of ocean surface temperature, amounting sometimes to 3 °C over areas up to 1000 km across, which occur from time to time (a) in the tropics, as a result of changes in the amount of cold water upwelling (under the influence of the winds at continental coasts), and (b) in high latitudes, when an ocean current boundary is displaced, change the rate of heating of the overlying atmosphere. The change in such cases is equivalent to a significant fraction of the solar heating available. Comparable changes, with an even bigger immediate effect on prevailing temperatures, occur when the area of snow or ice is extended or reduced, particularly when extensions of these surfaces into middle latitudes are involved: in those latitudes in the absence of ice and snow the intake of solar radiation is substantial. And the conversion of a desert or semi-desert area to a moist surface with grass cover, or the reverse change of savannah to desert, produces smaller but still significant changes of the heat absorption: the former increases it and the change to desert reduces it. These changes seem to introduce self-perpetuating (or 'positive feed-back') tendencies, in that there is more convection and therefore tends to be more rainfall over the vegetation-covered than over the desert surface. The effect of cloud cover in low latitudes is so great that over the Indian monsoon area in July there is actually a net loss of radiation – i.e. net outgoing radiation – from the Earth to space.

The effectiveness of extensive sea surface temperature anomalies of the scale mentioned above has been convincingly demonstrated both by theoretical modelling and observed correlations. Thus, J. Namias has shown that the prevailingly cold weather of the 1960s and again the cold winters of the late 1970s over the eastern two-thirds of North America and over Europe were associated with a distortion of the circumpolar upper wind vortex producing outbreaks of cold polar surface winds, apparently induced by anomalous sea surface temperatures in the central part of the North Pacific Ocean. A. Gilchrist and P. R. Rowntree of the United Kingdom Meteorological Office have shown that anomalously high sea surface temperatures in the tropical Atlantic near the Cape Verde Islands (latitude 17 °N) tend to produce patterns of the atmospheric circulation which give cold winter weather in Europe. Similar associations between anomalous warmth in the equatorial Pacific and cold winter weather over most of the United States were earlier demonstrated by theoretical modelling by Rowntree, following the observational studies of the late Professor Jakob Bjerknes.

IMPACTS ON THE CLIMATE OF VARIOUS HUMAN ACTIVITIES

We must now consider the range of effects, and possible effects, of human activities intruding upon the climatic regime. The greatest change in the terrestrial environment so far produced by man is the clearing of the northern hemisphere's temperate forest zone, which began on a small scale five thousand or more years ago, and its conversion to cultivation of the grasses which we use for grain crops and animal husbandry. This must have increased the prevailing wind strengths but may not have had a great effect on the heat absorption of the lands in the latitudes concerned. On the other hand, removal of the forest cover in low latitudes such as is now occurring in the Amazon basin, and either occurring or contemplated elsewhere, may be more serious: theoretical modelling studies suggest that the increase of surface albedo (the reflectivity of the Earth's surface) in this case would be likely to reduce heat absorption, and hence reduce convection and rainfall significantly. About 34 per cent of the equatorial zone between latitudes 5 °N and 5 °S is at present covered by tropical rain-forest. And it is estimated that complete deforestation might cool the Earth by 0.2–0.3 °C and by a larger fraction of a degree in parts of the zone concerned. One would also expect evaporation and rainfall to be reduced by several per cent in the tropical zone itself, possibly by 10 per cent in some part of the zone, with a smaller net decrease (about 1 per cent in the study reported) for the Earth as a whole.[9]

Afforestation and deforestation on a merely local scale, as within a single river catchment, are unlikely to have any significant effect on climate save within the area of the forest itself where moisture is retained within the forest canopy. Occasional exceptions may occur on showery and thundery days with light winds, such that the moisture is recycled and precipitated again within the same general region: but the general residence time of moisture in the atmosphere, reckoned to be about ten days, means that most rain deposits moisture far – even thousands of kilometres away – from the region where it was evaporated.

DEVICES TO ALTER THE CLIMATE AND ENVIRONMENT

Deliberate attempts to modify the climate began on a local scale with the planting of shelter belts of trees, to reduce wind speeds and protect light soils from blowing away, among the agricultural improvements pioneered in Norfolk in the eighteenth century. That practice has now been success-fully introduced in many other places, notably near the coasts of Denmark and north Norway and on the plains of Russia.

Where modern irrigation projects have sought to modify the conditions for agriculture, if not the climate itself, over larger areas it has often been hoped that a general increase of moisture in the atmosphere over the arti-ficially watered ground might result, particularly from the evaporation from reservoirs. But in naturally arid regions, particularly in the case of the Aswan dam which is more or less at the axis of the Saharan desert zone, the evap-orated moisture is likely to be dispersed in the atmosphere and carried far away by the winds, so representing a loss of water from the region. Even in the lower Volga basin, in latitudes near 50 °N, it seems that the exten-sive tapping off of the river water for irrigation has so increased the loss by evaporation as to reduce the flow into the Caspian Sea, contributing to the lowering of its level in recent decades.

The growth of population and industry in Soviet central Asia, and the need to cultivate the plains of that region so far as possible to grow grains and cotton, etc., have already for several decades placed excessive demands on the water resources of the region. The levels of the lakes and rivers, and of the water table in the subsoil, have been falling. The Aral Sea is liable to dry up completely and disappear by about the end of the century; and the Caspian Sea is becoming so much more saline, as its water level falls, that the supply of sturgeon – and hence the caviar – for which it is famed is threatened. The Soviet Union has therefore been driven to consider what could be done to supplement the natural water supply. The great rivers of Siberia and some in the northern part of European Russia flow north into the Arctic Ocean, and the needs of central Asia have given rise to a grandiose scheme which the Soviet authorities have often described as 'reversing the flow of the rivers'. What is contemplated is illustrated in fig. 113. The full project would not only provide irrigation for the areas opened up for cultivation under Kruschev's 'virgin lands' scheme in the 1950s and since, but could also include draining of the marshes in northwest Siberia. The scheme has been under consideration for half a century and would undoubtedly be a triumph for what twentieth-century engineering can do, including the use of atomic power to blast rocky barriers away, provided that the side-effects on climate did not turn out to be serious. This aspect has been the subject of much research and has caused hesitation. Later reports that work has begun on the scheme suggest that a good deal of caution will be

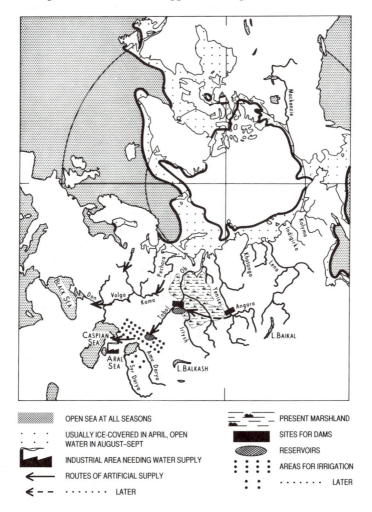

▓▓▓ OPEN SEA AT ALL SEASONS	═━═ PRESENT MARSHLAND
· · · USUALLY ICE-COVERED IN APRIL, OPEN WATER IN AUGUST–SEPT	▆▆ SITES FOR DAMS
▆▆ INDUSTRIAL AREA NEEDING WATER SUPPLY	⬭ RESERVOIRS
← ROUTES OF ARTIFICIAL SUPPLY	⁚⁚⁚ AREAS FOR IRRIGATION
← ─ ─ · · · · · · LATER	· · · · · · LATER

Fig. 113 Scheme for reversal of the flow of the Russian and Siberian rivers, the Pechora, Ob and Yenesei, to water the arid lands in Soviet central Asia.

331

exercised – beginning, at least, by tapping off a very limited proportion of the flow of the rivers for pumping south to the dry regions near the Caspian and Aral Seas.

The dangers foreseen in the project outlined arise from the fact that it is the fresh water from the rivers concerned which forms a large proportion of the thin layer of low-salinity water that covers the surface of the Arctic Ocean. The river Yenesei alone provides on average over 10 per cent of the total run-off of the northern continents into the Arctic Ocean. It is in this layer that the Arctic sea ice is formed. And if the layer were removed or seriously diminished, so that much of the polar sea became a salt-water ocean from the surface down, ice might not form on it – for the same reason that the Norwegian Sea and most of the Barents Sea remain open the year around. When cooled, water with the salinity normal in the world's oceans becomes denser and does not reach its maximum density until near its freezing point, at about –2 °C. By contrast, fresh water is densest at about +4 °C (39 °F), so that when the surface is cooled below that temperature the coldest water stays on top and at 0 °C ice is formed. The salt water of the deep oceans, when cooled at the surface, goes into patterns of convection, the coldest and densest portions gradually sinking to the depths. Hence the whole ocean – or a great depth of it – would need to be cooled to near its freezing point before ice formed and remained on its surface. In practice, inhomogeneities and salinity (and density) differences might allow some ice to form, as occurs seasonally on the Southern Ocean around Antarctica; but it would be easily disturbed and destroyed by vertical mixing in rough weather and would doubtless be limited in extent to the colder and shallower regions near coasts. It would probably always be much thinner than now, even in those regions where it still existed, and therefore likely to disappear in summer and be patchy in the transition seasons. Altogether the climate of the areas of the Arctic converted into an open ocean north of 70–75 °N would be, on average over the year, some 20–25 °C warmer than now. This huge change would be liable to shift the main thermal gradient of the hemisphere, and so alter the patterns of the large-scale wind circulation as to send the rain- and snow-giving cyclonic activity on new tracks, predominantly into the Arctic and sub-Arctic regions. The consequence might well be to reduce the rainfall both in central Asia – the region designed to benefit from the engineering scheme – and, to a less extent, also over most of Europe.

The scheme discussed in these paragraphs is an example of how man might be able – whether inadvertently or intentionally – to alter the climate in a big way by disturbing the global regime at some point where it is delicately balanced. In the case of most schemes, however, which have been suggested for deliberately modifying the climate, examination suggests that the global regime is extremely well buffered against upsets which might be caused by the relatively puny amounts of energy which are even now at human disposal. The energy released by a one megaton nuclear explosion is of the order of one-hundredth of that disposed of by a single cyclone/ depression over an hour or so or by a moderate-sized volcanic explosion. (The indirect effects of nuclear explosions at the ground through the screening off of solar radiation by the dust injected into the stratosphere, as after volcanic eruptions, are likely to be the only significant effects on weather far from the scene.)

Around the 1950s and early 1960s there was some discussion, and in Moscow a conference was held, on the possibilities of modifying world climate deliberately with aims such as to increase as far as possible the total cultivable area of the Earth. Since that time there has been a change of emphasis, probably due to a clearer understanding of the fact that the growth of population has already made the world community much more vulnerable to the dislocations that must result from any climatic shift and the wide-ranging year-to-year variations which would doubtless occur in the course of it. Nowadays, the chief concern is over the possibility of a large-scale shift of world climate being brought about inadvertently, as a side-effect of human activities and their increasing scale.

THE INCREASE OF CARBON DIOXIDE

The main worry about the impact that human activities are likely to have is related to the increase of the seemingly innocuous gas, carbon dioxide, in the global environment. Carbon dioxide (CO_2) is the end-product of the burning not only of wood but of all fossil fuels – coal, gas, oil, etc. It is a very minor constituent of the atmosphere, only about 350 parts per million (ppm) by volume, but it is important because of its effects on the radiation passing through the atmosphere. This applies particularly to the radiant energy going out from the Earth, because CO_2 is not transparent to radiation at some of the long wave-lengths most strongly represented in the emission from bodies at the temperatures prevailing at the Earth's surface and in the atmosphere. Hence, this radiation is absorbed on its way upward from the Earth by the CO_2 in the atmosphere and re-radiated in all directions, so partly back to the Earth. As a result of this, and the similar action of the water vapour in the atmosphere on radiation at a range of wavelengths partly overlapping those which CO_2 absorbs, the Earth's surface climate is some 35–40 °C warmer than would be expected on a planet at this distance from the sun. The action is reminiscent of a greenhouse, and the warming is sometimes spoken of as the greenhouse effect of carbon dioxide. Similarly, the temperatures prevailing on other planets with CO_2 in their atmospheres, notably Venus whose atmosphere consists largely of carbon dioxide, and also Mars, seem consistent with the magnitude of the expected carbon dioxide warming effect.

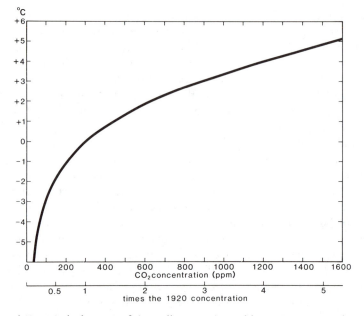

Fig. 114 Expected changes of (overall average) world temperature to be caused by different proportions of carbon dioxide in the atmosphere. (Derived from a theoretical modelling study by T. Augustsson and V. Ramanathan, 'A radiative-convective model study of the CO_2 climate problem', *Journal of Atmospheric Sciences*, vol. 34, pp. 448–51, 1977.)

One would expect that an increase in the amount of CO_2 in the Earth's atmosphere would increase the greenhouse effect. There is no doubt, from actual measurements, that the amount of carbon dioxide in the atmosphere has been increasing and that, with the increased rate of burning of fossil fuels,[10] the rate of increase has become greater. In 1880–90 the CO_2 seems to have been around 290 ppm in the atmosphere. Some who have studied the subject believe that before the massive clearing of forests for agriculture in the nineteenth century, the proportion (owing to assimilation of carbon from atmospheric CO_2 by the vegetation) may have been as low as 270 ppm. By 1950 the proportion had risen to about 310–15 ppm and by 1980 to 335–40 ppm. These figures mean that the proportion of carbon dioxide in the atmosphere had increased 9 per cent by 1950 and 15–17 per cent by 1980 above the 1890 level, and over the last 150 years the increase may be as much as 26 per cent. It has been calculated that this carbon-enriched atmosphere may have contributed to richer crop yields to the extent of a few per cent. Carefully calculated estimates have been made as to the proportion of the extra CO_2 generated by man's fuel-burning which stays in the atmosphere, how much is absorbed by the plant world (or elsewhere in the biosphere), and how much is dissolved in the oceans, where it may ultimately end up fixed as an increase of the calcium carbonate deposit on the ocean floor. From these studies it seems that about half of the output of CO_2 is remaining in the atmosphere, though the proportion varies slightly, as the oceans give up CO_2 to the atmosphere when they warm up either seasonally or over longer periods and absorb more when they become cooler. The most generally accepted calculations of the warming effect of increased carbon dioxide (through its increasing the effectiveness of the long-wave radiation trap) indicate that overall average temperatures at the surface of the globe should rise by about 1.9 °C[11] if the carbon dioxide concentration were doubled. A more complete coverage of the expected scale of temperature change from the 1920 average (when CO_2 was 300 ppm) for a wide range of concentrations is given in fig. 114.

There have been various other estimates of the warming of world climate to be expected from a doubling of the atmospheric carbon dioxide, since the case was first thoroughly argued through from physical principles by G. N. Plass[12] in the 1950s. Plass suggested a figure of 3.6 °C temperature rise for a doubling of the CO_2 and a 3.8 °C fall if the CO_2 were halved: on this basis it seemed that the entire warming of world climate (by about 1 °C) since the industrial revolution in the eighteenth and nineteenth centuries might be explained in this way. The CO_2 warming thesis has always had a specially strong appeal to physical meteorologists as one element in the complex problems of atmospheric science which should be entirely predictable, since the effects of CO_2 on radiation are clearly demonstrable and well understood in theory. Nevertheless little was heard about the thesis in the 1960s, when it was discovered that world temperature was falling despite the more rapid increase of CO_2 in the atmosphere than ever before. Others professionally concerned with the CO_2 problem suggested that the warming effect might be no more than one-tenth to one-fifteenth of Plass's figure; but recent studies with the most sophisticated models, which not only allow for atmospheric transport of the CO_2 and heat about the world but take at least some formal account of exchanges with the top layer of the ocean and possible effects on cloud cover, have pointed once more to a greater warming, between 2.0 and 3.5 °C, for a doubling of the atmospheric carbon dioxide to 600ppm.[13] These figures imply a steeper curve than that shown here in fig. 114; but they continue to be treated with scepticism by some climatologists and atmospheric modellers, because it has not yet been possible to include in a realistic dynamical way in the theoretical models the exchanges with the ocean (and to deeper layers within the ocean) or the effects on cloud and atmospheric humidity. It is pointed out that a 1 per cent increase in mean cloudiness over the Earth, if such an increase occurred, would probably completely counteract the proposed CO_2 warming effect – at least as regards the net effect over the globe. Such an increase of cloud might come about through a real warming, of the ocean surface in low latitudes, leading to increased evaporation and input of moisture into the atmosphere, and hence more cloud over middle latitudes, producing lower temperatures than before there.

OTHER EFFLUENTS FROM HUMAN ACTIVITIES

Besides carbon dioxide there are other substances which human activities are putting into the atmosphere, mostly in increasing amounts, which must also be supposed to affect the radiation balance. The most discussed of these items has been the solid particulates in the form of dust and smoke, the latter from industrial and domestic fires and from agriculture, particularly the primitive slash and burn agriculture of the tropics. At the same time the greatly increased area of the world that is tilled, often large-scale tillage of open prairies by tractor-drawn ploughs and other implements, seems to have increased the amount of dust in the lower atmosphere. Atmospheric turbidity measurements at Washington, DC, in the United States and on the heights of the Alps at Davos agree in indicating substantial increases (by 50–80 per cent) from the beginning of this century. The dust deposit measured in the ice-layers of glaciers on the high Caucasus shows an even greater increase but suggests that little change had occurred from the eighteenth century until around 1930. Some have thought that the increase of dust haze should cool the Earth's surface (through interfering more with the incoming solar beam than with the Earth's outgoing radiation) and may cancel the effect of carbon dioxide warming entirely. Professor R. A. Bryson has likened the effect to that of volcanic dust – although the latter is in the stratosphere in those cases where its effect lasts for more than a few weeks, whereas the dust that man produces is maintained in the lower atmosphere – and has written of the 'human volcano'. One calculation suggested that increasing the suspended particulate matter by a factor of four should lower the Earth's mean surface temperature by 3.5 °C, but later modelling studies show that the effect must vary greatly with the prevailing size and absorptive properties of the particles and may in some cases be in the direction of warming.[14]

Smoke trails, or 'plumes' as they are more usually called, from industrial or thickly inhabited areas commonly remain identifiable and reduce visibility over long distances down-wind from their source, particularly in air that is cool near the ground and warmer above so that convection is checked. The industrial haze from the Ruhr and Belgium has long been liable to produce murky conditions in England in light easterly winds in winter (though less in recent decades with more efficient firing of domestic and industrial hearths, as also applies to the haze generated in British centres of population). At the time of the great Fire of London in 1666 John Locke observed in Oxford some 80 km away that the unusual colour of the air 'made the sunbeams of a strange red dim light', later noting that they had heard nothing of the fire at that point, though it now seemed that it must have been due to the smoke. And in the late sixteenth century the smoke from large-scale moorland fires in England was said to have ruined French vine crops in the bud[15] – evidently indicating northerly winds in spring.

Other substances put into the atmosphere, which seem important in the same connection, are (a) the nitrous oxide produced by breakdown of nitrogen fertilizers in the soil and (b) methane and chlorofluoromethanes (more widely known as 'freons') – a range of chemical substances in which one or more chlorine or fluorine atoms replace some of the hydrogen atoms in methane – used in aerosol sprays and also by refrigerators. If they get into the stratosphere, the latter substances, besides selectively absorbing long-wave radiation like carbon dioxide, destroy some of the ozone there. The stratospheric ozone is important because it absorbs solar short-wave radiation, including some wave-lengths which would have lethal effects on living organisms, and thereby warms the stratosphere at the expense of the Earth's surface and lower atmosphere. The ozone therefore has a cooling effect so far as the surface climate is concerned. There may also be significant contributions to the human disturbance of the radiation balance from sulphur dioxide (which is also harmful to human health, to vegetation and buildings, etc.), hydrogen sulphide, carbon monoxide and ammonia, even though the usual residence time of these in the atmosphere before removal by chemical action is much shorter. The quantities of water vapour added by man (except very locally) are trivial compared to those naturally occurring.

THE NET EFFECT

The net effect of the increase of all these substances in the atmosphere as a result of man's activities is apparently in the direction of warming, and may *in toto* add about 50 per cent to the CO_2 effect. Flohn[16] suggests that the simplest way of dealing with all these intrusions into the atmosphere is to consider a 'virtual CO_2 concentration', which should have the same theoretical effect on temperature as the combined greenhouse effect of all the substances actually involved.

The reason most commonly advanced for why the carbon dioxide, or combined greenhouse effect, warming is not obvious at the present time is that it is not yet big enough to go beyond the range of the climatic fluctuations – sometimes, of course, in the opposite direction – produced by natural causes. This range of natural climatic fluctuation is sometimes described as the 'noise level', which must of course make it difficult to identify any new trend – whether or not the trend were produced by man's impact – before it had already reached a substantial amplitude. Efforts have therefore been made to decide how soon the (assumed) further increase of carbon dioxide will produce a warming too strong to be offset or obscured by the natural variability of climate. In such writing the natural variability is dismissed as unforecastable and therefore to be treated as random. Those putting forward this view of the matter have taken +1 °C as the approximate range of variation of the long-term temperature average produced by natural causes in the post-glacial world. In consequence of this, they expect the warming by carbon dioxide, combined with the other substances contributing to an intensification of the greenhouse effect, to gain the upper hand and 'swamp' all other elements of climatic variation from the end of this century onwards and possibly from the 1980s on. This view was strongly put in a statement approved by the executive committee of the World Meteorological Organization in 1976 (reported in *The Times*, London, 22 June 1976).[17]

There is a fallacy in this part of the case, however, since it is impossible to define a figure for the range of natural variation of climate which is meaningful in this connection. The record of prevailing temperatures, whether over the past few centuries or over the much longer-term record of ice ages and interglacial periods, shows that the range of variation is itself subject to variation. We know, both from early thermometer records and from the indirect indications of tree ring sequences in Europe, that in the Little Ice Age climate, specifically in its later stages for some decades around AD 1700 and again between about 1760 and 1850 or later, the year-to-year variability (measured by the standard deviation) of these items was from 30 to 60 per cent greater than in the earlier twentieth century. There are similar indications from the European tree ring studies regarding the last decades of the warmer climate of the high Middle Ages between AD 1280 and 1350. And it is clear from isotope studies of the Greenland ice and other evidence that some much sharper changes took place in the later part of the last warm interglacial period.

One surely implausible suggestion that has been put forward on the basis of serious scientific arguments is that the undoubted strong warming that took place in high southern latitudes, south of about 45 °S, between the 1950s and 1970s (amounting to rather more than 1 °C in the overall average for the Antarctic south of 60 °S) may be the first direct sign heralding the dominance of carbon dioxide warming. The argument that the effect should be first noticed there, so far from all the man-made carbon dioxide sources, depends partly on heat storage and transfer by the oceans and partly on the freedom of high southern latitudes from the (increasing) contamination of the lower atmosphere by dust. The authors of this suggestion take no cognizance of the evidence that high southern latitudes have a record of a partly antiphase relationship to the temperature variations affecting the rest of the world and enjoyed somewhat milder conditions during some of the sharpest phases of the recent Little Ice Age. We shall return to some consideration of the distribution of temperature changes over the world, and their climatic consequences, in the next chapter.

There are other serious enigmas and difficulties which remain to be sorted out before we can be sure that the thermal effect of increasing carbon dioxide on the passage of radiation of different wave-lengths through the atmosphere, which appears straightforward in theory, should emerge clearly in the complexity of the world environment. The effects involved in the oceans, and in the atmospheric moisture content and cloudiness, need further observation and study. And proper allowance needs to be made for the reactions of the biosphere, since vegetation (including the plant life in the sea) growing more luxuriantly in a carbon-rich atmosphere may change colour sufficiently to affect the reflection and absorption of solar radiation. It can be deduced that there was a much bigger change of the atmosphere's CO_2 content far back in the geological past between the conditions before the development of the first vegetation cover[18] and afterwards, but neither in this case nor with the later changes of vegetation extent does there seem to be the expected correlation with the development of climatic changes. In theory the presumed changes of CO_2 should have tended to stabilize the respective climatic regimes. During the long warm periods of geological time, when there was little or no polar ice, the oceans must have had a smaller capacity to hold carbon dioxide dissolved and so the atmosphere should have been richer in CO_2. And during the ice ages the colder seas could dissolve so much more CO_2 that its proportion in the atmosphere may have been much reduced. A figure of 200 ppm has been suggested for the time of the ice age climax. Contrary effects would, however, be implied by the changes of vegetation extent between warm periods and ice ages.

The most successful mathematical simulation of the variation of world temperature since AD 1600, and more specifically over the last hundred years (as in fig. 91), has been by an equation involving just three variables:[19]

1 an index of the amount of volcanic material in the atmosphere;
2 warming latterly introduced, and increasing, through the continual addition of carbon dioxide to the atmosphere through the burning of fossil fuels; and
3 an index of solar disturbance.

The fit was improved by adjusting the equation so as to double the effect of volcanic dust. In a preliminary draft of their work the authors of the equation added a caution, which could apply equally to much more elaborate theoretical modelling work: 'We are hesitant to try to improve the fit of our calculations to the observations by 'tuning' the model. . . . With so many free parameters to vary one could fit almost anything to anything . . .'. They also punctiliously added that they had played down

339

on performed of a correlation coefficient between their results
hundred year record of global surface temperatures used
ertainty about the reliability of that record.

of Dansgaard and Hammer, and of Bryson and his co-
ed earlier in this chapter seem to reinforce the lesson of
ocnneider's experience and suggest that the effect of volcanic matter in the
stratosphere in cooling the surface climate may bulk larger – and possibly
a good deal larger – *vis-à-vis* the carbon dioxide warming effect than is
commonly assumed today.

EFFECTS IN INDUSTRIAL AND URBAN AREAS

This chapter would not be complete without some further notice of man's
activities as a cause of unintentional modification of climate on a local
scale, particularly in cities and industrial areas and down-wind from them,
and also in enclosed valleys and in waters with restricted circulation and
outlet. The artificial temperature rise in cities and in enclosed waters has
been mentioned in earlier chapters. These effects are sometimes referred to
as urbanization and thermal pollution. The artificially maintained warmth
of some coastal inlets and backwaters near electric power stations and oil
refineries, or other industrial complexes, may be able to support an exotic
fauna and flora. It was reported some years ago that a warm-water species
of crayfish had established itself in Southampton Water, near the Fawley
oil refinery, presumably introduced in bilge discarded by ocean liners
approaching the port.

Despite some controversy it seems established that 'urban heat islands',
the artificially warmed central parts of cities, tend to increase the activity
of convection clouds, showers and thunderstorms over them and some way
downwind from them. This is liable to produce an increase of up to 10
per cent in the average yearly rainfall totals in the part most affected in
big cities.

These conditions, and the relative freedom from frost and shelter from
strong winds, may facilitate the cultivation of exotic plants in urban
environments.

Perhaps the most important local effects produced by the activity in
towns and industrial areas, and in a few other places (e.g. motorways, railways), are those due to smoke, steam and chemical pollution. In this class
perhaps the most widespread damage has been caused (to human health
and to buildings and other property – cars, etc.) by sulphurous gases and
sulphuric acid in the atmosphere, although in some places carbon
monoxide, ozone and other gases, and in rural areas the substances in
crop sprays and nitrogenous fertilizers, may also have serious effects.
Buildings and machines as well as human lungs are subject to corrosion
and decay caused by these chemicals in the atmosphere. Because lichens

are particularly sensitive, their growth or absence may serve as an indicator of the cleanness of the air.

P. Brimblecombe of the School of Environmental Sciences in the University of East Anglia has traced the history of air pollution in London since the thirteenth century, and to some extent in other European cities, and of public attitudes to it, in a series of publications.[20] It is probable indeed that the smoke pollution and smells produced by industries such as the tanning of leather, pottery and lime production caused local complaint even in much earlier times in places where these industries were carried on in light winds and sheltered areas in and near towns, particularly where inversions of the usual vertical lapse of temperature with height developed in cold winter weather and prevented the escape and dispersal of the pollution upward into the atmosphere. Such situations probably sometimes affected the choice of sites for industry, and the growth of the urban settlements near by, from ancient times. Brimblecombe reports that coal was introduced to London for lime burning and smelting soon after AD 1200; and the results, together with the sewage problem associated with the building of privies over ditches and gutters, soon gave rise to unbearable stenches and many complaints. In 1257 King Henry III's queen was among the complainers, and from that time on commissions were appointed to consider the problem. Matters became worse in the time of Queen Elizabeth I with the further introduction of coal, initially smoky Tyneside coals brought by sea from Newcastle (and therefore known as 'sea-coals'), when wood was becoming difficult to secure, for domestic fires. At first nice people refused to enter rooms where coal had been burnt. And during the course of the seventeenth century the incidence of rickets, a children's disease associated with deficiency of sunlight, increased sharply. This disease ultimately became so rife that a survey in Leeds in 1902 found that half the children in the poorest districts were suffering from it. And a similar incidence of it was characteristic in other cities of industrialized Europe. But it was inevitable that coal should replace wood as fuel.

Much more complaint arose and a smokeless stove was invented before the seventeenth century was out, but nothing much seems to have been done about it. Sir Christopher Wren's new St Paul's Cathedral, built in the last part of the century after the fire of London in 1666, is said to have been badly soiled already before completion. The hanging of tapestries on the walls of rooms was largely given up by the early eighteenth century because they became so dirty and spotted, as reported in 1658 by Sir Kenelme Digby who became one of the early Fellows of the Royal Society. More interestingly in the matters with which we are basically concerned, Brimblecombe has found it possible to illustrate the long-continued increase of pollution in the air over London (and neighbouring parts of the most densely inhabited and increasingly industrialized Europe) from a progressive change in the colour of the background skies painted

341

by landscape painters, changing from the early dominance of blue to an increasing dominance of pinks and muddy shades of yellow-brown. London fogs became widely known from this colour as 'pea-soupers'. Visibility at its worst sometimes fell so low that it was difficult for a person walking to see his feet. And the darkness entailed using lamps throughout the day. Dickens called these fogs 'London particular'. For as long as nothing was done to improve the efficiency of hearths and furnaces, right up to the present century, the growth of pollution of the city air was neatly paralleled – and could effectively be measured by – the increase of coal consumption. Whatever measure we study, whether from the evidence of artists or the frequency of reports of fog[21] or damage to buildings and measurement of the soot deposit on them, the pollution of London's air was increasing all the time from before 1600 to about the beginning of the present century, the increases being most rapid in the seventeenth century and again after 1800.

Scientists at least from the time of Benjamin Franklin onwards pointed out that smoke from chimneys was unburnt fuel going out into the air to waste. And smoke abatement legislation was first proposed in 1843 by a

Fig. 115 Atmospheric pollution at its source: smoke in its densest form, emitted by a coal-fired tilery in England's 'Black Country' before the Clean Air Act of 1956. (Reproduced here by kind permission of the National Society for Clean Air, Brighton.)

committee which reported to the British government, in the Mackinnon Report, but was not acted upon. Improvement began in London and elsewhere with the adoption of more efficient furnaces in industry from early in the twentieth century, but the pollution from house fires continued. It had earned Edinburgh the name of Auld Reekie (meaning the old smoky one). Under the dark skies and surrounded by the smoke-soiled grasses, trees and buildings of the most industrialized areas of England (fig. 115) continental Europe and North America, some species of moth whose survival during the day depends on camouflage developed all-black forms. (This melanism, first investigated in England, is a remarkable example of quick biological adaptation by the processes of mutation and selection.) Substantial improvement was achieved in many more centres in Britain by the Clean Air Act, which at last became law in 1956. Since that time the amounts of sunshine registered in British cities have increased. (Rickets has now become much rarer, partly through advances in medical science but probably largely through better living conditions, including healthier air in cities.) But the Act was against visible smoke, and the noxious fumes of sulphur dioxide (SO_2) harmful both to lungs and to the stonework of buildings have continued. These have been heavily implicated in several 'smog' disasters, in Belgium and the United States as well as in London, when in fogs formed under an inversion of temperature in winter, and with a lack of wind, sulphur dioxide concentration rose to lethal levels.[22] In the great London smog in December 1952 deaths in the Greater London district rose from 2062 in the week ended 6 December to 4703 in the following week. Deaths from bronchitis and pneumonia showed a sevenfold increase. There had been similar occurrences in 'Cattle Show week', 7–13 December, in 1873 and again in 1880, 1892 and 1948; and there was a recurrence in December 1962 which was somewhat less serious, perhaps already thanks to the new Clean Air legislation. A similar great fog, 'with a very sensible effect on the eyes' and an acrid smell, extending all around London between 27 December 1813 and 2 January 1814 is recorded in the *Annals of Philosophy*.

If the situation has improved in London and Britain's industrial areas thanks to legislation and the independent actions of industry and householders at last to burn fuel more efficiently, and elsewhere in the advanced countries in the temperate zone by similar moves, other pollution problems remain and not all are localized. The sulphur dioxide from industrial and domestic chimneys continues as a threat to life, and seems to have increased. Measurements of SO_2 concentration in the air in Epping Forest (about 30 km northeast of London) show an increase from an average of 30 micrograms per cubic metre in 1784–96 to about 60 of the same units a century later, from 60 to 70 in 1909–19 and between 70 and 120 around 1970. Not only human lungs are damaged, particularly in the old, but the animal world suffers and the leaves of vegetation. For many years now there has been alarm and complaint in Scandinavia at the increasing acidity of the lakes and rivers there, and decline of the valued fish stocks, because of sulphur dioxide whose origin is attributed to the industrial areas and electrical power stations of Britain, the Benelux countries and Germany.

Los Angeles, despite the sunny warm climate general in California, has its own special smog problem. The cool air off the sea tends to underlie warmer air overhead, and the inversion of temperature stops convection which might disperse the pollutants put into the city air; the situation is particularly common and at its worst in the warm seasons, summer and autumn. The biggest contribution to the pollution seems to be the exhaust gases of motor vehicles, though industry is also implicated. Concentrations of noxious fumes enough to make eyes smart are common, and there has been damage to vegetation a considerable distance inland. A warning system and special controls on days of high pollution have been instituted.

New experiences of local pollution, sometimes of a serious order, arise in newly industrialized areas in the Third World, especially in enclosed valleys (fig. 116) and still air situations. Even on camps and outstations,

Fig. 116 Smog in the valley of Mexico. Formerly the area was famed for its views of the mountain peaks 50–100 km to the east, especially the beauty of the light of the setting sun on the volcanoes. Now, owing to the fumes of industry and transport and the dust commonly blown up from the drained lake areas, such views are seldom seen. (Photograph kindly supplied by Dr Humberto Bravo of the University of Mexico.)

expedition sites and airfields established in the Arctic and Antarctic, smoke concentrations on some cold days with strong inversions of temperature may halt some activities. At these places also the moisture put into the very cold air by aircraft and other vehicle exhausts on the ground may produce dense fog – a case of pollution by excess water in air which becomes saturated at very low water vapour concentrations. Similarly fogs are occasionally formed or thickened by moisture from vehicle exhausts along motorways and from steam engines on railways in temperate latitudes in winter. And it is a not uncommon sight in still, fair winter weather to see a layer of low stratus cloud formed by the steam from industrial cooling towers.

17

FORECASTING

THE DEVELOPMENT OF DAILY WEATHER FORECASTING

When we come to consider the possibilities of forecasting weather and climate to guide our planning for the future, we must take account of many different time-scales. The processes involved in the short and long-term developments, and the influences chiefly at work in each, the amount of detail which we may be able to foresee, and the degree of reliability attainable, are so different for each time-scale that different handling methods and different ways of stating our conclusions are inevitable.

The first daily weather maps had been drawn by H. W. Brandes in 1820 from the observational data for 1783 then available after thirty-seven years! It was the invention of the electric telegraph that by 1850 first made it possible to track the movement of storms before they arrived at the area of interest, and so made gale warnings and daily weather forecasts possible. By the late nineteenth century these had attained a degree of success not altogether dissimilar from today's, though the recognition of fronts and the characteristics of different air-masses in the 1920s and 1930s introduced some details that were not understood earlier. As upper air observations from balloon and aircraft became increasingly available, so it became possible to include adequate details of the upper winds and cloud conditions some hours ahead in forecasts for flying. But it was the development of instrumentation capable of reporting the winds and temperatures in the upper air above the clouds and in all weathers – balloons carrying automatic radio-sounding apparatus and tracked by radar – from about 1940 onwards that first brought the circumpolar vortex under daily survey.

Up to this point forecasting had been more or less limited to plotting the travel and development of individual weather systems from their first appearance on the surface map day by day, steered by the winds aloft (as indicated by the surface winds in the warm sectors of the frontal cyclones) to their ultimate decay or absorption in another system. Only when a slow-moving anticyclone settled over an area could 'outlooks' for two to three days fine weather ahead – e.g. for hay making and harvesting – be issued with satisfactory reliability. But once the circumpolar vortex, and soon afterwards the jet stream, had been recognized, the principles governing the development and locus of formation of new weather systems – i.e. individual travelling cyclones and anticyclones on the surface weather map – could be better formulated and in ways adapted to the computer age and the numerically calculated forecasts for one, two, three and more days ahead. It is in this realm that the main advance of daily weather forecasting since the 1930–50 period has come, in the greatly improved ability to indicate the weather development over several days ahead.

LONGER-TERM FORECASTING

This type of forecasting, in all the detail that handling of the individual weather systems can give, runs into its ultimate limit as the errors resulting from any weaknesses of theory and coarseness of the specification of the situation existing at the outset on day zero, by observations at a necessarily limited network of points, build up when one computes the situation at successive intervals ahead. Moreover, the atmospheric circulation uses up its existing store of energy in about five days, and the heating patterns by which the store is continually renewed themselves depend upon the weather that the circulation itself has produced. Generally, it is found that the time taken for errors in the forecast situation to become doubled is from three to eight days. And so at some number of days ahead – at present it looks like being on average around five to ten days ahead – the forecast map gradually ceases to bear a useful resemblance to the situation which actually emerges.

Forecasts for longer periods of the order of a season ahead depend on trying to spot by one technique or another the broad characteristics of that season, determined by the prevailing steering of the warm and cold surface winds and of the weather systems by the lay-out of the patterns of the circumpolar vortex. This leads to statements about the probable departure of prevailing temperatures from the long-term average and about rainfall totals, rain or snow frequency, and storminess in relation to what is normally expected, and so on. Because of the sharp local differences of rain and snowfall caused by most terrains, forecasts of these items are best expressed as a percentage (or range of percentages) of the long-term average. At most only the boldest features of any time sequence of events within the season are likely to be specified, either on the basis of the gradual attainment of a climax of the regime characterizing that season and its subsequent weakening, or through recognition of some (often recurring) oscillation within the season. Most success has been achieved by statistical approaches based on understanding (or perhaps just successful hypotheses) regarding the physical controls of the seasonal development of the atmospheric circulation in the particular year and those that determine the average course of the general run of years. Rules have been recognized regarding developments associated with long-lasting anomalies of temperature in sequels to various developments of sea ice and snow cover distribution, to the formation of volcanic dust veils, and changes of solar activity, and so on. Mistakes have probably been made through some 'blind' use of statistical associations not linked to correctly recognized physical processes, and through failing to recognize when such rules were merely duplicating each other (and so adding no weight to the argument for a particular forecast). There seems to be room for the development of further statistical rules guided by the findings of theoretical modelling, but based, of course, on the use of observations from the real world.

Some help has been obtained from recognizing that the year tends to divide itself up into natural seasons, within which long spells of set weather type often prevail: towards the end of each natural season the controlling pattern of the circumpolar upper winds adjusted to the particular global heating pattern tends to break down, and the long spell of weather ends with it. It is in high latitudes that the dates of the year which mark off these seasons seem to be most nearly fixed. (With the declining sun in autumn within the Arctic circle nothing can stop the general freeze-up except a south wind, which cannot possibly occur everywhere.) In middle latitudes there is rather more variation of the critical dates from year to year, but alike in the Soviet Union and at the Atlantic fringe of Europe it is recognized that there is a season from around the beginning of July through most of August when a persistent weather character tends to prevail. Similarly, long spells of one character or another often develop within the winter season after Christmas or New Year until a change of pattern some time in mid-February or March. The preceding late autumn or 'forewinter' period is different and is commonly marked by a quite different spell of weather. In low latitudes, within the range of the seasonal migration of the equatorial rains, the spells of weather are much more alike in character from one year to another – as is recognized in the well-known monsoons and rainy and dry seasons, associated more or less with particular months of the year – but the dates of onset and ending of these seasons, and the occurrence or not of breaks within them, may differ widely from year to year.

Some use can also be made of the more or less biennial (two-year) cycle, which is a quite marked tendency in many (though by no means all) series of weather observations around the world. Its effect is most marked in the summer and winter temperatures in some places, and rainfall in others, and is accompanied by a corresponding tendency for shifts of the large-scale wind circulation pattern. Indeed, it seems that more or less regularly associated changes in the winds in the stratosphere – over both low and high latitudes – which, however, vary significantly in their timing from

348

year to year, may serve as signals of the progress of the cycle and therefore of the surface weather likely to dominate the ensuing season. Similarly, there is a suggestion of what amounts to a roughly 5½ year cycle, which may also be used by the seasonal forecaster. It was. discovered, and tentatively explained, by the German pioneer of weather forecasting on this time-scale, the late Professor Franz Baur, whose judicious use of it contributed to his remarkable record of seasonal forecasting successes in the 1950s. It seems that the global wind circulation tends to produce maxima of the westerly winds in middle latitudes in the intermediate phases of the (rather variable) eleven-year sunspot cycles – i.e. when the sunspots are declining after a maximum and when they are rising towards the next maximum – such that Europe in particular gets more mild winters and rather dry (anticyclonic westerly) summers at these times. Again it is important in forecasting practice to monitor the progress of the cycle.

How far it may ultimately be possible to exploit these tendencies to forecast more than one season, or more than one year, ahead – taking whatever account may be necessary of other influences working on that time-scale – cannot be adequately judged in the present state of knowledge. There are suggestions that tidal forces, which affect the atmosphere and the Earth's crust itself as well as the oceans, and other forces associated with the alignment and occasional conjunctions of the planets, as well as the Chandler wobble of the Earth's axis, may play some part. Some have suggested that the last two items point to particularly disturbed years, with a climax of blocking, around the 1980s. What is certain is that, in any attempt to foresee the prevailing weather development several years ahead, the reliability of specific detail must decline as the range of the forecast is extended. It may, for instance, be – or it may become – possible to foresee a tendency for severe drought or the likelihood of a notably severe winter in a certain decade, or even about some particular year or years, but necessary to allow for a probable error of plus or minus one year (or, in some cases, several years) in its arrival.

CLUSTERS OF LIKE YEARS

Some mention must be made at this point of a so-far unexplained phenomenon, operating on the time-scale of a few years together, which is occasionally very marked and may for the time being override the two-year and 5½-year (and other short-term) fluctuations. This is the clustering of several years – not always in unbroken succession – with some similar point of character. An early example is enshrined in the Fimbulvinter legend quoted on pp. 147–8. Other runs of three severe winters in a row in Europe with very similar wind circulation patterns occurred in 1878–81 and 1939–42. Sometimes the similarity concerns surprising detail. One example was the sequence of three 'skating Christmases' in England in

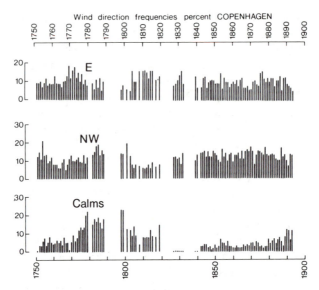

Fig. 117 Observed frequencies of winds from east and northwest, and of calms, at Copenhagen, Denmark, yearly from 1752 to 1893. The variations show groups, or clusters, of years with similar character.

1961, 1962 and 1963, with very severe frosts beginning on or just before 25 December with a strong north European anticyclone and easterly winds blowing right across the European plain. In a moderated degree, with snowy weather or a dry frost sufficient for skating in southern England between 25 and 28 December, the sequence continued for two more years in 1964 and 1965. (This sequence must be viewed against the background of only seven to ten Christmases in the first fifty years of the century in southern England with any claim at all, on grounds of white frost or snow, to be classed as a 'white Christmas'.) Another, similarly precise cluster was shown in the years 1965–71 by a repeated high frequency of northerly winds in the first five days of January over the British Isles: the winds were northerly on 30 per cent of the days, while only 10 per cent had westerly winds. No other run of years for which we have daily weather maps (i.e. between 1781 and 1786 or from 1861 to date) shows this feature. The overall average for the first five days of January over this long period was westerly 32 per cent, northerly 7 per cent; and from 1921 to 1932 (another cluster) the frequencies were westerly 73 per cent, northerly 8 per cent. Other clusters may be seen in the long record of yearly total frequencies of winds from east and northwest, and of calms, at Copenhagen from 1752 to 1893, here shown in fig. 117. The explanation must presumably lie in some strong feature of the surface heating pattern persisting over the years concerned, probably an anomalous position of a major ocean current boundary such as Namias identified in the North Pacific Ocean in the 1960s[1] and such as we have pointed out (see fig. 23, p. 60) in the northeast Atlantic in the 1690s.

THE DEVELOPMENT OF CLIMATE

Any advice on future weather, over any time-span whatever, must be accompanied by a proviso that the forecast would have to be changed – to a colder regime with a different wind pattern and a different distribution of rain – if volcanic outbursts should occur sufficient to create (and, perhaps, maintain) major dust veils in the stratosphere. Essentially the same proviso would be needed to provide for the case of possible further volcanic activity adding to any stratospheric veils already in existence at the time of issue of the forecast advice. The need for these provisos is perhaps most obvious in the case of any forecasts to cover one to ten years ahead, since the whole (or a large part of) the forecast period could be affected by the dust veil from a single great eruption. But, in fact, the caution is necessary in relation to all the longer time-spans, since changes in the frequency of volcanic explosions such as are known to have taken place in the past could alter the whole prospect, the effect being generally in the direction of cooling if volcanic dust veils should occur.

In looking next at longer time-scales, it will be convenient to consider first the climatic variations produced by nature and not attributable to man.

CYCLICAL DEVELOPMENTS

In the first place, some apparently cyclic oscillations of longer period than those already mentioned deserve notice. In all these cases neither their origin nor the outlines of their working in terms of the global wind patterns, which would make it possible to keep a running watch on the evolution of the current round of the oscillation, have yet been properly identified. Yet variations such as the recurrences of drought in the United States Middle West at roughly 20–23-year intervals must engage the attention – and perhaps precautionary actions – of people (in this case farmers) in the area, whose activities are vulnerable. The central England temperature record for most seasons of the year (fig. 28, pp. 80–1) shows variations tending to occur on this same time-scale, which recur in other data around the world and may be triggered by alternate sunspot periods affecting some natural 'resonance' period in the atmosphere. There have been suggestions that the same periodicity is one element in the variations in the occurrence of blocking, specifically of high latitude anticyclones in the Greenland-Iceland-Scandinavia region: but these also show longer-term variations, including perhaps a roughly fifty-year periodicity of which hints were found in the analysis of the many historical manuscript references to seasonal weather over the last nine hundred years in Europe. Anticyclonic conditions over northern and central Europe appear to have been more frequent in and around the thirties and eighties of most centuries than in the other decades. The accompanying high frequency of easterly winds, and of calms or light winds, near latitude 50 °N, make for a 'continental' tendency of the climate in central and western Europe with warm summers and cold winters at such times. This tendency, however is plainly liable to come into conflict with the effect of any oscillation that is close to twenty years in length.

This brings us to notice the evidence for a cyclical tendency which is very close to a hundred years in length.[2] Some of the evidence suggests, as in the case of the clustering phenomenon, a surprising approach to precision. The data in tables 8, 9 and 10 give a survey of some relevant items. We notice that, whether we consider what is known however roughly of the record since the third century AD or just the last 320 years, for which we have thermometer readings, severe winters seem to have been commoner in the forties and the sixties to nineties of each century than in the other decades.[3] The bunching of cold winters in and around the sixties and nineties looks interesting and the extraordinary record of the years ending in ninety-five or within one year or so of that.[4] Although the sample of mild winters is confined to the extremest cases, with winter temperatures at the level normally expected in March, the sequence 1734, 1834 and 1935 accounts for nearly half of them (and there is reason to suppose that 1634 – as well as 1648 – should be added to the table). Bunching is less obvious in our treatment of the summers, but the concentrations around the years ending in 15–17, 27–31 and 56–63 include over 58 per cent of the very wet summers in England. Even so, some opposite experiences (summer 1718 had only 56 per cent of the twentieth-century average rainfall, 1818 37 per cent, 1921 57 per cent) occurred close to these dates. As noted earlier, the 'blocking' tendency which is conducive to cold winters in middle latitudes is liable, by a small shift of the controlling anticyclone from one year to another, to produce an extremely mild winter in close proximity to some of the coldest winters: thus the record is dangerous to use in forecasting.

The cyclical tendencies considered in these paragraphs – notably the 2.0- or 2.2-year, roughly 5½-year, 20–23-year, 50-year and 100-year periods and some others including longer periods, such as 200 years – appear in this or that series of observation data from most parts of the world. So they are presumably manifestations of certain evolutions in the wind and ocean circulations of world-wide range (whatever external influences, if any, may trigger them off). However, in many observation series most of them are of only modest amplitude (even if they are to be seen at all) and explain only a small proportion of the variability. This, and their failure as a guide to the specific year (particularly when opposite extremes can occur), has led most meteorologists to discount their possible value as a forecasting

Table 8 Coldest winters[a] of each century in western Europe, from the compilation by C. Easton (*Les hivers dans l'Europe occidentale*, Leyden, Brill, 1928)[b]

Century	3rd	4th	5th	6th	7th	8th	9th	10th	11th	12th	13th	14th	15th	16th	17th	18th	19th	20th
	296 (10)	359 (10)	401 (21)	545 (21)	695 (10)	760 (21)	822 (10)	913 (10)	1033 (21)	1125 (10)	1205 (10)	1303 (21)	1408 (4)	1511 (21)	1608 (4)	1709 (4)	1830 (4)	1917 (1.5)
			411 (10)	554 (10)		764 (10)	845 (21)	928 (21)	1044 (21)	1143 (21)	1210 (10)	1306 (10)	1423 (10)	1514 (10)	1621 (10)	1716 (21)	1838 (20)	1929 (1.7)
			432 (21)	566 (21)			856 (21)	940 (21)	1068 (21)	1150 (10)	1217 (21)	1316 (17)	1432 (17)	1544 (21)	1656 (17)	1740 (8)	1845 (20)	1940 (1.5)
			462 (10)	593 (21)			860 (10)	975 (21)	1074 (10)	1179 (21)	1219 (17)	1363 (21)	1435 (5)	1546 (21)	1658 (10)	1784 (11)	1871 (19)	1947 (1.1)
							874 (21)		1077 (10)		1225 (10)	1364 (10)	1443 (17)	1565 (4)	1667 (21)	1789 (10)	1880 (12)	1963 (−0.3)
							881 (21)				1236 (10)	1394 (21)	1458 (17)	1569 (17)	1672 (21)	1795 (11)	1891 (8)	1979 (14)
											1270 (10)	1399 (21)	1465 (21)	1571 (10)	1677 (21)	1799 (21)	1895 (16)	
													1481 (17)	1573 (17)	1684 (17)			
														1587 (21)	1695 (21)			
														1591 (21)				
														1595 (10)				

Notes:

[a] The winters are numbered according to the year in which the January falls. Much of the sparseness of entries in the earlier centuries of the table must of course be attributed to lack of information.

[b] Easton's ratings of the temperatures implied by the manuscript descriptions are given (in brackets) according to his index, which increases with the temperature level so that the mildest winters are rated over 80 on the scale. Since in the twentieth century the Easton index is lacking, the central England temperatures averaged over December, January and February are given instead.

tool. The position is a very clear warning against 'juggling with figures' without a known physical basis. If these phenomena are to be handled successfully, it is essential to identify their physical origins and acquire some capacity to monitor (and interpret correctly) the unfolding of the phases of the current cycle of the evolution. There is no denying that some of these cyclical elements in the course of climate development at some times and in some places acquire an importance (e.g. to farmers) not to be ignored. Some continuing effort to improve the forecasting position is surely demanded.

FORECASTS OF THE NATURAL CLIMATE

The most accepted forecast of the broad tendency of the natural climate so far issued rests partly on this sort of basis. In 1974 a specially appointed panel of the National Science Foundation in the United States produced the analysis of the position summarized here in table 11. The net outcome is a suggestion that the natural climate is at present cooling at an average rate of about 0.15 °C per decade. On this analysis, the net cooling rate would be expected to decline to zero by about the year 2015 and be followed by two or three decades of slight warming, the peak rate being about 0.08 °C per decade around AD 2030, and thereafter little change before a further decline a century later. The variations considered in this treatment over periods of about 100, 200 and 2000 years (or, as some writers would have it, 250 and 2500 years) are perhaps generally thought to be solar in origin, although some variations of the tidal forces may be involved. Traces

Table 9 Numbers of coldest winters in western Europe falling in the different decades of each century (averaged)

Decade	00–09	10–19	20–29	30–39	40–49	50–59	60–69	70–79	80–89	90–99
Percentage of the winters with Easton index under 22 listed in table 8	7.6	10.9	7.6	7.6	13.0	7.6	14.1	12.0	7.6	12.0
Average number of freezing months (mean temperature below 0 °C) in central England since 1659	0.3	0.7	0.7	0.7	1.3	0.7	1.2	1.5	1.5	2.7

Table 10 Mildest winters[a] in central England (mean temperature over December, January and February over 6 °C) and wettest summers in England and Wales (June, July, August rainfall totals over 140 per cent of the 1916–50 average)

	Winters over 6 °C since 1659				Summers with over 140 per cent of the average rainfall since 1697 (record 1697–1726 from Kew, London only)		
Century	17th	18th	19th	20th	18th	19th	20th
							1912 (186)
					1715 (194)	1817 (149)	1917 (138)
					1729 (169)	1828 (148)	1927 (155)
						1829 (168)	1931 (147)
		1734	1834	1935			
						1839 (148)	
						1848 (157)	1946 (138)
						1852 (166)	
					1758 (143)		1956 (145)
						1860 (169)	1958 (147)
					1763 (181)		
			1869		1768 (164)		
				1975	1775 (144)	1872 (140)	
						1879 (186)	
	1686					1882 (141)	
		1796			(1797 140)		

Note: a Numbered according to the year in which January falls.

of a periodic variation of about 200 years period-length have also been reported in the (global total) volcanic activity; if real, these could contribute to the climatic swings and perhaps help in the forecasting problem.

When we come to the longest periods of variation here mentioned, it is no longer necessary to limit any forecast entirely to a statistical statement. The amplitudes of the temperature changes are bigger, and they apparently rest on the well-understood changes in the Earth's orbital arrangements, which – like other astronomical variations – can be predicted with some precision. The associated changes in radiation available to heat the Earth at different seasons can be similarly calculated. Nevertheless, the effects in terms of temperature (and consequential changes in the wind circulation and rain and snow distribution) are amplified presumably by the reflectivity of an increasing area of snow and ice. And the changes are at times greatly sharpened in some way, perhaps by volcanic activity and the dust veils produced. Another suggestion – put forward by Professor A. T. Wilson of the University of Waikato, New Zealand – is that towards the end of each warm interglacial period the remainder of the great inland ice on Antarctica tends to become unstable. And it is thought that, aided by melting at the base, virtually the whole ice dome covering West Antarctica – the Pacific Ocean sector of the continent – where the bed-rock is far

Table 11 Estimated characteristics of the principal fluctuations of the natural climate[a]

Characteristic period length of the fluctuation	Estimated range of the prevailing temperature	Date of last warm peak	Present stage of the current cycle	Rates of temperature change in °C per decade	
				Fastest stage	Around the 1970s
(years)	(°C)	(years ago)			
100,000	8.0	10,000	Very high	0.0025	−0.0015
20,000	3.0	800	High	0.0045	−0.003
2000	2.0	1750	Middle	0.025	+0.024
200	0.5	75[b]	High	0.075	−0.053
100	0.5	35[c]	Middle	0.13	−0.121

Notes:

a   After figure given in *Report of the Ad Hoc Panel on the Present Interglacial*, Washington, DC, National Science Foundation, 1974, 22pp. The figures seem to relate to middle latitudes of the northern hemisphere.
b   Around 1900.
c   Around 1940.
    Net predicted outcome around the 1970s: cooling at c. 0.15 °C per decade.

below sea level is liable to surge out into the surrounding sea (and perhaps parts of the bigger East Antarctic dome as well). This should so broaden the floating pack-ice belt as to cool the entire southern hemisphere climate and ultimately cool the oceans all over the world. Hence, the magnitude and the timing of the more abrupt steps in the climatic progression are subject to influences, such as the incidence of volcanic dust in the atmosphere, and perhaps Antarctic ice surges, which may have to be treated as random. This means that details of timing of the progression towards another ice age can probably only be stated in some sort of statistical terms based on comparisons with the declining stages of previous interglacials.

The most thorough refinement of the calculation of the Earth's orbital variations, extending back (for each month of the year at thousand-year intervals) over the last million years and forward sixty thousand years into the future, has been carried out by Professor A. Berger of the Institute of Astronomy and Geophysics at the Catholic University of Louvain-la-Neuve, Belgium. Berger has also been able to demonstrate a convincing (statistically significant) association between these variations and past climatic effects on the scale of ice ages and interglacial periods. This was done by studying with Dr G. Kukla of the Lamont-Doherty Geological Observatory, New York, the significance of the climatic response to the radiation changes in different months of the year and at various latitudes.[5] Two different models were used for examining the association between the radiation balance variations and the climatic response. One model examined the incidence of warm and cold climate periods separately; the other amounted to a single integrated expression of the regime, including a persistence effect from the condition of the climate three thousand years before. The results from the two models agreed so well, and they explained such a high proportion (in one case 87 per cent) of the past climatic variation (as known from the oxygen isotope variations in cores from the bed of the deep oceans), that in Berger's words they 'authorize the prediction of the future natural climate'. The result is seen in the righthand portion of fig. 118. The key points are that:

1 Unless counter-effects due to man's impact on the climate supervene, the descent of prevailing temperatures towards the next ice age is due to steepen in the next millennia.
2 The first (modest) climax of colder, more or less glacial climate appears to be only around three thousand to seven thousand years from now.
3 Despite some recovery peaking about fifteen thousand years hence, return to climates as warm as today's is not expected until after a full glacial climax about sixty thousand years hence. According to one of the models used, 114,000 years of glacial climates lie ahead.

This outline of the development of climate over all these thousands of years ahead may be regarded as the most guaranteeable part of our capacity to foresee the future, because its basis is of similar nature to the succession of night and day and the yearly round of the seasons.

These findings explain the importance attached in some quarters to studies of the declining stages of the last warm interglacial period after its peak about 120,000–125,000 years ago. The detailed curve from northwest Greenland which we have shown in fig. 35 (p. 92) looks less drastic between 120,000 and 90,000 years ago than the past record as represented by fig. 118, though there are features 5000–8000 years after the peak of the last interglacial which could signify effects that would be alarming today. It behoves us to look into the evidence from other parts of the world, and this commonly produces a sharper feature around the time mentioned, more like that seen in fig. 118.

The most detailed record we have for the period of interest is the pollen record from a peat-bog at Grand Pile in the Vosges mountains in northeast France, examined by Dr Geneviève Woillard of Louvain-la-Neuve.[6] This is the longest continuous pollen record so far obtained anywhere in the world, going back right through the last ice age and the interglacial before it to 140,000 years ago (at which date an early post-glacial type of vegetation was present). Only for 11,000 years around 125,000 years ago does the pollen from the surrounding vegetation indicate a climate as warm as in the current post-glacial times up to the present. Then, as with the ending of the temperate stages of other interglacials examined elsewhere, a

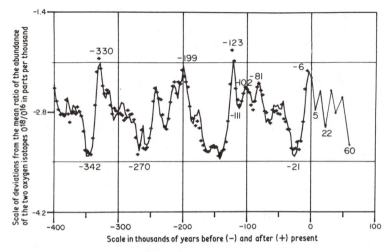

Fig. 118 Long-term variations of climate (effectively of world temperature) over the past 400,000 years and calculation for the next 60,000 years, based on the orbital variables. Crosses show how far the part of the curve relating to the past agrees with oxygen isotope measurements from deep ocean bed cores. The numbers give the dates in thousands of years before and after the present time of key points on the curve. (Reproduced by kind permission from Professor Berger's article in *Vistas in Astronomy* (1980) – see note 5, p. 406.)

series of abrupt changes of the vegetation character followed. The changes at the end of the warmest part of the interglacial, which it was possible at Grande Pile to follow in close detail,[7] indicate that the change-over from a temperate fir-spruce forest mixed with alder, box, hornbeam and oak to a typical boreal forest dominated by pine, birch and spruce, as in Scandinavia today, took only about 150 years. (The error margin on this estimate is not thought likely to exceed seventy-five years either way.) There seem to have been three quite abrupt stages in the transition, the first marked by decline of all the broad-leafed trees but most notably by a sharp decline of the fir (*Abies*) which had been present. The most drastic change took place 150 years later, when within 20 years the remaining fir and the broad-leafed trees virtually disappeared. It is suggested by those engaged in this work that the first sharp decline of fir may have been due to a very dry hot summer – the case of 1976 had a similar effect – but that most details of the transition point to a net cooling of the climate. (Perhaps the current European and North American elm disease is part of a similar picture.) And it is claimed that, although it has never been possible to indicate the detailed timing before, these changes are typical for the corresponding stage of all interglacial periods that have been examined. As Dr Woillard warned, we cannot exclude the possibility that we are already living today in the beginning stages of the corresponding vegetation changes and the fact may be masked from our perception by the extent of artificial management of forests.

Meteorologists engaged in climatic research have thought it best to treat the forecasting of the next ice age development in statistical terms. Whether referring to the orbital variations or to a random variation of volcanic activity as the supposed cause, they have rated the probability of ice age onset within the next hundred years as of the order of 1 or 2 per cent and so as a risk that may be ignored. However, a change in middle latitudes from an oak- to a birch- and pine-forest climate must come much sooner; the lessons of previous warm interglacial times suggest that that change should be expected to have several abrupt stages, and its beginning could quite well be imminent. A specific forecast, giving perhaps excessive weight to the not adequately explained 200 and 2000–2500 year cyclical tendencies, would probably expect the change to a birch- and pine-forest regime between about 3300 and 4300 AD, and one such pronouncement has in fact appeared in print from an internationally respected scientist working in the field. But, if we were to take the view that most of the recovery in the last hundred years or so from the Little Ice Age climate of recent centuries is attributable to man's output of carbon dioxide into the atmosphere, it may be that the unmodified natural climate would already be nearly at that change-over stage. This suggests that if we had a physical basis for making a statistical estimate, the probability of the required further (sharp?) cooling of the natural climate to a pine-forest regime occurring within the next 20–200 years could be around 10 per cent or higher.

POSSIBLE EFFECTS OF HUMAN ACTIVITY AND POLICY DECISIONS

Having thus completed our review of how far the present state of knowledge enables us to make useful statements about the current and future tendencies of the natural climate, we must consider how man's activities may modify the prospect. This is more difficult even than attempting to forecast the natural climate, because it involves forecasting what mankind will do. We will assume here that man will refrain from blowing himself up and making the planet uninhabitable with nuclear fall-out. Past history suggests some pessimism about the likelihood of blame for climatically induced difficulties and changes of land-use (particularly any change in food resources) being imputed to this or that class or nation. One may also expect fierce competition to grab any dwindling resources, quite apart from the immediate political contentions of the present day. Mankind is doubtless capable of continuing to inhabit the Earth and survive through the next ice age, and of doing so far better than our primitive forebears who survived the last one. And we should be at least equally able to adapt to a much warmer Earth with the productive crop belts and the deserts

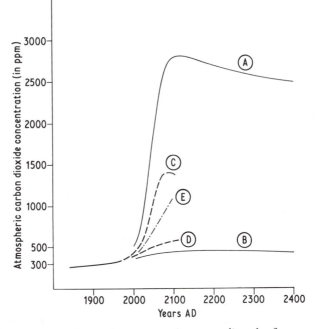

Fig. 119 Curves representing various expectations regarding the future proportion of carbon dioxide (CO_2) in the atmosphere. (Curves A and B were given by Siegenthaler and Oeschger in their article in *Experientia* (1980) – see footnote 11, curves C and D are from J. Williams's article in *Multidisciplinary Research Related to the Atmospheric Sciences* – see footnote 9; curve E was included by F. Niehaus in a research report, 'A non-linear eight-level tandem model to calculate the future CO_2 burden to the atmosphere', Laxenburg, International Institute for Applied Systems Analysis (IIASA), 1976.)

shifted poleward. But there is no warrant for the unfulfillable hopes – in extraordinarily many quarters today *the basic assumption* – that a constant, or even a steadily rising, standard of living can be achieved.

In what follows we shall also assume – as is common practice, though seldom specifically said – that the solution to the riddle of the lack of any demonstrable effect from the increase of carbon dioxide so far is that the expected warming has been offset by tendencies of the natural climate working in the opposite direction.[8] It is agreed by most of those actively engaged in climate modelling that the main threat from human activities to the stability of the existing natural climate regime, to which our present-day international order is adapted, is the warming – possibly inconveniently large – to be expected from the continuing increase of carbon dioxide. And this, as we saw in the last chapter, may be boosted as much as 50 per cent by other pollutants which have a similar action upon the radiation balance.

Later in the twenty-first century, at some point which will depend on how much power is generated from nuclear or other fuels, the output of artificially produced heat may itself begin to have effects on a global scale. This is certainly the major effect on climate to be expected from the large-scale use of nuclear energy. In some ways it is analogous to the unsolved problem of disposal of the nuclear waste itself (a problem to which there may be no solution on an Earth where no part of the crust can be guaranteed earthquakefree over the periods of continuing dangerous radioactivity). There may be climatic troubles arising from the emission of great quantities of heat whatever the locations chosen for electrical power generation. Some studies[9] have already been directed towards discovering what sort of effects on the world climate pattern might be expected to result from the

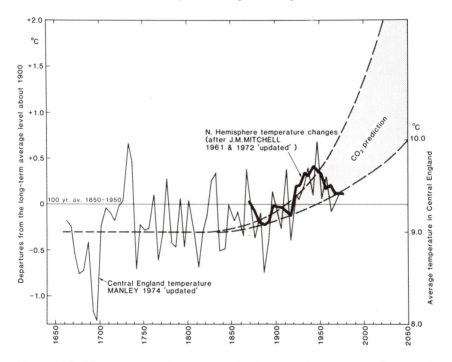

Fig. 120 World temperature: the past record and predicted future course as affected by increasing carbon dioxide in the atmosphere. The bold line shows the varying level of temperature averaged over the northern hemisphere in successive five-year periods since 1870 (as originally published by J. M. Mitchell, 'On the world-wide pattern of secular temperature change', *Changes of Climate – Proceedings of the Rome Symposium 1961*, Paris, UNESCO, 1963 and updated to 1978 by the present author). Successive five-year averages of temperature in central England, as given by Manley, and reproduced more fully in fig. 28a in this book (p. 80), are shown by the thin line in order to carry the record back to 1660, before CO_2 changes began to be significant. The various predictions of future temperature, based on CO_2 effect, are explained in the text.

disposal of the waste heat from nuclear power generation in various parts of the world's oceans.

Various alternative potential sources of energy open to man are under consideration: absorption of solar energy, either directly to be stored as heat in water systems or on a vastly larger scale in the world's deserts to produce electricity, tapping of thermal energy from the oceans, harnessing the power of the tides, growing fuel to burn in the shape of either wood or oil producing crops, use of wind power, and so on. Not all of these are free from awkward side-effects. According to one estimate, the heliostat arrays required for conversion of solar energy to satisfy the projected demand for the expected population of the Earth – assumed to have doubled – before the middle of the next century would need to cover nine million square kilometres, or about 6 per cent of the total land area of the globe, if only this source of energy were used. The tapping of potential energy from the oceans would change their temperature distribution, affecting the wind and ocean circulation and hence the climate, and causing them to release carbon dioxide to the atmosphere. Perhaps only the large-scale use of power from the winds and the tides, for which the technology still needs to be mastered, would be free of major objections. Nevertheless, all these alternative sources of energy are probably preferable to either fossil fuel or nuclear energy. Strategies for their most effective use should certainly be explored vigorously and put into practice wherever appropriate. In the case of most of these types of resource, the proportion of their world total availability which could in fact be used may be rather small, but in some cases the world totals are very large. And with each type local circumstances govern strongly how much advantage is to be obtained, most obviously so with hydroelectricity or harnessing the power of wind and tides. Overall, their potential contribution to our energy needs is by no means negligible.[10] Whatever values we accept for the effects on world climate of the exploitation of various fossil fuels and other sources of energy, the outcome to be expected will depend to an important extent on decisions as to how far to develop each.

Fig. 119 indicates the range of curves currently being put forward for the probable course of the carbon dioxide proportion in the atmosphere over the next centuries.[11] Curve A is based on the proposition that all the Earth's readily exploitable fossil fuel will be burnt within the next two hundred years. As a result the CO_2 proportion in the atmosphere would rise to more than eleven times the natural level that existed before the nineteenth-century acceleration of the Industrial Revolution. At the other extreme, curve B presents the least change from the pre-existing natural conditions that seems in any way conceivable, the proposition being that power production can be so managed that the CO_2 level shall not come to exceed one and a half times the nineteenth-century level. Curves C and D are taken from the work by Jill Williams and colleagues at the

International Institute for Applied Systems Analysis at Schloss Laxenburg, near Vienna, already cited above. Curve C represents the outlook if artificial energy production continues to grow, but at a more modest rate than in curve A, and nuclear power is not used; curve D represents an 'optimistic' energy strategy which keeps down the consumption of oil and coal. Curve E illustrates the results of another strategy of decisions studied by F. Niehaus of the International Atomic Energy Agency, allowing more use of coal than curves B or D. These curves certainly identify the increase of carbon dioxide in the atmosphere as one of the most alarming changes of the natural environment due to man and require its potential impact on the climate to be taken with the utmost seriousness.

Discussion of which of the curves in fig. 119 is likely to prove the most realistic has included the – possibly too optimistic – conception that decisions affecting the bulk of the world's energy production may come to be aimed at reducing the dangers inherent in such a drastic modification of nature. At the other extreme, questions arise of how rapidly the world's oil and coal reserves could in fact be exploited[12] as the portions easiest of access get used up, as well as the problems of how rapidly a switch from one type of fuel and fuel policy to another could be implemented. The question whether the world's population will really double again also comes into it.[13] Birth rates are now falling in almost every country in the world (e.g. in China by 41 per cent between 1970 and 1975); and although improving availability of medical services in the Third World may be expected to lower the death rates from disease, rising death rates were in fact reported in the 1970s in a number of poorer countries – particularly in the Indian sub-continent – due to starvation following harvest failures and other natural disasters. The energy growth situation has already begun to be affected by dawning appreciation of the squandering of limited resources, leading to price rises, and the first beginnings of the idea of strategies to reduce energy demand. The latter ultimately entails rethinking even in architecture and clothing habits and will undoubtedly require a few decades to evolve. At present consumption rates there will be no more fossil fuels after about AD 2200, though the enchanced CO_2 in the atmosphere will be with our descendants for long after that: political decisions to reduce consumption and so spin the process out can, of course, be expected long before exhaustion of the fossil fuels is reached. Thus the first part of the projection into the future in fig. 119 is at least on fairly firm ground in expecting less change of present growth rates than may later on be achieved.

What is clear is that the possible warming of world climates due to carbon dioxide chiefly concerns the next few centuries as the CO_2 builds up and, on the present showing of human behaviour, reaches a peak with the exhaustion of fossil fuels sometime between AD 2100 and 2600. Thereafter the atmospheric CO_2 proportion is expected to decline slowly over many hundreds of years. Theoretical modelling suggests that in AD 3500 the proportion will still be 70 per cent above the pre-industrial level.[14]

Fig. 120 shows the range of various publicized forecasts of world temperature change based on the carbon dioxide warming effect resulting from different assumed fuel policies. To put the prospect in perspective against the net outcome when natural climatic variations are also included, what is known in outline of the history of world temperature from the seventeenth century to date is indicated by the most widely accepted variations of five-year mean temperature level in central England and over the northern hemisphere (the latter available only since 1870). This makes it obvious that the CO_2 climate theory is not doing very well as the sole explanation of the changes and that other causes of climatic variation are also important. The common decision to treat the natural climatic variations as unforecastable 'noise' (i.e. random events) is plainly not satisfactory. Research effort must be continued, aimed at improving our capacity to foresee the variations of the natural climate. Nevertheless, widely publicized expert opinion from the leading theoretical modelling laboratories in climatic research expects the increase of CO_2 to have raised world temperature by between 1 and 2 °C by the period AD 2050–2100 in the case of the most restrictive strategies on fossil fuel consumption and by from 4 to 9 °C (refer to fig. 114 and text on pp. 334–6) on the basis of what are thought to be likelier developments. These figures may even be increased by the contributions from other pollutants.

Changes of this magnitude imply bringing world temperature to a level which has not occurred in the last two million years, since the Tertiary geological period. The polar sea ice would be expected to melt and disappear, but the great masses of inland ice covering Greenland and Antarctica would take a long time to go. This is just as well, since melting of the Greenland ice-cap alone would raise world sea level by 6 or 7 m, while the addition of water from the Antarctic inland ice would ultimately – after a delay of some centuries as melting proceeded – raise the oceans by between 50 and 100 m, submerging lowland plains in every continent and drowning nearly all the world's great cities. Clearly the proposed change of world temperature level would also shift all the vegetation and crop belts poleward by many degrees of latitude, and this would take more or less immediate effect, dislocating the existing economies of nations.

Research aimed at studying the geographical distribution of expected climatic effects at each stage of the progression towards this artificial world of the twenty-first century is therefore seen as urgent. Many exploratory studies have already been done. Some use theoretical models of the climate system. Others proceed by studying the climate patterns of various warm periods in the past.[15] These 'scenarios', as they are called, start with the regime in the earlier part of the present century and/or the patterns of individual warm years, a level commonly expected (on the basis discussed) to return within a decade or two. The next stage is likened to the medieval warm epoch, with average temperature over much of the northern hemisphere around 1 °C above present. Thereafter, stages equivalent to the warmest post-glacial times, to the warmest part of the last interglacial periods, and to the late Tertiary geological epoch, are supposedly reached in succession, within at most six hundred years: the conditions of those times may therefore be relevant studies. The general tenor of the conclusions may be summarized as follows:

1 The temperature rise over the Arctic regions generally is expected to be several times as great as the world average. For the first doubling of the CO_2 level a warming by 8–10 °C near latitude 80 °N is suggested.

2 With so big a change of temperature gradients, and of their position, and of that of such ice surface boundaries as would still be present, the patterns (and intensity) of the world's wind and ocean circulations would be shifted and changed.

3 The changes of the wind and ocean circulations would alter the distribution and amounts of rain and snowfall.

The expected temperature changes have been widely announced in an ever-increasing volume of meteorological literature, notably at the World Climate Conference organized by the World Meteorological Organization in February 1979 and at other scientific conferences before and since. The alarm that has been raised over the dislocation which such great changes would be liable to cause is entirely proper, even though the actual net outcome when the natural climatic variations also have their effect (and even our view of the CO_2 effect itself when the theoretical modelling has been improved) is by no means certain. There is not very much time to acquire the necessary further knowledge to resolve this question and only too little time to adapt national and international habits and policy in the use of energy to minimize the dangers ahead – especially since some radical changes may be called for. The dilemma is a very difficult one. Despite the uncertain reputation of even short-term weather forecasts, and the uncertainties involved in this rather different problem, the potential for disorganization and disaster is so great that the meteorologists' warnings must be taken as a very serious matter.

Nor is it only the temperature changes that look serious. The changes of precipitation, and of the balance between down-put and evaporation, would also be important.[16] Although precipitation would be expected to increase at most latitudes because of the extra water vapour picked up from warmer seas, it is only in high latitudes and the monsoon regions of Asia that a general increase above the increased rate of evaporation would be expected. Over most of the northern hemisphere's land-masses conditions could turn out significantly drier than today's. And, as the warming should move the belts of cyclonic activity polewards, the Mediterranean winter rains would be expected to fail. Indeed, at that latitude (35–40 °N) total rainfall would probably decrease; with more evaporation there, as elsewhere, the aridity of the desert would presumably advance over the region.

The patterns that have to be considered if and when the generation of nuclear power, and the waste heat from this and from the cities of the future besides, begin to affect the climate on a global scale differ from those arising in the CO_2 problem. The artificial generation of heat is now, and presumably always will be, concentrated in limited areas. Globally, the heat artificially generated today is only about one ten-thousandth part of the energy absorbed at the Earth's surface from the sun. And it seems unlikely to rise above one half of one per cent within the next century or two, possibly implying a rise of world temperature by about 1 °C. But already in some great urban and industrial areas the artificial production of heat is more than a thousand times the world average, and in certain cases exceeds locally the average heat absorbed directly from the sun. For the possible generation of much larger amounts of heat from nuclear power production to meet future demand, the impact of heat input concentrated in various specially chosen 'energy park regions' in the ocean has been investigated. This has been done by theoretical work using a 'general circulation model', in a collaboration between the United Kingdom Meteorological Office and the International Institute for Applied Systems Analysis (IIASA).

The possible energy parks considered were (a) just southwest of the British Isles, (b) in the region of the Cape Verde Islands near 17 °N in the eastern Atlantic, and (c) east of Japan, as well as various combinations of these. In view of inadequacies of present modelling capacity to explore interactions between atmosphere and ocean, huge inputs of heat of prob- ably unrealistic magnitude – to supply a world population five times as big as at present and with a per capita energy consumption ten times the present average – were considered in the theoretical modelling experiments. This was done in order to make sure of getting an identifiable response, standing out above the random variations. It was found that there were effects on the large-scale atmospheric circulation which varied according to where the heat was put in, and how much heat, and what proportion went into the ocean and what was allowed to escape into the atmosphere. But for the smaller heat inputs that might in fact be realized it is suggested that there might be no significant effect on the climate system.[17] This aspect of nuclear energy seems therefore to entail much less difficulty or danger than the carbon dioxide produced from burning traditional fuels (but, of course, this has nothing to do with the problem of radioactivity of the nuclear waste).

CONCLUDING SUMMARY

To conclude this chapter, we must return once again to an attempt to see the matter whole: the possibility of global warming, even drastic warming with dislocation of other elements of the climate pattern as a consequence, has to be balanced against the possibility of cooling, even drastic cooling, as the natural climate develops over the same period. Neither side of the balance is yet adequately known and understood. The effect of CO_2 increase itself, although clear in the laboratory and in theory, is not proven as applicable in the global environment context where feedback (i.e. consequential) effects operating through the oceans and water vapour in the atmosphere may greatly alter the outcome. Nor is the net effect of the global increase of turbidity (particulate matter) in the atmosphere as yet certain, since (a) the sizes and distribution of the suspended particles in the world atmosphere may make the difference between a net warming and a net cooling effect, and (b) in this matter also complications may arise through some of the substances facilitating the condensation of water vapour to form clouds.[18] All forecasts must in any case be subject to the proviso that volcanic activity does not produce so much aerosol in the atmosphere as to impose cooling, as may indeed have been happening since 1950. So there are many reasons for scepticism about the confident forecasts based on present theoretical models, even though their warnings of what may happen must be taken so seriously as to guide policy decisions which have to be taken very early if the dangers are to be averted.

In many of his papers published in the last decade or more, Dr J. Murray Mitchell of the United States weather service, who was universally respected as one of the most cautious, as well as one of the most widely knowledgeable, research workers in this field of science, indicated that the global climate regime which we know and take for granted may be subject at this time to influences tending to push it far off course in either direction.[19] On the one hand, the Little Ice Age of recent centuries, which must be seen as just the latest in a series of 'neoglacial events', looks appreciably shorter than previous events of the series, and may not be over but only interrupted or disguised in this century, perhaps by the side-effects of man's activity. On the other hand, the effects of man's activity are presumably becoming much stronger than before, but may not all tend in the same direction.

We may mention at this point that suggestions have already been made that man may be obliged in the future either to seek to avert, or slow down, the onset of a new ice age by deliberately increasing the CO_2 in the atmosphere or, on the other hand, to offset the effects of his own excessive heat production by using aircraft to spread dust in the stratosphere in order to screen off the sunshine. This latter suggestion was made by Professor M. I. Budyko in the Soviet Union as long ago as 1960.

The analysis of our present climatic situation certainly reveals basic reasons for instability of the existing climatic regime. And in a heavily populated world where it is difficult to produce enough food, climatic instability, fluctuations and change in any direction threaten all the perils of disappointed hopes leading to conflict and, in some large areas, carry directly the threat of mass starvation. It is vital therefore to pursue all lines of research which are likely to bring better understanding and some capacity to forecast the tendencies of both the natural climate and human impact upon it.

The instability already apparent in the climatic situation over the past twenty years has led to a position, bewildering to the public and its leaders faced with decisions affected by climate, in which the leaders of meteorological and climatic research have given conflicting advice about probable future trends. To some extent the confusion has been due to a failure to distinguish between tendencies operating on different time-scales. There is no necessary conflict in diagnoses which identify:

1 a cooling, especially in the northern hemisphere, since 1950 and which might be expected to continue (with shorter-term fluctuations super-posed) for some decades further;
2 warming attributable to the increase of CO_2 – and other pollutants with similar effects – in the atmosphere due to human activities: this effect to become stronger over the next century or two and reach a peak around AD 2100 or some time after;
3 the progression towards the next ice age, with an expectation of some abrupt cooling phases such as to change the vegetation character in Europe and temperate North America within one to two thousand years.

Our present uncertainties about the overlaps between these tendencies do, however, frustrate forecasting attempts. They also make it imperative to learn to identify as early as possible the signals of change when they come.

18

WHAT CAN WE DO
ABOUT IT?

PERCEPTION AND RESPONSE

So, what can we do if the climate does fluctuate and change? And what can we learn from the past? To answer these questions briefly: the main requirement is realism about our situation. We must seek to know and understand enough about the behaviour of climate and its effects upon our environment and resources to cast off illusions and false expectations. And to be realistic also demands humility about what man can do in the face of climatic shifts, even today, other than adapt his ways. It may well be that mankind has, and perhaps always has had, an exaggerated impression of his power to alter the climate, intentionally or otherwise, for good or ill – except on a quite local scale. Numerous global budget calculations, covering many aspects of the atmospheric system, have been aimed in recent years at producing what are hoped to be realistic numerical estimates of the effects of human activity. Yet our theoretical modelling is still (and may continue to be) inadequate to reveal the full power and means at nature's disposal to buffer the climate against such interference as man produces. Nor can we be sure that the natural causes of climatic change will not overmaster the side-effects of even our enormously increasing energy production.

It is in any case among the remaining mysteries of the planet Earth which is our home – mysteries in the sense that we all find difficulty in fully fathoming and adjusting to them – that the scenery surrounding our lives is always changing. Sometimes the changes are slow and hard to notice. Sometimes they are fast. Some of the changes are due to man. Others are due to the climate and to slowly evolving successions in the natural vegetation and soils. The rapid changes sometimes shock us and confront us with difficulties and disasters before which we still feel helpless, although modern technology has certainly enabled us to do far more than ever before in rushing aid to the scene of immediate calamity and in many cases to reduce the toll of suffering and death by short-term forewarnings issued a little before the event. It is doubtful, however, whether we are any more capable than

369

our forefathers of coping with long-term change – especially if it happens quickly and affects areas inhabited by millions. Perhaps the difficulty is greatest when events come in the shape of occasional, irregularly spaced disasters – as, for instance, by drought or flood or sea storm – which give the economy time between whiles to resume its previous pattern and the population to reoccupy threatened areas. The same human psychology that builds residential areas on active geological faults, as soon as one or two generations have lived in the area since the last great earthquake disaster, equally dulls the response of human planning to a climatic threat.

How to respond is rendered still more difficult and doubtful by uncertainties in the scientific predictions. Yet some heed must be taken of the magnitude of the difficulties that mistaken development planning may pile up for the future. If we decide to concentrate on what we know we have achieved and can deliver, then we must note the triumphs of our times in the reductions of loss of life attributable to coastal defences against sea flood, to hurricane and gale warnings, to highway management in frost and snow, and various forms of protection of aircraft and shipping against ice accretion, and to the interventions of medical science where water supply and hygiene are disrupted by droughts and flooding. The successes of clean air control and legislation in industrial areas also deserve a place in the list. The effectiveness of shelter belts and irrigation in agriculture are also beyond doubt; but with irrigation, as with coastal defences, there are obvious limits to the natural situations that can be coped with – and, perhaps, some less obvious restrictions, if unwanted or disastrous side-effects are to be avoided, as in the case of the Siberian rivers scheme discussed in earlier chapters. Such questions demand the widest possible knowledge, understanding and caution.

The more grandiose schemes for 'altering the face of nature' – plans such as diversion of the Gulf Stream or the Siberian rivers or abolition of the Arctic ice – should be approached not only with caution but with scepticism. As long as our capacity for forecasting the weather is limited and sometimes marred by gross errors affecting large areas, our ability to foresee the consequences of any deliberate manipulation of the climate system that might be attempted must be subject to the same danger. Our world economy is geared to the existing climate, and any major change – even one aimed at increasing the overall cultivable area – would entail grave dislocation, quite apart from the likely short-term vagaries of weather and failures of forecast that would have to be expected, let alone the possibility of long-term deviations from the result planned. These could obviously affect some areas and even whole countries adversely. And it seems certain that fully international agreement to accept the hazards involved could never be obtained.

There is already a demand – and a need – for international agreements on a modest and surely attainable scale to control and avoid activities which might have, or in some cases are known to have, adverse effects on environment and climate. Cases in point range from the emission of sulphurous gases from chimneys to the unlimited use of aerosol sprays and nitrogen fertilizers. And it is clear that national and international policy with regard to future fuel development, which involves great unsolved problems ranging from the effect of increasing carbon dioxide and waste heat output to the disposal of nuclear residues, demands a continuing search for knowledge and, in the meantime, caution and flexibility.

Turning once more to matters within our present capabilities, Walter Orr Roberts and Henry Lansford[1] have made the valid point that

> in the absence of forecasts or outlooks precise enough to satisfy a meteorologist, the farmer is likely to make some hard and important decisions on the basis of intelligent guesses . . . about future weather and climate that fall far short of the atmospheric scientist's rigorous standards of acceptability. For example, every dryland farmer in the high plains [of the United States] probably knows by now that severe drought has struck the region about every 20 to 22 years for the last 160 years. . . . Even conjectures can be of some use in making climate-related decisions in the real world, provided they are not completely wild.

APPROACHES TO CLIMATE FORECASTING AND THEIR USEFULNESS

Another point which should already affect decisions today comes from studies of the aftermath of the world-wide stresses of the early–mid-1970s by Michael Glantz.[2] Officials of the governments and others concerned in the countries in the Sahel were asked what they would have done if a reliable climate forecast had been available before the worst phase of the Sahel drought around 1972–3. A common answer was that the cattle-carrying capacity of the rangelands should have been assessed and cattle-herders required to keep down the size of their herds to prevent overgrazing. A policy of culling the herds to improve them by keeping only the best beasts could have been enjoined upon the owners at the same time.

There has also been already a more general pay-off from the increased activity in climate research over the last ten to twenty years in an awareness – however little acted upon so far – that climate is not as constant as it appeared to be in the most benign decades of the present century. Even the most extreme and divergent forecasts of future climate, put forward in this period prematurely by scientists who were expert in only this or that part of the enormously wide fields of relevant knowledge, may have done some good by undermining complacency and alerting the world

371

community to what can happen. Nevertheless, this is a situation which cannot be allowed to continue. The daunted decision-makers, who must have been confused and disillusioned about the value of 'experts', should perhaps see it as a stage that had to be gone through after the long neglect of investigation of the history and development of climate. The need is for research to improve knowledge and, particularly, to understand the limitations of each kind of approach to forecasting. And for the planners the lesson already is to allow somewhat wider margins for the possibility of climatic change.

There are two main problems, to extend our knowledge of (a) the behaviour of the natural climate, and (b) the effects of the intrusion of human activities and pollutants, both those now occurring and those implied as the situation is planned to develop, or may develop, in the future.

There are also two lines of advance needed:

1 To reconstruct an ever fuller and more extensive past record of the global climate. This is the essential observation base of climatology, without which some of the processes and phenomena we have to deal with in forecasting may remain undiscovered and our theoretical concepts and models may remain incomplete and untested.
2 To achieve fuller understanding of the controls and mechanisms of climatic behaviour, and their range of variation, by physical and mathematical climate theory.

The theoretical models may be of various kinds. Their range includes physical models such as experiments with fluids in rotating dishpans, simulating in a simplified way the flow of the atmosphere when the heating is varied; and it extends through mathematical models from quite simple ones, which express only the mean state of the atmosphere and oceans and explore the balance of energy received and heat transported by the winds and ocean currents, to the most elaborate models of the general wind circulation (as used in numerical daily weather forecasting). Either type of model may also be used to consider the budgets of heat, momentum and water vapour transfer. The simplest models may be designed to consider only the situation averaged around the world for each latitude. The most elaborate models offer some insight into regional patterns and make it possible to consider the effects of mountain or hill barriers and other local disturbances upon the winds, all necessarily simulated in simplified form.

All models need calibrating and testing by comparisons with results observed in the real world. The climatic situations reconstructed from the past, provided the job has been reliably done, are needed also by the theoretical modeller. General circulation models are conventionally 'run' – i.e. integration of the equations is continued, as if for forecasting – for periods of eighty days to at most (on grounds of cost) a year or two. The maps produced for this period are then used to provide a statistical picture of the 'climate' – for example, maps of the frequency of anticyclones and depressions, of rainfall and different wind directions – of the period covered. This theoretical climate can be compared with the observed climate of any epoch which it was intended to simulate. The effects of altering the ground conditions and heating pattern, or of putting more or less water vapour and other substances into the air, and other changes, can be similarly explored. By repeating runs of the model, from slightly different starting conditions specified for day zero, an idea of the stability of the statistics derived from the runs to represent a given climatic regime can be derived – or, to put this another way, one is enabled to see how big a random element there is in the result.

So far the models do not incorporate fully the exchanges with the ocean, and effects within the ocean, and how these react upon the atmosphere. A more serious uncertainty affects the theoretical results. This is because the complexity of the climate system, the more fully and elegantly it is represented, provides opportunities at so many points to adjust this or that component and obtain at least some sort of match with the climatic regime to be explained. This is a matter of giving more weight to this or that and making compensatory adjustments elsewhere. As Schneider put it in the case described in chapter 16 (pp. 339–40), one can match anything to any-thing in this way, but the question of whether the set-up then expressed by the equations corresponds to the mechanisms of the real climate regime remains open. (There may even be a variety of hypothetical (modelled) set-ups by which the characteristics of the actual regime could be reproduced.)

Modelling in a realm as complex, and with as many interactive variables, as the climatic system is primarily an aid to thought and to conceiving the patterns of the real world rather than an automatic provider of accurate or reliable answers. It can suggest probable linkages of cause and effect in the climate system and often the probable order of magnitude of some of the effects. And it is obviously the main way of exploring the possible consequences of human activities which introduce new elements or changed conditions into the climate system, whether by pollutants in the air or extra heat or alterations of the face of the Earth – as in the creation of artificial lakes or clearing of the tropical rain-forests (and proposals like removal of the Arctic sea ice).

Two quite different types of forecast, whether for a season ahead or of the climate in the longer term, can also be attempted. One is specific, stating that the prevailing weather will be warmer, or colder, or perhaps even specifying a temperature range (and correspondingly for rainfall). The forecasts of carbon dioxide warming, and of the next ice age some thousands of years ahead, are in this category. (This seems also to be the style preferred for all occasions by amateurs and quacks.) The other type of forecast takes the form of a statistical statement of the probability of this or that range of conditions. The modelling approach can be used to produce forecasts in either form. Forecasts based on analysis of the past record of climate can logically only be made in the statistical type of statement of the probability of certain outcomes following the known initial conditions.

It is arguable that the statistical form of statement is most helpful to the recipient, especially when great risks (economic risks or human lives and sufferings) depend upon the decisions he has to make. But the statement of probabilities only has meaning in relation to the range of thinking, of items known to be relevant, and of the reference material surveyed, in making the forecast. Those items which constitute the basis of the probability statement must be made clear to the recipient: for without them the alleged probability is no more than a guess, which the recipient cannot evaluate, and which may be quite unrelated to the realities of the situation. It seems highly desirable that forecasts based on insight gained from modelling studies should also be produced as statistical statements of probabilities which are similarly made understandable – i.e. assessable – to the recipient.

In the present state of knowledge the basically empirical approach to forecasting resting on the past record of climate will commend itself to most recipients. The probabilities of various future developments of the natural climate can be clearly and explicitly assessed on this basis by consideration of suitable numbers of previous occurrences of an apparently similar climatic situation and what followed in those cases. The contribution of theoretical modelling can best come in by illuminating whether and in what ways the previous occurrences were really similar to the existing situation. In the case of any new climatic trend or developments which may result from man's activities, however, theoretical modelling may be the only way of predicting the outcome and its probable order of magnitude, apart from such additional information as may be gleaned from study of seemingly relevant 'scenarios' chosen from climates which did occur in the past. In connection with the possibility of drastic warming resulting from the prospective further increase of carbon dioxide, comparative studies have been made of the world climate patterns of the warmest of past interglacials and the still warmer climates of the Tertiary geological period (more than two million years ago), as a guide to the patterns of warmth and rainfall which might arise and dislocate the economy of the world as we know it.

INTERNATIONAL EFFORTS NEEDED TO IMPROVE KNOWLEDGE

While our knowledge of climate development processes remains far from complete, the immediate needs are that research continue and that a running watch be kept on the state of world climate. In the latter connection, identification of a few items (e.g. the Arctic temperatures, and extent of sea ice, or perhaps the occurrence of the westerly winds near the British Isles, as discussed in chapter 14) which could serve as a quickly responding, economical index of world climate may be of value. But in relation to the vulnerability of our economy and international arrangements to climatic changes, the bald assertion in a recent British government report[3] that no big natural changes are likely soon has no value at all. And the corresponding assertion of one leading scientist that it is a waste of money now to support any research into climatic change other than changes likely to be produced by man's impact is equally without foundation and likely to lead to a vital element of the problem being overlooked.

In fact, the increasing concern in recent years over climatic change led to the inauguration in 1979 of an international programme of climatic research under the World Meteorological Organization, known as the World Climate Programme, and national programmes in several countries (e.g. in the United States and in the European Community). It seems unfortunate that, according to report,[4] the Climate Impacts Assessment side proposed for the World Climate Programme was left to be taken care of by the United Nations Environment Programme. Studies of how to reduce the vulnerability of our food supply systems to climatic variability must be one of the most vital practical problems confronting mankind, affecting the whole economy. It has been pointed out[5] also that some schemes, successful or otherwise, to modify the climate or the environment of certain areas – as, for example, diversion of the Siberian rivers for irrigation in central Asia, with possibly serious repercussions through diminishing the Arctic sea ice, or the seeding of clouds in many areas to extract rain from them – may have damaging effects on the climate and economic interests of other countries beyond the borders of the region concerned.

> As populations and demands on resources continue to increase, governments will be under mounting domestic pressure to put national requirements first. . . . If the world is not to relapse into anarchy, with states warring over use and abuse of natural resources, some sort of international agreement . . . a self-denying ordinance and commitment to consult will be essential.

The author of these sentences, C. Tickell, formerly of the Office of the President of the European Community, goes on to suggest that an international organization – a World Climate Organization, perhaps – will be needed to monitor and take appropriate action on such matters.

There is no doubt that all the problems of adaptation to climatic fluctuations and change are made harder by the high level and continuing increase of world population. In an interesting article in the *Yale Review* (vol. 64, pp. 357–69, 1975) on 'An Ecologist's View of History', Paul Colinvaux has argued that all poverty (on a mass scale) in every age is caused by the continued growth of population and that behind all the great aggressive conquests of history will be found a rising population who have seen for a while hopes of a rising standard of life. He believes that the 'brooding about the possibilities of nuclear war' between the great continental states which are the superpowers of today may be misplaced, that the real threat comes from island peoples or other nations with teeming populations living in a confined space and with an aggressive nature evidenced in their history. He sees hope for the future in the likelihood that technology can continue to find raw materials and even energy for manufacturing almost without limit. But he remarks that, even so, we are clearly going to force people to live in uncongenial ways, with rationing of space and few outlets for adventure: 'for a time at least we are going to deny them the right to aggressive war'. Parts of this case may be plainly overstated, but its main themes are assuredly partly true. What the statement does not include is that the pressure towards such an outlook for mankind will be further intensified by any reduction of resources and living space such as climatic fluctuations and change are liable to bring at least temporarily and in some cases for the longer term.

THE LESSONS OF HISTORY

In the preparations for the World Climate Conference in 1979 Professor B. Bolin of the University of Stockholm suggested the following points as common ground, namely that:

1   the variability of climate, as experienced during the last few centuries, has had a marked effect on man's activities and well-being;
2   the variability of the natural climate will continue during the next hundred years . . . and that there is some possibility that a more extreme and probably cooler climate, as during the seventeenth to nineteeth centuries, may develop;
3   man is already influencing climate on a local scale to an extent which is significant when compared with the natural variability of climate;
4   man's activities may come during the next hundred years to induce global climatic changes as great as, or even significantly larger than, the climatic changes experienced in the last few centuries;
5   the effects of man's activities will probably be to produce a warmer climate and significant changes of the world rainfall pattern;
6   for mankind to adapt better in the future to the variability of climate, even to bigger changes than those experienced in the recent historical past, will demand more effective use of climatic data and continued research effort to improve our capability of forecasting.

No doubt some will regard these anxieties about climate as the least of our worries in a world troubled by sharply rising energy costs and concentrations of wealth in oil-producing states, by increasing violence every-where and the threat of nuclear war. After all, the climate does not seem to have changed, many will say. And anyway we have always had to cope with climatic extremes from time to time. But this is to overlook the in-built trap in the nature of the climate problem, that the wide range of year-to-year variations will always make it hard to recognize any new trend until this is already strongly established. It is true that in some recent years India has been able to spare some food for export. But the increasing population pressure on food resources increases vulnerability to even one bad year. In the midst of the better-known symptoms of tension, it may be overlooked that already in the 1970s, even in the United States, with increased acreage sown, yields of grain per acre dropped sharply and that monsoon failures in India and Bangladesh twice in the early and mid-1970s seem to have caused over a million deaths.[6] And should we see in the tragedy of the emigration of the 'boat people' (with countless drownings) from southeast Asia in the late 1970s a (possibly not new) twist in the problem of food shortages, caused by weather as well as the ravages of war, whereby political prejudices choose the victims, the classes of the popula-tion on whom the main brunt falls.

This is close to one of the lessons of history, that in troubled times and periods of scarcity scapegoats are usually found – and often illogically chosen – to take the blame and become the targets of vengeful acts, of riots and war, or else that it is merely the weakest sections of society – the poor, the old and the children – who are made to suffer the worst consequences.

THE NEED FOR FLEXIBILITY, DIVERSIFICATION AND MARGINS OF SAFETY IN AGRICULTURE AND ENERGY POLICY

We live in a precarious world overshadowed by threats of food and energy shortages as well as nuclear holocaust. In the most productive countries it now takes two calories of fuel on the farms to produce one calorie of food, and when transport and storage costs are added the ratio may rise to ten or twelve calories of fuel for one of food. So there is no mistaking the fact that the destabilizing effects of both short-term climatic fluctu-ations and any long-term change are deeply involved with the more obvious threats named. We must be prepared to develop our technology in ways that decrease, not increase, the risks. This may mean avoiding that degree of rationalization in agriculture, aimed at maximum production, which would concentrate too much of the production of one crop in a few areas or which would concentrate the production from one area too much on one crop. In this respect the very practices which are used today to maximize food production increase the risks of various kinds of disasters.

Apart from the direct effect of adverse weather beyond the supposed extreme occurring in one, two or more years in succession,[7] the possibilities of disaster from plant parasites, which may be encouraged by a certain type of weather, are more serious where monoculture is practised. An object lesson in this was provided by a newly developed strain of wheat in the Netherlands in the 1950s. It was a product of scientific plant breeding, which had been carefully tested and found to be resistant to all the then known forms of yellow rust disease. In 1955, three years after its introduction, over 80 per cent of the wheat sown in Holland was of this variety, Heines VII. A new variety of the yellow rust appeared and, as a result of its attack, over two-thirds of all the winter wheat sown in this country for the 1956 harvest was destroyed.[8] The risks of similarly wholesale crop failures – possibly over much wider regions – that would accompany a global warming of the magnitude that some current scientific work suggests could come from the increase of carbon dioxide through increased use of coal, oil and other non-nuclear fuels are no less than the dangers in nuclear waste. They may even include the same kind of risks, if and when melting of the world's glaciers raises the sea level to the point where nuclear power stations on the often-favoured coastal sites become flooded.

This chapter began with the need for realism and humility about our situation. We must now stress the needs for diversification, flexibility, and margins of safety in our energy, agriculture, food and population policies. As Schneider has written[9]

> coal is environmentally damaging, air polluting and may ... alter the global climate. Nuclear power advocates have not fully solved the radioactive waste disposal problem ... they have yet to make a reliable assessment of the problem of serious accidents which could release lethal quantities of radioactivity into populated areas. Wind, hydro and solar power are promising renewable energy alternatives, but each has difficulties. ... The only safe projection for energy system planning is ... that surprises are sure to come. This requires flexibility. One step in that direction is the parallel development of many energy alternatives. ... A massive and disproportionate investment in one energy resource ... is likely to create an inertia in special interests which will restrict our readiness to react to new information about risks and benefits. The implications of all this may further require the return of our fashions in architecture and clothing to styles which diminish the demand for fuel of whatever kind.

In the realm of agriculture we must recognize what may be implied by allowing one country to become the world's sole producer of exportable surpluses of basic foods. For here we glimpse the possible emergence of a new *Realpolitik*, whereby the producer and holder of available food surpluses in a hungry world could exercise an overwhelming power. Doubtless there are other risks too, for instance of armed attempts to seize the stocks and dispute that power.

And just as we need margins of safety in the form of planned grain surpluses,[10] and the storing of them safely against the lean years or for emergency aid anywhere, so also grazing lands should not be occupied to the limit – herds should not be allowed to build up to the maximum that can be supported in the best years. Or, if the land is so occupied, a policy of culling may have to be instituted. If we wish for a stable world, we must hope to control events so as to break out of the historical cycle of drought and starvation, followed by a build-up of cattle and population during the recovery years to a level which makes it certain that there will be starvation again in the next dry period. And as Bryson[11] has put a related point:

> of course, efforts to increase agricultural production, and distribute food where needed most, and to make human lives more important than profit and power, are worthwhile. But such measures may only increase the number of people who will starve to death . . . if the population does not stay below the level of the least food supply that will be all that is available in some years.

We are surrounded by many dangers, but one of the most hopeful things is the progress this century has seen in understanding how human beings and human societies function and how they must be expected to react to their neighbour's, and to neighbouring nations', doings. Nevertheless, the development of an international moral sense lags somewhat behind this growth of knowledge. In Tickell's words, 'no responsible and, still less, elected government could lightly sacrifice a short-term and direct advantage in . . . wealth and employment for its people to avoid a long-term . . . and uncertain disadvantage for the human race . . . as a whole'.[12] It has been wisely observed that there are already strong incentives for most nations to reduce the consumption of fossil fuels. Conservation and the development of solar, tidal and wind energy should be given priority over nuclear energy. Despite the fact that the CO_2 emitted per unit of energy produced is 50 per cent higher for coal than oil, there is bound to be a decision to go over to coal from oil wherever possible, as oil reserves dwindle.[13] So adaption to whatever effects on the climate result is sure to be needed.

It is important now that we also obtain a better understanding of the physical world and its climate and what that should be expected to do, or may do, to the circumstances of our life on this planet. We should note the words of Lord Zuckerman, OM, FRS on this subject:[14] in the long history of the Earth we see

> the shapes of continents and oceans . . . continually changing; moun-tain ranges have thrust up to the skies and then disappeared; and ice

has covered the land. We . . . have a sense of the physical forces that have been at work, but we certainly do not wind the clock which triggers major changes in geography and in climate . . . these forces are still there; . . . our Earth is still changing; . . . the axis on which the globe spins every twenty-four hours is not immutably stable; the orbit in which we move annually around the sun is not constant; . . . the sun . . . is itself subject to change; the climate we know – the winds, the rains, the seasons – is also changing from year to year . . . nature itself has been responsible for far more significant changes in the physical world . . . than any for which we, the human species, have been or are likely to be responsible. We should begin to organize, on a world-wide scale, to monitor what is happening. It would be too late to do anything if, to take an extreme example, part of the ice which covers Greenland . . . were to break away. And undoubtedly we would be slow, and even reluctant, to recognize the first signs that anything like that was happening.

THE CHALLENGES OF TODAY AND THE FUTURE: WATCHFULNESS, UNDERSTANDING AND REALISM

The difficulty of recognition of a new trend or a lasting change in the general performance of the climate is real and constitutes an important difficulty for policy-makers. Such developments are always obscured by the wide range and suddenness of the short-term variations. It may be that particularly at times of long-term change, the weather 'slaps about' from one extreme to the opposite extreme from year to year. The situation is analogous to the familiar course of seasonal changes in middle and higher latitudes. As the autumn progresses, the sudden onset of winds from lower latitudes may bring an interval of mild, even summer-like weather. But the season does change nevertheless and sooner or later makes itself known.

Perhaps, in one regard the smallness of man's powers in relation to those of the natural world even today is a matter to be thankful for. The list of ideas for climatic and other types of warfare by altering the environment is a frightening one. But the experiments of the 1970s in defoliating forests, interfering with the monsoons of Asia, and ruining crops, seem to have had results which fell far short of expectation. A fringe activity of a more positive kind, which has long been considered as a possible source of supply of fresh water for arid lands near the sea, is to tow icebergs from the Antarctic. Small bergs were occasionally towed north from southern Chile to the drier parts of the country in the 1890s and even as far as Callao in Peru (latitude 12 °S). But more recently operation on a bigger scale has been advocated, towing some of the huge tabular icebergs which calve off from the Antarctic ice-cap, sometimes individually as much as 10–100 km long and deep enough to strand in 40–60 m of water, to the desert lands in the Middle East. However, indications so far are that 50 per cent of such a berg would melt away on a journey to even the easiest destinations, while the cracks and crevasses that are nearly always present would threaten break-up and total loss of the berg on the way and its overturning (as is usual in such developments) would endanger the towing vessels.[15] The effectiveness of artificial seeding of clouds to produce rain, to forestall hail or clear the cloud, has remained debatable or at most has had success on only a local scale.

Man's history has been played out in an ever-changing world, the changes sometimes slow, sometimes fast, the nature of the long-term ones always obscured by the bigger swings that distinguish the individual years. The environment will continue to change, partly due to human activities with their effects both intentional and unintentional, and partly due to natural causes. There is certainly no warrant in this for expecting that either a constant or an ever-rising standard of living will in the long run be possible. But we can with good reason continue to seek a juster world in which the poor and vulnerable – both individuals and nations – are less and less disadvantaged. There is encouragement, too, in that people in every generation, even amidst the discomforts and hardships of primitive times, have found their joys and happiness. Those in middle latitudes have thanked their gods for the green Earth, the lilies of the field and the golden corn, those in other latitudes for the beauty of the polar and mountain snows, the shelter of the northern forests, the great arch of the desert sky, or the big trees and flowers of the equatorial forest. How many of our present problems arise from not understanding our environment and making unrealistic demands upon it?

This book has presented human history as a climatologist sees it. The climate seems to have had many effects; though seldom a determinant of human history, its influence on the overall picture of society may be great. Again and again the development of climate seems to enter in as a destabilizer and catalyst of change.

Adaptability and flexibility in our planning in the face of climatic and environmental change may in extreme situations be the price of survival. It has even been argued that the demise of the old European colony in Greenland in the late Middle Ages was due not so much to the increasing difficulty of the climate as to the colonists' failure either to go over to an Eskimo way of life or to evacuate the country. But it is important also to note the lessons in history that by the time when climatic stresses become severe the people at risk tend to lose their power, or their willingness, to adapt and with it lose their resilience.

The world of human idealism, the faith of the Christian believer and other devoted people, and the sympathies of the humanist will for ever be engaged with a changing scene and must rise to meet ever new challenges. And our actions need the best assessment of the development of the physical world about us, and the likely effects of any course of human action impinging on the natural world, that science can bring.

19

RECENT DEVELOPMENTS AND THE OUTLOOK

THE MID-TWENTIETH-CENTURY COOLING

The lowering of world temperatures from the 1930s and 1940s to some time between 1975–80 and 1985[20] has been less than the rise over the previous fifty years. Hence, the twentieth century has been warmer than the previous two centuries. This is a broad summary statement that is generally true the world over. However, estimating average temperatures for the whole Earth to something approximating the degree of precision claimed (or, at least, generally implied) by figures now commonly published can surely never be realistic because of the huge ocean spaces – about 70 per cent of the surface of the globe – and the local diversity of soils and drainage, etc. on land, not to mention the error margins to which all sensing methods are liable. In central England the 1900–93 average temperature is 0.8 °C above that which Manley derived from the late seventeenth century (1659–99) and 0.3 °C above the figures for the eighteenth and nineteenth centuries.

Despite the remarkable warmth in 1989 to 1990 (years which in England were however not significantly warmer than 1948 and 1949), no later decade has so far equalled the average for the 1940s or for the 1930s and 1940s combined. This position may not be true for Scotland and Scandinavia, where heavily predominant west and southwest winds made the years from about 1987 to 1991 outstandingly warm. We still need more knowledge and understanding of the variations, which in some cases evidently last up to several decades, in the correspondence (or lack of it) between the temperature trends in different latitude zones and other fairly large regions. As pointed out on p. 39, there were times during the Little Ice Age period in recent past centuries when the climate of the Antarctic became relatively milder, at least round a wide fringe zone, just when some of the coldest phases were being experienced in the middle and higher latitudes of the northern hemisphere. From 1950 to the 1990s the world situation has approximated to the reverse of that pattern. The North Polar basin and, especially, a region extending south from there as far as Iceland, has tended to be out of step with much – perhaps most – of the rest of the world. A count of the number of months each year when most of the polar cap north of 70 ° was warmer, or colder, than the 1931–60 average showed a preponderance of cold months in every year from 1960 to 1986, mostly a fourfold preponderance (see fig. 96, p. 270).

Taking the world baldly as a whole, there is no doubt that the twentieth century has been a warm time. We have to take note that at times some large areas may be out of step, as in the recent examples we have quoted in these paragraphs and as, indeed, China seems to have been (see p. 171) during much of the Middle Ages, when Europe and much of the Arctic had a warm epoch. It has been suggested that such asymmetric patterns around a hemisphere may be linked to the known wanderings of the magnetic poles of the Earth as a result of changed targeting of the corpuscular streams of radiation from the sun. No physically complete explanation seems, however, to have been presented.

Another asymmetry, in the development of the seasonal round, has been noticeable in the later half of the twentieth century: the autumns in the northern hemisphere, particularly the Octobers, continued at, or near, their warmest level through the 1960s, 1970s and 1980s, even when the other seasons, particularly the springs, became colder. This last item is mentioned here chiefly as a warning against expecting too simple patterns. No doubt a physical explanation would involve understanding changes in the latitudes occupied by the jet stream and the average strength of the wind circulation, as well as more localized controls of where the main energy sources and the strongest flow lies, and perhaps such external matters as the solar constant.

Since around 1980, international concern about the environment and the possibility of disastrous changes to the climate, to the atmosphere which we breathe and to the Earth's surface, as an outcome of the ever-increasing scale of mankind's intrusions and pollution of many kinds, has been continually increasing. Prospects of global warming are now spoken of on every side and are treated by many, including people whose decisions affect millions, as if the more alarming forecasts were already established fact.

Let us consider some facts which lie deep in the framework of what we have to consider.

THE UNRELENTING GROWTH OF THE WORLD'S HUMAN POPULATION

This is the biggest of all the threats to life on Earth and is responsible for many of the other threats. And it ensures that the consequences of climatic changes and variations, which occur all the time, will thrust far deeper into the lives of people and be harder to adjust to.

The Irish potato famine in the 1840s, already referred to (see pp. 252–3) provides an object lesson. It was surely the most horrifying example in Europe of a well-documented climatic disaster – in this case simply caused by a run of warm, moist summers – which had become certain to happen whenever the appropriate weather occurred. Its consequences were greatly aggravated by the fast-growing overpopulation. Ireland's rural population had been multiplying, having probably doubled from 1820 to the mid-1840s. By that time the potato was the only crop that could produce the bulk of food needed to fill the people's bellies. And the cheap 'lumper' variety, which was inevitably mostly chosen, proved particularly vulnerable to the blight. The farmers' holdings were commonly only one to 1½ or 2 hectares, as a result of repeated subdivision among the inheritors in successive generations: this process could go no further. The historian, Robert Kee,[21] writing of this famine, tells of the pitiful scenes that followed of disease and death, as well as the packed emigrant ships on which more died, and, probably inevitably, the lasting sense of outrage and resentment that it bred. He adds: 'it is easy now to say that the accusations of genocide made by some Irish writers at that time and since were unjust . . . the government [in London] was the prisoner of the economic philosophy of the day which taught that economic laws had a natural operation' and to interfere . . . would bring chaos. In fact, 'the government looked on with . . . increasing dismay at what it regarded as its helplessness before irresistible economic and social forces'. And he goes on, in the end, 'by what seemed a superhuman effort at the time, it succeeded in abandoning . . . some of the principles it held most sacred and brought itself to distribute government charity'. Such are the dilemmas faced whenever natural trends bring new, but urgent and distressful, situations and such are the hesitations and delays that characteristically hinder helpful action by the authorities. The divisive human reactions among those affected are also predictable, as we can now see from more recent events in other parts of the world.

Other possible climatic developments now before us threaten difficulties of no less magnitude.

THE MORE OR LESS WORLD-WIDE WARMTH OF THE TWENTIETH CENTURY, EXCEEDING THAT OF MOST RECENT CENTURIES

The warming generally seems to have begun around 1700 and has gone through a number of rapid phases as well as some sharp setbacks, one of which between about 1780 and 1850 brought things back to more or less as they were before it. It seems that a number of different factors have contributed to the sequence, among them variations in the amount of volcanic activity which loads the atmosphere with dust, gases and vapours that may still be carried and veil the sun's radiation for some years after the greatest eruptions, as well as the larger sizes of debris that soon fall out. The strength and constitution of the solar beam itself are also subject to some variations. And changes in the amount of carbon dioxide and other 'greenhouse gases' in the atmosphere, as well as variations of the water vapour content, and of cloudiness, must also be expected to affect the climate.

It now seems necessary to admit – though this is seldom mentioned in recent literature – that none of these variations explains the timing of the general warming and cooling phases altogether satisfactorily, certainly not as well as widely claimed. In particular, the onset of the sharp warming phase around 1700, and the mid-twentieth-century cooling from about the 1940s to the early 1980s, are not well accounted for. Nor is the magnitude, nor the distribution, of warming and cooling over the Earth in good agreement with most global warming model predictions. Even the great warmth of the years 1989–91, hailed in some quarters as proof of the reality of the predicted global warming due to the enhancement of the greenhouse effect by increasing carbon dioxide and other effluents, requires the usual adjustments. But it may also have a surprising analogy in the past to the remarkable warmth – well attested in Europe – of the year 1540, shortly before the sharpest onset of the so-called Little Ice Age. Pfister[22] records that for several decades before 1564 the climate in Switzerland – and this seems to be in line with the implications of other European chronicles – was on average about 0.4 °C warmer, and slightly drier, than today. The summers in the 1530s were at least as warm as in the warmest ten years of the present century, between 1943 and 1952. And the year 1540 outdid the warm dry year 1947 appreciably. From February till mid-December 1540 rain fell in Basle on only ten days. And young people were still bathing in the Rhine on the Swiss-German border at Schaffhausen in the first week of January 1541 after a ten-months-long bathing season. The warm anomaly of 1540 is the more remarkable because the weather then became severely wintry, and spring came late in 1541. Moreover, only twenty-four years later the 1564–5 winter was one of the longest and severest in the whole millennium in most parts of Europe and marked the arrival of the most notable cold climate period of the Little Ice Age, with ten to twenty historic winters, very late springs, cool summers and advancing glaciers.

THE DEVELOPING OZONE HOLES

Ozone is created in the upper atmosphere at heights near 50 km above the Earth's surface by the action of the sun's ultra-violet radiation on the oxygen molecules and free oxygen atoms at those heights. The process absorbs this lethal (UV) constituent of the sun's rays. The ozone diffuses downwards and reaches its greatest concentration in the layers between 15 and 50 km, especially between 15 and 30 km, at one to ten parts per million. At lower levels the ozone (O_3) molecules become dissociated as they oxidize things they come in contact with, and this action leaves ordinary oxygen (O_2) molecules once more in the atmosphere.

The ozone in the stratosphere, like the volcanic eruption products that occasionally reach those levels, is gradually carried polewards. It reaches its greatest concentration over high latitudes, but near the jet stream it leaks downwards into the lower atmosphere and is destroyed. The absorption of some of the sun's short-wave radiation in the process of forming the ozone warms the stratosphere at the levels where it takes place.

Destruction of the ozone in the atmosphere by chemically active agents released in the exhausts of high-flying aircraft and rockets or space-ships, as well as from household use of aerosol sprays and from refrigerators, which are – however surprisingly – found to reach the stratosphere, is now a major cause of anxiety, no less than other forms of pollution. It was first noticed in 1984 over the Antarctic, where substantial depletion of the ozone was discovered in the core region of the stratospheric circulation during the winter night. This has been observed again and again in every year since, and the area affected has grown bigger. By 1992, the extent of the 'hole' in the ozone layer over the Antarctic at its seasonal maximum was about four and half times what it was in 1984. A similar feature has since appeared over the Arctic as well during the northern winters. And when the winter stratospheric circulation regime is approaching its end, these features wander far enough from their origin to expose some areas in the inhabited temperate latitudes for a time to solar radiation from which the harmful ultra-violet rays have not been filtered out by passing through the ozone layer.

This now regularly repeated destruction of the protective ozone layer demands modification of fashionable and popular habits that almost universally have become part of the twentieth-century way of life. Exposure to the sun must be severely limited, and dress modified accordingly, if skin cancers and other undesirable effects are to be avoided.

Another issue is that loss of the ozone layer may be expected to reduce heating of the stratosphere and contribute to warming the lower atmosphere layers that we inhabit.

POLLUTION

Humanity is continually polluting the atmosphere in various ways. The sulphur dioxide (SO_2) released by burning coal, gas and oil in industrial processes and in domestic fires is the cause of many anxieties, from creating breathing problems to turning the rain acid. Acid rain,[23] reported now in many regions of the world, damages plant life – sometimes killing the trees (even whole forests) – ruining the soil and poisoning lakes and other water bodies. Sometimes all the fish die.

The carbon dioxide (CO_2), which is the chief product of burning all fossil fuels, wood and cut vegetation, is added to the atmosphere. The small proportion of CO_2 in the air has increased from 260 to 280 parts per million in the mid-nineteenth century to about 350 ppm today. The carbon dioxide in the atmosphere is the basic food of vegetation, which may be expected to grow more luxuriantly in a more carbon-rich environment. If climates get warmer, it should benefit from that too.[24] The atmosphere's carbon dioxide is expected to increase to about 600 ppm by the year 2100, which, if it occurs, must be expected to alter the balance of radiation passing through the atmosphere and is generally expected to warm the climate significantly – according to some forecasts to a temperature level that has not occurred for many millions of years. Much research published in recent years has been directed at anticipating the increased crop yields that might be expected in a warmer world and the possibilities which might open up of growing warmth-demanding crops in new areas. But against these advantages must be set the likely extended ranges of insect pests and diseases from warmer latitudes. However, the match between past periods of increasing CO_2 and climatic warming seems not to be as close as expected and widely claimed.

Worst of all the types of pollution is the accumulation of nuclear waste materials on or in the ground and the decay products in the atmosphere and terrestrial environment. DDT that must have come from insect sprays used in the main inhabited countries of the world has been found in the snows of Antarctica, and radioactive caesium from the nuclear plants at Sellafield (Windscale) in northwest England has been found in the sea surface water in the polar ocean current moving south off the coast of East Greenland. The range of consequential threats has been illustrated by the contamination of sheep in England, Wales and Scotland with wind-borne radioactive matter from the accident at the nuclear electricity plants at Chernobyl in the Ukraine in 1986. Over eight years later the contamination still persists in some of the areas across Europe reached by this fall-out. Similar nuclear accidents have occurred elsewhere, as at Three Mile Island in the eastern United States in 1980, and more must be expected in the future. There is nowhere near the surface of the Earth where radioactive nuclear wastes can be stored indefinitely without risk of dispersal at some later date by earthquake, volcanic activity or war.

WINDINESS AND STORMS

There have been indications in many regions, seemingly representative of most of the Earth, that windiness – as shown, for instance, by average wind speeds and the frequency of storm winds – has been increasing since about 1950.[25] Flohn and others have argued, surely soundly, that this can logically be associated with the undoubted rise of ocean surface temperatures over these years, principally in the tropics and in the higher southern latitudes. This has been accompanied by an increase of the area in the warmest tropical oceans with surface water temperature above 27.5 °C, which seems to be critical for tropical hurricane formation. These authors also mention the doubt entertained by several leading investigators about attributing the twentieth-century warming mainly to the increase of 'greenhouse gases' in the atmosphere. Care must certainly be exercised before attributing the increase of general windiness to global warmth, since there seem to have been stormier and less stormy periods in the past which cannot simply be aligned with warmer and colder periods respectively. (A possible counter-argument would be available if it could be shown that in past cold episodes which were, or are, regarded as stormy, the storminess was narrowly restricted to some particular zone.)

The present (1990s) warmth of the tropical oceans, on average warmed by about 0.3 °C over the last thirty to thirty-five years, can safely be presumed to have been accompanied by increased evaporation from their surface. There is indeed indirect evidence of an increase of rainfall over those oceans in a measured reduction of the salinity of the surface waters. (Rainfall at sea cannot yet be directly measured reliably.) And it must be safe to conclude that the overall water vapour content of the Earth's atmosphere has been increased in consequence. More latent heat must therefore have been released by condensation of this water vapour in clouds. In this way, the total energy of the world's winds and weather systems must have been increased. Flohn estimates the overall intensification of the general wind circulation by these processes over recent years at probably 10–12 per cent of the energy budget. And he points to the implication that more extreme weather systems and events should therefore be expected: 'more severe cyclones and hurricanes . . . very heavy precipitation and intense hailstorms, but also (at least to a more minor extent) more active subsiding air motion in anticyclones'.

The semi-permanent centres of low pressure over the northern North Atlantic and northern North Pacific have shown this expected intensification. But there is doubt as to how far this can be due to global warming from the man-made increase of carbon dioxide and other 'greenhouse gases'. Doubts arise because the geographical distribution of the warming differs markedly from model predictions. It is not as marked in the Arctic as in some other regions. And the prolonged cooling period between about the 1940s and 1975–80 or after coincided with a time when the increase of carbon dioxide, etc., was more rapid than ever before.

The warmth noted over much of the northern hemisphere, particularly Europe, the European Arctic, Greenland and North America during the high Middle Ages does seem to have led to a very notably stormy period (see pp. 191–4), at least in the North Atlantic and European sector, around its closing stages, particularly in the 1200s AD. But we have noticed else-where in this book (e.g. pp. 218–19) evidence of another climax of storms and blowing sand at the coasts of Europe coinciding more or less with the coldest climate period between about 1550 and the 1720s or after.[26]

In Britain, as in other places near the Atlantic fringe of the continent of Europe from northwest France to Norway, the storm which struck on 15–16 October 1987 was certainly one of the severest in the last three hundred years or more, generally thought to be comparable with the famous storm in 1703 (see p. 219) which was very fully described by Daniel Defoe. The strongest gusts of wind in the 1987 storm ranged up to 119 knots (220 km/hr) at the coast of Brittany. There was enormous damage to forests and woodland. Traffic was halted for many hours in all the countries near the path of the storm. Insurance losses from damage, chiefly to buildings and trees, in England alone was estimated at £1,000 million at 1988 prices, but the number of people killed (eighteen in England) was not very great, as the worst of the storm was in the night hours.

The trend over recent decades to increasing storminess[27] has produced new records for low pressure over the Atlantic. Older texts on climate reported that barometric pressures over the North Atlantic corrected to sea level had been known to go down to about 925 mb, although values below 940 mb were very rare. Nevertheless, in the winter of 1982–3 three depressions deepened to between 930 and 934 mb and these figures were repeated by two more North Atlantic lows in November 1992. On 15 December 1986 a centre between southernmost Greenland and Iceland had a pressure value of 916 mb and on 10 January 1993 another case, with central pressure as low as 912 to 915 mb, occurred close to southeast Iceland, near 62 °N 15 °W. A count made in the Deutsche Seewarte, Hamburg, of the numbers of lows attaining depths below 950 mb on the North Atlantic, winter by winter since the late 1950s, produced the following average figures per winter: 1956–9 (4 winters) average 5 to 6 per winter; 1960–4 average 3 to 4; 1965–9 average 1.2; 1970–4 average 5.2; 1975–9 average 5.6; 1980–4 average 5.4; 1985–9 average 5.6; 1990–1 (2 winters) average 15.5. Some remarkable wind strengths have been experi-enced. Indeed, an extreme gust of about 174 knots was allegedly measured just north of Shetland on 1 January 1992 in another storm that produced a measured gust strength of 119 knots at Ålesund on the west coast of Norway and £35–40 million damage in that country with great destruc-tion in the forests as well as to shipping and coastal installations.

RECURRING OSCILLATIONS IN LARGE-SCALE WEATHER PATTERNS

The most important of these, both introduced in chapter 3, are the Southern Oscillation (see p. 48) and the development of 'blocking' of the middle latitudes westerlies by slow-moving or stationary high pressure systems (pp. 36, 55). Neither is a regular oscillation with constant period.

We reported on p. 306 the wide-scale regional anomalies that develop in the world's climate with, and after, the greater than usual magnitude of the swing of the ocean currents and sea temperatures in the broad Pacific Ocean in 1972, a familiar pattern to the coast dwellers and fishermen of Peru as 'El Niño'. This occurs at intervals of about two to seven years. It affects the very cool Humboldt Current that normally brings water from southern temperate latitudes north all the way along the coast of South America to Peru, where it further draws up cold bottom water as it begins to turn away from the coast to proceed west across the ocean near the equator. The El Niño, interrupting this normal regime, most characteristically begins to be noticeable about Christmas time – the Spanish name means 'the baby' – and develops during the months that follow. It is now understood to be an integral part of the great Southern Oscillation (q.v.) so that some meteorologists (and accounts of it in the literature) now prefer to rename both – in the regrettably obscurantist fashion of these times – as ENSO (El Niño – Southern Oscillation). The normally prevailing ocean currents in the region make the waters near the Peru coast, and some way from there across the Pacific, the coldest sea surface in the world at such latitudes, with temperatures around, or below, 20 °C and occasionally as low as 16 °. During El Niño events, this pattern is replaced by warmer water with temperatures normal for the equator spreading from the north. The fisheries are, of course, affected. The meteorological consequences include very much higher rainfall than in other years in Peru and the Pacific islands. But, through the much wider range of the Southern Oscillation, this is linked with anomalies extending to the occurrence of blocking anti-cyclones (and easterly winds) over the higher middle latitudes (sometimes) in both hemispheres.

Our account on pp. 306–8 mentions the economic and cultural disasters that were associated with the great 1972 El Niño, when sea surface temperatures west of Peru, which had been 2–2.5 °C below normal about New Year, rose to 3.5–4 °C above normal from June to the following December. There was another, briefer and less intense, El Niño in 1976–7. But the next one, in 1982–3, was an even greater and longer-lasting El Niño, perhaps matching that reported in 1877. (No other comparable case is known from the period for which we have instrument measurements.) Sea surface temperatures off the coast of Peru rose up to 7 °C above normal in June 1983.[28] The impacts on human affairs in many countries were much as in 1972. One place in northern Peru had 3950 mm of rain between November 1982 and June 1983, compared with 25 mm twelve months earlier. This situation was followed by other extremes. The most striking of these was the most extreme phase of the drought (and consequent famine) that has affected the Sahel–Ethiopian zone of Africa since the late 1960s. This has made its mark on enormous numbers of people all over the world through the emergency appeals and relief work in Africa by all the leading charities, and popularized by the 'Band Aid' and 'Food Aid' activities in 1985 and since.

There is a 'Southern Oscillation Index', defined as the difference of monthly mean barometric pressure between Tahiti in the Pacific and Darwin in northern Australia. It indicates the relative strengths of the South Pacific subtropical anticyclone and the winds in the equatorial convergence zone over the Indian Ocean. In the normal climate situation this Index has positive values, but during El Niño events the values are consistently negative. In 1983 the greatest negative values of the century occurred. Extreme occurrences such as this may damp out the trend of climate or may even be able to switch it into a new course. Handler[29] maintains that El Niño and Southern Oscillation events are liable to be linked to the distribution of volcanic aerosols in the atmosphere over the northern and southern hemispheres after great eruptions.

NOTES

2 THE CLIMATE PROBLEM

1 T. Bergeron, 'Richtlinien einer dynamischen Klimatologie', *Meteorologische Zeitschrift*, vol. 47, pp. 246–62, Berlin, 1930.
2 *Geographical Journal*, vol. 157, no. 3, pp. 326–9.
3 *Geographical Journal*, vol. 159, no. 2, pp. 219–26.
4 A. Bourke and H. Lamb, *The Spread of Potato Blight in Europe in 1845–6 and the Accompanying Wind and Weather Patterns*, Dublin, Meteorological Service, 1993, 66 pp.
5 The wind-vane is of much greater antiquity. Earlier rain-gauges are also known to have existed in the Far East and according to some reports existed in parts of the ancient Roman world, but their records have been lost.

4 HOW CLIMATE COMES TO FLUCTUATE AND CHANGE

1 R. R. Dickson, J. Meincke, S. A. Malmberg and A. J. Lee, 'The "Great Salinity Anomaly" in the northern North Atlantic 1968–82', *Progress in Oceanography*, vol. 20, pp. 103–51, 1988.
2 L. A. Mysak, D. K. Manak and R. F. Marsden, 'Sea-ice anomalies observed in the Greenland and Labrador Seas during 1901–1984 and their relation to an interdecadal Arctic climate cycle', *Climate Dynamics*, vol. 5, pp. 111–33, 1990. L. A. Mysak and S. B. Powers, 'Greenland Sea ice and salinity anomalies and interdecadal climate variability', *Climatological Bulletin*, vol. 25, no. 2, pp. 81–91, 1991.
3 J. L. Knox, K. Higuchi, A. Shabbar and N. Sargent, 'Secular variation of the northern hemisphere 50kPa (equivalent to 500 millibars) geopotential height', *Journal of Climatology*, vol. 1, pp. 500–11, 1988.
4 P. M. Kelly, P. D. Jones, C. B. Sear, B. S. C. Cherry and R. K. Tavakol, 'Variations in surface air temperatures: Part 2. Arctic regions, 1881–1980', *Monthly Weather Review*, vol. 110, pp. 71–83, 1982.
5 R. Reiter, H. Jäger, W. Carnuth and W. Funk, 'The stratospheric aerosol layer observed by lidar since October 1976. A contribution to the problem of hemispheric climate', *Archiv für Meteorologie, Geophysik und Bioklimatologie*, B27, pp. 121–49, 1979. R. Reiter and H. Jäger, 'Results of 8-year continuous measurements of aerosol profiles in the stratosphere with discussion of

the importance of stratospheric aerosols to an estimate of effects on the global climate', *Meteorology and Atmospheric Physics*, vol. 35, pp. 19–48, 1986.

6 *Climate Monitor*, vol. 8, no. 3, 1979.

7 We shall return to this point and the evidence for it in chapter 17.

8 On the argument that, if a change in the overall annual mean temperature resulted from just random variations within the system, one would not expect any correlation between the trends shown by the individual months or seasons and that emerging for the year as a whole, a recent statistical study by Dr C. D. Schönwiese of Munich *(Meteorologische Rundschau,* vol. 32, pp. 73–81, 1979, and *Meteorological Magazine,* vol. 109, pp. 101–13, 1980) explored the longest temperature record, the 320-year record of temperatures in central England, from this point of view. It came out that there were strong correlations, embracing all the months of the year, in the case of the long-term trend. This certainly makes it appear that the general warming of the climate since the seventeenth century (particularly its first, very strong phase from the 1690s to the 1730s) could be attributed to external causes affecting the whole climate system. A similar conclusion seemed to apply to the evidence of cyclic elements in the data with period lengths around 100 years and 2.2 years.

5 HOW WE CAN RECONSTRUCT THE PAST RECORD OF CLIMATE

1 The general reliability of the temperature variations here derived seems to be indicated by comparisons with the results obtained by other workers who have used different sets of historical data. A correlation coefficient of +0.77 was produced by comparing the basic Summer Wetness index values for England here used for the fifteen decades between AD 1200 and 1350 with J. Z. Titow's rating of the summer and autumn weather (June to October) from the manorial accounts of the bishopric of Winchester in the south of England. And the successive fifty-year mean values of the present author's Winter Severity index for England from AD 1100 to 1600, when compared with the assessments by P. Alexandre of Liège of the winters in the records for southeast Belgium and northeast France, gave a correlation coefficient of +0.74. Both these figures appear statistically significant. (See J. Z. Titow, 'Evidence of weather in the Account Rolls of the bishopric of Winchester 1200–1350', *Economic History Review* (second series), vol. 12, no. 3, pp. 360–407, 1960. Also J. Z. Titow, 'Le climat à travers les rôles de comptabilité de l'évêché de Winchester (1350–1450)' *Annales: Economies, Sociétés, Civilisations,* no. 2, Paris, Armand Colin, 1970.)

2 D. W. Moodie and A. J. W. Catchpole, 'Environmental data from historical documents by content analysis: freeze-up and break-up of estuaries on Hudson Bay 1714–1871', *Manitoba Geographical Studies 5*, Winnipeg, University of Manitoba, 1975.

3 J. Iversen, '*Viscum, Hedera* and *Ilex* as climate indicators', *Geologiska Föreningens Förhandlingar*, vol. 66, pp. 463–83, Stockholm, 1944.

4 The information in this paragraph comes largely from P. V. Glob's book *Mosefolket: jernalderens mennesker bevaret i 2000 år* (Copenhagen, Gyldendal, 1965), which has been published in English by Faber & Faber, in paperback, as *The Bog People* (London, 1969).

6 CLIMATE AT THE DAWN OF HISTORY

1 There seems to have been too little moisture in the polar atmosphere to provide the substance to build up an ice-sheet over northern Alaska. A somewhat analogous situation exists over northern Greenland today.

2 For more on this, see R. F. Flint. *Glacial and Quaternary Geology*, p. 785, New York, Wiley, 1971, 893 pp.; D. M. Hopkins (ed.). *The Bering Land Bridge*, Stanford University Press, 1967, 495 pp.; H. H. Lamb, *Climate: Present, Past and Future – Volume 2: Climatic History and the Future*, London, Methuen, 1977, 835 pp.; P. Woldstedt, 'Die Beringstrasse und die Einwanderung des Menschen von Asien nach Amerika', in *Das Eiszeitalter*, Band 3, 2, Auflage, pp. 220-4, Stuttgart, Enke, 1965, 328 pp.

3 For the reader's convenience the dates in this book have been converted to calendar dates, even when they are based on radiocarbon or other approximate dating methods. Where there are margins of uncertainty, these are indicated by quoting the dates in rounded figures and by appropriate wording of the text.

4 M. R. Bloch, 'Zur Entwicklung der vom Salz abhängigen Technologien: Auswirkung von postglazialen Veränderungen der Ozeanküsten', in *Saeculum*, Band 21, Heft l, pp. 1–33, Munich, 1970. See also Bloch's 'Salt in human history', *Interdisciplinary Science Reviews*, vol. 1, no. 4, 1976.

5 The snowline on the Taurus and Zagros mountains from southern Turkey to the Iran–Iraq border seems to have been lowered by 1200–1800 m below its present altitude during the last ice age. So great a lowering cannot be explained by the lowering of temperature alone, which probably amounted to no more than 4 or 5 °C in that area: the difference must be explained by the accumulation of ice and snow owing to a much greater precipitation than now. This in itself is interesting as evidence that the area was close to the main zone of cyclonic activity passing south of the European ice-caps. By contrast, in the Alps the upper tree line was depressed much more than the snowline. (These details are given by H. E. Wright, 'The late Pleistocene climate of Europe: a review', *Geological Survey of America Bulletin*, vol. 72, pp. 933–84, Rochester, N.Y., 1961; and H. Firbas, 'The Late Glacial vegetation of central Europe', *New Phytologist*, vol. 49, pp. 163–73, Oxford and Edinburgh, Blackwell, 1950.)

6 H. E. Wright, 'Natural environment of early food production north of Mesopotamia', *Science*, vol. 161, pp. 334–9, Washington, DC, American Association for the Advancement of Science, 1968.

7 W. M. Wendland and R. A. Bryson, 'Dating climatic episodes of the Holocene', *Quaternary Research*, vol. 4, pp. 9–24, 1974.

7 IN THE TIMES OF THE EARLY CIVILIZATIONS

1 We may have a more or less direct account of these experiences, though no doubt first written down after generations of oral memory, in the first book of the Bible (Genesis 11–13, 19, 26, 41 and 46), telling of the successive migrations of the ancestors of the Israelites, which are now dated between about 2100 and 1700 BC, in the generally drying landscapes of the region between Mesopotamia and Egypt. The account, albeit personalized in the names of the leaders, as is common in such early writings, starts with Abram's departure from Ur in Chaldea. The people wandered, accompanied by their flocks and herds which the various lands they stopped in proved unable to

support, and at various times they were either sold or taken captive into slavery. Although several droughts of limited duration feature in the account, the impression is given of the necessity of permanent abandonment of previously occupied lands and so of a general drying of the region. Other details which may now seem to be related to this are the tale of Lot's wife being turned into a pillar of salt and the earlier account, in chapter 4 of the same book, of the rivalry and conflict between Cain the agriculturist and Abel the herdsman.

2   The post-glacial period is known to geology as the 'Holocene' or the 'Flandrian Interglacial'.

3   K. Aaris-Sørensen, 'Atlantic fish, reptile and bird remains from the Mesolithic settlement at Vedbæk, north Zealand,' *Videnskabelige Meddelelser*, vol. 142, pp. 139–49, Copenhagen, Danish Natural History Society, 1980.

8 TIMES OF DISTURBANCE AND DECLINE IN THE ANCIENT WORLD

1   The history of the vegetation in this region was worked out by H. E. Wright of the University of Minnesota. The vegetation distribution over the entire region of the United States east of the Rockies has now been carefully mapped, species by species and stage by stage, through postglacial time by Bernabo and Thompson Webb of Brown University, Providence, Rhode Island.

2   A useful global survey of current knowledge in this field is given by J. M. Grove, 'The glacial history of the Holocene', *Progress in Physical Geography*, vol. 3, no. 1, pp. 3–54, London, Arnold, 1979.

3   C. E. P. Brooks. *Climate Through the Ages*, 2nd edition, p, 300, London, Ernest Benn, 1949. (It is to be noted that some of the dates then given may need adjusting in the light of later knowledge and more precise dating techniques.)

4   H. Gams, 'Aus der Geschichte der Alpenwälder'. *Zeitschrift des deutschen und Osterreichischen Alpenvereins*, vol 68 (yearbook for 1937), pp. 157–70, Stuttgart.

5   The version here quoted is compounded of the clearest sections of the translations given in the Revised Standard version of 1881 and the Knox Bible.

9 ROMAN TIMES AND AFTER

1   I am indebted to Drs M. Ryckaert and F. Verhaeghe and Professor A. Verhulst of the University of Ghent for the information about the situation at the times here mentioned on the Belgian–Dutch coastal plain.

2   H. Salvesen, *Jord i Jemtland*, Östersund, AB Wisenska bokhandelens förlag, 1979, 187 pp.

3   Published in a special issue of *Die Alpen,* the journal of the Swiss Alpine Club, in November 1976.

4   Several other advances of these glaciers in about the last four thousand years, including one around AD 1200–1300, were somewhat less extensive.

5   In a valuable summary by Dr Jean Grove, 'The glacial history of the Holocene', *Progress in Physical Geography*, vol. 3, no. 1. London, Arnold, 1979.

6   The world survey of glacier advances by J. M. Grove (op. cit.) cites six main phases of glacier advance in the last six thousand years.

7   C. E. P. Brooks, *Climate Through the Ages,* 2nd edition, London, Ernest Benn, 1949.

10 THROUGH VIKING TIMES TO THE HIGH MIDDLE AGES

1 In his treatise *Liber de Mensura Orbis Terrae*.
2 Lauge Koch, '*The East Greenland ice*', *Meddelelser om Grønland*, band 130, nr 3, Copenhagen, 1945.
3 A. Holmsen, *Norges historie*, Oslo and Bergen, Universitetsforlaget, 1961.
4 Kenneth Clark, *Civilisation*, London, BBC and John Murray, 1969.
5 Hugh Trevor-Roper, *The Rise of Christian Europe*, London, Thames & Hudson, 1965.
6 Information kindly supplied by Professor A. Verhulst of the University of Ghent.
7 J. L. Anderson in a paper given at the Climate and History Conference, University of East Anglia, Norwich, July 1979.

11 DECLINE AGAIN IN THE LATE MIDDLE AGES

1 As cited by Vilhjalmur Stefansson in his *Greenland*, London, Toronto, Bombay and Sydney, George Harrap, 1943, 240 pp.
2 See A. A. Ruddock, 'John Day of Bristol and the English voyages across the Atlantic before 1497', *Geographical Journal*, vol. 132, pp. 225–33, 1966. See also A. A. Ruddock, 'Columbus and Iceland', *Geographical Journal*, vol. 136, pp. 177–89, 1970.
3 See L. P. Kirwan, *A History of Polar Exploration*, London, Penguin Books, 1962, 408 pp.
4 Miscellaneous Papers No. 14, Reykjavik, Museum of Natural History, Department of Geology and Geography, 52 pp.
5 The work by Professor M. K. E. Gottschalk of Amsterdam, *Stormvloeden en rivieroverstromingen in Nederland*, Deel I (voor 1400), Deel II (1400–1600), Deel III (1600–1700), Assen, van Gorkum, 1971, 1975 and 1977, is a model of comprehensive and critical compilation.
6 G. M. Trevelyan, *History of England*, London, Longmans, Green & Co., 1928, 723 pp.
7 H. H. Lamb, 'What can historical records tell us about the breakdown of the medieval warm climate in Europe in the fourteenth and fifteenth centuries – an experiment', *Contributions to Atmospheric Physics*, vol. 60, no. 2, pp. 131–43, Wiesbaden, Vieweg, 1970. Chr. Pfister, 'Veränderungen der Sommerwitterung im südlichen Mitteleuropa von 1270–1400 als Auftakt zum Getschershochstand der Neuzeit', *Geographica Helvetica*, no. 4, pp. 186–95, 1985.
8 See A. Comfort, *Nature and Human Nature*, London, Penguin Books, 1969.
9 Quoted from M. L. Parry, *Climatic Change, Agriculture and Settlement*, Folkestone, Dawson, and Hamden, Connecticut, Archon Books, 1978.
10 W. Abel, *Die Wustungen des ausgehenden Mittelalters*, 3rd edition, Stuttgart, 1976.
11 The fact that there was a notably high percentage of deserted villages in the east of England, e.g. in Norfolk, and particularly in marshy places and places on high ground with northeast aspects in the east, though none on the extremely well drained chalk soils of the Chiltern Hills, suggests that an important factor was a marked increase, in and around the fifteenth century, of rainfall on the east side of the country. This points to a substantial increase in the frequency of east winds.

12   A. Holmsen, *Norges Historie*, Oslo and Bergen, Universitetsforlaget, 1961. Also for details and listing of sources see A. Holmsen, *Hva kan vi vite om Agrarkatastrofen i Norge i Middelalderen*, Oslo, Bergen and Tromsø, Universitetsforlaget, 1978.

13   J. Sandnes and H. Salvesen, *Ødegards tid i Norge*, Oslo, Bergen and Tromsø, Universitetsforlaget, 1978.

14   The southerly foehn wind, a warm blustery wind of the northern (i.e. leeside) alpine valleys in central Europe, is warmed like other winds that blow across mountain ranges elsewhere by the latent heat of condensation gained by the air in the formation of the abundant clouds, and rain and drizzle, on the upslope side of the mountains.

15   Kåre Lunden, '*Norge under Sverrætten 1177–1319*', Bind 3, *Norges Historie*, ed. Knut Mykland, Oslo, J. W. Cappelens Forlag, 1976.

16   A. E. Christensen, 'Danmarks befolkning og bebyggelse i Middelalderen', *Nordisk Kultur*, vol. 2, pp. 1–57, Copenhagen, Oslo, Stockholm, 1938.

17   Sv. Gissel, 'Forskningsrapport for Danmark', *Nasjonale forskningsoversikter – Det Nordiske Ødegardsprosjekt, Publikasjon Nr. 1*, Copenhagen, Landbohistorisk Selskab, 1972, 223 pp.

18   W. G. Hoskins, BBC broadcast talk, 24 November 1964.

19   Investigated by M. L. Parry. See his *Climatic Change, Agriculture and Settlement*, Folkestone, Dawson-Archon Books, 1978, 214 pp.

20   B. Huber, 'Durehschnittliche Schwankung und Periodenlänge von Jahresring-Breitenkurven als Klima-Indikatoren', *Geologische Rundschau*, vol. 54, no. 1, pp. 441–8, Stuttgart, Enke.

21   E. Le Roy Ladurie, *Times of Feast, Times of Famine*, New York, Doubleday, 1971, 426 pp.

22   There is a difficulty about the eighteenth-century vintage dates, however, owing to the adoption in France of a new policy of later harvesting of the grapes to secure a stronger wine, enforced by decree.

23   K. Müller, '*Geschichte des Badischen Weinbaus*', Lahr in Baden, von Moritz Schauenburg 1953, 283 pp.

24   Trevelyan, *op. cit.*

25   See M. Beresford, *The Lost Villages of England*, London, Lutterworth Press, 1954, 445 pp.

26   I. E. Buchinsky, *The Past Climate of the Russian Plain*, Leningrad, Gidrometeoizdat (in Russian), 1957. This is a valuable collection of relevant excerpts from these sources.

27   K. Pejml, 'A contribution to the historical climatology of Morocco and Mauretania', *Studia geophysica et geodetica*, vol. 6, pp. 257–9, Prague.

28   K. S. Lal, *Growth of the Muslim Population in Medieval India*, Delhi, Research Publications, 1973.

29   The story of these investigations is attractively told in R. A. Bryson and T. J. Murray, *Climates of Hunger*, Madison, University of Wisconsin Press, 1977, 171 pp. For a further account and list of references see also R. A. Bryson, D. A. Baerreis and W. M. Wendland, 'The character of late glacial and postglacial climatic changes', in *Pleistocene and Recent Environments of the Central Great Plains*, pp. 53–74. Special Publication No. 3, Department of Geology, University of Kansas, 1970.

30   R. C. Euler, G. J. Gumerman. T. N. V. Karlstrom, J. S. Dean and R. H. Hevly, 'The Colorado plateaus – cultural dynamics and paleoenvironment', *Science*, vol. 205, pp. 1089–101, 1979.

12 THE LITTLE ICE AGE

1 Work by A. T. Wilson and C. H. Hendry reported in *Nature*, vol. 279, pp. 315–17, London, 24 May 1979.

2 Wetness and coolness of the summers in the odd-numbered years 1529, 1531, 1533 . . . 1541 is indicated by on average 16 days later vintage dates and one-third wider tree rings than in the even-numbered years 1530, 1532 . . . 1540.

3 Chr. Pfister, *Agrarkonjunktur und Witterungsverlauf im westlichen Schweizer Mittelland 1755–1797*, University of Bern, Geographical Institute, 1975, 279 pp.

4 Sigurdur Thorarinsson, 'The thousand years struggle against ice and fire', Miscellaneous Papers No. 14, Reykjavik, Museum of Natural History, Department of Geology and Geography, 1956, 52 pp.

5 I am indebted to Gisli Gunnarsson of the Economic History Institution, University of Lund, Sweden, for this information.

6 Daniel Defoe, *The Storm*, published in London, 1704.

7 S. G. E. Lythe, *The Economy of Scotland 1550–1625*, Edinburgh, Oliver & Boyd, 1960; T. C. Smout, *Scottish Trade on the Eve of Union 1660–1707*, Edinburgh, Oliver & Boyd, 1963, 320 pp.

8 J . M. Grove, 'The incidence of landslides, avalanches and floods in western Norway during the Little Ice Age', *Arctic and Alpine Research*, vol. 4, pp. 131–8, Boulder, Colorado, 1972. I am indebted to Dr Jean Grove of Cambridge, England, for many unpublished details from the original records kindly supplied by the Norwegian archives.

9 See, for example, Kari Lundbekk. *Lofoten og Vesteralens Historie 1500–1700*, Stokmarknes, Lofoten and Vesterålen Communes, 1978, 335 pp.

10 The landscape, the whole way of life largely based on the sea, the failing fisheries, the crops inland, and the poverty of the northern counties of Norway in the last decades of the seventeenth century have been immortalized in a book of poems by the Lutheran poet-priest, Petter Dass, *Nordlands Trompet*, Oslo, reprinted by H. Aschehoug & Co. (W. Nygaard), 1974.

11 E. Österberg, 'Kolonisation och kriser – bebyggelse, skattetryck, odling och agrarstruktur i västra Värmland ca. 1300–1600', *Det nordiska ödegardsprojektet publikation nr 3*, Lund, Gleerups, 1977, 308 pp. See also H. Salvesen, *Jord i Jemtland*, Östersund, 1979, 187 pp.

12 G. M. Trevelyan, *English Social History*, London, Longmans, Green & Co., 1944, 628 pp.

13 W. G. Hoskins, 'Harvest fluctuations and English economic history', *Agricultural History Review*, vol. 12, pp. 28–46, 1964, and vol. 16, pp. 15–31, 1968.

14 W. G. Hoskins, BBC broadcast, 24 November 1964.

15 E. A. Wrigley, *Population and History*, World University Library, London, Weidenfeld & Nicolson, 1969, 254 pp.

16 G. Manley, 'Central England temperatures: monthly means 1659 to 1973', *Quarterly Journal of the Royal Meteorological Society*, vol. 100, pp. 389–405, London, 1974.

17 See H. M. van den Dool, H. J. Krijnen and C. J. E. Schuurmans, 'Average winter temperatures at De Bilt (the Netherlands) 1634–1977', *Climatic Change*, vol. 1, pp. 319–30, Dordrecht, Reidel.

18 See E. Le Roy Ladurie, *Times of Feast, Times of Famine*, New York, Doubleday, 1971.

19 These data are collected in a thesis by J. Maley, 'Études palinologiques dans le bassin du Tchad et paléoclimatologie de l'Afrique nord-tropicale de 30,000 ans à l'époque actuelle', Académie de Montpellier, Université des Sciences et Techniques du Languedoc, 1980.

20 A remarkably high level of lakes and rivers in Ethiopia, reported by the Portuguese missionary Manoel de Almeida in 1628, is referred to by A. T. Grove, Alayne Street and A. S. Goudie in 'Former lake levels and climatic change in the rift valley of southern Ethiopia', *Geographical Journal*, vol. 141, no. 2, pp. 177–202, London, 1975. See also A. T. Grove, 'Geographical introduction to the Sahel', *Geographical Journal*, vol. 144, no. 3, pp. 407–15, 1978.

21 A study by J. Chang ('*Climatic change and its causes*', Peking Scientific Publications, 1976, in Chinese, reported by Professor M. M. Yoshino in *Climatic Change and Food Production*, University of Tokyo Press, 1978) listed four main cold periods in China in the last 500 years: 1470–1520, 1620–1720 (especially the decades between 1650 and 1700), 1840–90, and after 1945 (especially since 1963). All these periods – though the last one only weakly so far – have some title of a similar kind in Europe, espccially the main one in the seventeenth century. But of the main warm periods in China as listed by Chang (1550–1600, 1720–1830 and 1916–45) the first saw the sharpest cooling in Europe as the main Little Ice Age regime set in 'and the warmth of the eighteenth century in Europe was subject to many interruptions' e.g. by cold winters and by the run of cool wet summers in the 1760s and by all seasons of the year turning cold in the decade from 1810.

22 H. H. Lamb, *Weather, Climate and Human Affairs*, esp. pp. 141–63, London, Routledge, 1988, 364 pp.

13 THE RECOVERY, 1700 TO AROUND 1950

1 *Evelyn's Diary*, 10 September (Old Style) 1677: 'The Travelling Sands . . . have so damaged the country, rouling from place to place . . . like the Sands in the Deserts of Lybia, quite overwhelmed some gentlemen's whole estates.' It seems that great moving sand-dunes were a feature of this inland countryside near Thetford in the driest sector of England 50–70 km from the sea.

2 Winters similarly dominated in western Europe by persistent, bitterly cold continental east winds which froze the rivers occurred again in 1940 and 1947, as in 1830 and some other years, but many cold winters in the Little Ice Age period were characterized more by northerly winds which were less persistent so that at least in the west of Europe the snow and ice came and went in a succession of frequent changes. The long winter of 1963 belonged to an intermediate pattern with Arctic cold air from the north entering Europe over Scandinavia and the Baltic, and often reaching western Europe as an east wind: this is a pattern which dominated some of the great winters of the seventeenth century, as in 1684, and which has hardly done so since 1712. Many of Europe's other cold winters have been marked by weaker patterns and less air movement than usual in the European region.

3 For further details on the introduction and impact of crops from the Americas into Europe and of other exchanges between the Old and New Worlds, see A. W. Crosby, '*The Columbian Exchange: Biological and Cultural Consequences of 1492*', Westport, Connecticut, 1972, 268 pp.

4 Chr. Pfister, *Agrarkonjunktur und Witterungsverlauf im westlichen Schweizer Mittelland, 1755–1797*, University of Bern, Geographical Institute, 1975, 279 pp.

5 Luke Howard was a Quaker pharmacist living at Plaistow (now in east London) whose studies and classification of clouds not only interested John Constable but came later to form the basis of the modern *International Cloud Atlas.*

6 A great deal of detail of the strange summer monsoon of 1816, with the breaks that occurred in it, is given in the Introduction to a *Report on the Epidemick Cholera Morbus, As It Visited the Territories Subject to the Presidency of Bengal in the Years 1817, 1818 and 1819,* by James Jameson, published Calcutta, 1820.

7 See J. D. Post, 'Meteorological historiography', *Journal of Interdisciplinary History,* vol. 3, no. 4, pp. 721–32, Cambridge, Massachusetts, MIT, 1973.

8 This cautious description is necessary because the difficulty of homogenizing long rainfall records is greater than for temperature records. It is also more difficult to cover the country in a representative manner with fewer rain measurement sites than nowadays.

9 G. Lefebvre, *Études sur la Revolution Française,* Paris, Presses Universitaires de France, 326 pp. See also the detailed summary by J . Neumann, 'Great historical events that were significantly affected by the weather – 2. The year leading to the Revolution of 1789 in France', *Bulletin of the American Meteorological Society,* vol. 58, no. 2, pp. 163–8, 1977.

10 Details of the surveys are given in H. H. Lamb, 'Britain's changing climate', *Geographical Journal,* vol. 133, no. 4, pp. 445–68, London, 1967; and H. Neuberger, 'Climate in art', *Weather,* vol. 25, no. 2, pp. 46–56, London, 1970.

11 P. Brimblecombe, University of East Anglia seminar 1978 (unpublished).

12 Lt-Col. E. Sabine, 'On the cause of remarkably mild winters which occasionally occur in England', *Philosophical Magazine and Journal of Science,* London, Edinburgh and Dublin, April 1846.

13 E. Wahl, 'A comparison of the climate of the eastern United States during the 1830s with current normals', *Monthly Weather Review,* vol. 96, no. 2, pp. 73–82, Washington, DC, 1968. See also *Monthly Weather Review,* vol, 98, no. 4, pp. 259–65, 1970.

14 Intergovernmental Panel on Climate Change (Chairman Dr J. T. Houghton) *Scientific Assessment of Climate Change,* World Meteorological Organization, Geneva and United Nations Environment Programme, 1990. J. T. Houghton *et al.* (eds), *Climatic Change: The I.P.C.C. Scientific Assessment,* Cambridge University Press, 1990, 365 pp. J. T. Houghton (ed.), *Climatic Change (The Supplementary Report to the I.P.C.C. Scientific Assessment),* Cambridge, Cambridge University Press, 1992.

15 H. H. Lamb, *Weather, Climate and Human Affairs,* p. 160, London, Routledge, 1988, 364 pp.

16 Within the last hundred years or so the decade average length of the growing season in England has varied by as much as twenty days (shortest around 1880) and the length in individual years by nearly forty days (shortest in 1879), when there were barely 6½ months with temperatures above 6 °C in central England.

14 CLIMATE SINCE 1950

1 A simple index of global variability of climate devised by R. K. Tavakol and P. D. Jones of the Climatic Research Unit, University of East Anglia, consists of the normalized departures of atmospheric pressure or temperature averaged over all the 10° latitude and longitude intersections. For the northern hemisphere, for both these elements of the climate, the index of variability had

generally low values from about 1920 to 1960 (except for pressure in the 1940s, when there was an enhanced frequency of blocking). The index then rose sharply to high values in the mid 1960s and after. The variability index can be expressed as

$$\frac{1}{n}\sum_{1}^{n}\frac{\left|x-\bar{x}\right|}{\sigma_x}$$

for the deviations of the observed values x of an element from its average value $\bar{x}$ at each of n points on the map, σ_x being the standard deviation of x at each point.

2 The insurance industry, which already in the mid-1950s encountered heavy losses on shipping claims because rates for the tropical hurricane hazard had been kept at the level ruling from 1900 to 1930 and the frequency of such storms had significantly increased, found it necessary by 1980 to impose a more than 10 per cent increase on house insurance in the British Isles for landslip, subsidence and frost heave.

3 Reported, for instance, by K. E. Trenberth of the New Zealand Meteorological service in the Quarterly Journal of the Royal Meteorological Society, vol. 102, pp. 65–75, 1976.

4 I. Weiss and H. H. Lamb, 'Die Zunahme der Wellenhöhen in jüngster Zeit in den Operationsgebieten der Bundesmarine, ihre vermutlichen Ursachen und ihre voraussichtlichen weitere Entwicklung', *Fachliche Mitteilungen Nr 160*, Porz-Wahn, Geophysikalischer Beratungsdienst der Bundeswehr, 1970. See also *Fachliche Mitteilungen Nr. 194*, Traben-Trarbach, 1979.

5 M. Petersen and H. Rohde, '*Sturmflut*', p. 52, Neumunster, Wachholtz, 1977, 148 pp.

6 A series of useful surveys of the changes in sea temperatures and salinity, and of the fish stocks, in the waters about Greenland and Iceland has been published by Dr Svend-Aage Malmberg, most recently in *Ægis*, Reykjavik, 1979–80.

7 H. Dronia, 'Der Stadteinfluss auf dem weltweitem Tcmperaturtrend', Meteorologisch Abhandlungen. Band 74, Nr. 4. Berlin, Inst. für Meteorologie der Freien Universität, 1967.

8 However, Luke Howard noted already around 1810–20 that the temperatures in London tended to be more than 2.0 °F (1.1 °C) higher than those in the open country outside. We have noted (p. 181) indications in the beetle faunas studied in the archaeology of York that the densely built-up, though by modern standards tiny, medieval city may have been significantly warmer than its surroundings.

9 Cf. the study by N. E. Davis, 'The variability of the onset of spring in Britain', *Quarterly Journal of the Royal Meteorological Society*, vol. 98, pp. 763–77, 1972.

10 M. Tanaka, 'Synoptic study on the recent change in Monsoon Asia and its influence on agricultural production', in K. Takahashi and M. M. Yoshino (eds), *Climatic Change and Food Production*, pp. 81–100, Tokyo, University of Tokyo Press, 1978.

15 THE IMPACT OF CLIMATIC DEVELOPMENTS ON HUMAN AFFAIRS AND HUMAN HISTORY

1 See S. F. Markham, *Climate and the Energy of Nations*, Oxford, Oxford University Press, 1942, 144 pp.

2 The good housing of animals is known to make a very great contribution to their health.

3 B. H. Slicher van Bath, *The Agrarian History of Western Europe AD 500–1850*, trans. O. Ordish, London, Arnold, 1963.

4 A. W. Ruddock, 'John Day Of Bristol and the English voyages across thc Atlantic before 1497', *Geographical Journal*, vol. 132, pp. 225–33, London, Royal Geographical Society, 1966.

5 A. Bugge, F. Scheel, R. Tank and J. S. Worm-Müller, *Den norske sjøfarts historie*, vol. i, Kristiania, Steenske Forlag, 1923.

6 I am indebted to Cdr Ph. M. Bosscher of the Royal Netherlands Navy for this information.

7 Many interesting further details are given by G. Gunnarsson in 'A study of causal relations in climate and history, with an emphasis on Icelandic experience', *Meddelande från Ekonomisk Historiska Institutionen*, nr. 17, Lund University, Lund (Sweden), 1980.

8 *North-South: A Programme for Survival. The Report of the Independent Commission on International Development Issues under the Chairmanship of Dr Willy Brandt*, pp. 81–2, London and Sydney, Pan Books, 1980.

9 I am indebted to Dr P. M. A. Bourke, formerly President of the World Meteorological Organization's Commission for Agricultural Meteorology, for this information.

10 Some of the information in this paragraph and many other interesting details are collected in A. W. Crosby, *The Columbian Exchange: Biological and Cultural Consequences of 1492*, ed . R. H. Walker, Contributions in American Studies No. 2, Westport, Connecticut, Greenwood Publishing Co., 1972.

11 M. L. Parry, *Climatic Change, Agriculture and Settlement*, Studies in Historical Geography series, Folkestone, Dawson, and Hamden, Connecticut, Archon Books, 1978, 214 pp.

12 M. L. Parry, 'Secular climatic change and marginal agriculture', pp. 1–13, Publication No. 64 of the Transactions of the Institute of British Geographers, 1975.

13 I. T. Lyall, 'The growth of barley and the effect of climate', *Weather*, vol. 35, no. 9, pp. 271–8, 1980.

14 T. M. L. Wigley and T. C. Atkinson, 'Dry years in south-east England since 1698', Nature, vol. 265, pp. 431–4, London, 3 February 1977.

15 Christian Pfister. *Agrarkonjunktur und Witterungsverlauf im westlichen Schweizer Mittelland 1755–1797*, Bern, Geographisches Institut der Universität, 1975.

16 J. D. Post, 'Meteorological historiography' (a review of E. Le Roy Ladurie's *Times of Feast, Times of Famine*, New York. Doubleday, 1971), *Journal of Interdisciplinary History*, vol. 3, no. 4, pp. 721–32, 1973).

17 Statement by three Chinese workers, Chang Chia-cheng, Wang Shao-wu and Cheng Szuchung, in a paper given to the World Climate Conference 1979, World Meteorological Organization, Geneva.

18 H. Arakawa, 'Meteorological conditions of the great famines in the last half of the Tokugawa period, Japan', *Papers in Meteorology and Geophysics*, vol. 6, no. 2, pp. 101–15, Tokyo, 1955. Also in a number of more recent papers, e.g. by Professor Takeo Yamamoto.

19 J. D. McQuigg, 'Climatic change and world food production', lecture given at the University of Florida, 23 April 1975.

20 I have to thank Dr P. M. A. Bourke, formerly Director of the Irish Meteorological scrvice, for this information.

21 *Financial Times*, 16 July 1980.

22 A. Noye, 'Soviet grain – problems and prospects', *Food Policy*, vol. 1, no. 1, pp. 32–40, Guildford, IPC Science and Technology Press. See also the article by W. Schneider, 'Agricultural exports as an instrument of diplomacy,' pp. 23–31, in the same issue.

23 Brazil has succeeded in greatly increasing its production of corn (maize) and so has lately become an exporter of grain, and by the late 1970s it is reported that the European Community was once more producing a surplus of grain, including wheat.

24 Peru was said to have the biggest fishing fleet in the world at that time, bringing in annual catches of 12–14 million tonnes and supplying a large canning industry. Since that year catches have been under 2 million tonnes.

25 Quoted by S. G. Wheatcroft, 'The significance of climatic and weather change on Soviet agriculture with particular reference to the 1920s and the 1930s', *CREES Discussion Papers Series SIPS No. 11*, Centre for Russian and East European Studies, University of Birmingham, 1977.

26 Droughts are known to have affected Nebraska in a few years about 1800, from about 1822 to the early 1830s, a few years around 1840, from about 1859 through most of the 1860s, the later 1880s to early 1890s, around 1910, 1932–40, the mid-1950s and a briefer occurrence in the 1970s.

27 Quoted by W. O. Roberts and H. Lansford, in *The Climate Mandate*, San Francisco, W. H. Freeman & Co., 1979, 197 pp.

28 J. G. Charney, 'Dynamics of deserts and drought in the Sahel', *Quarterly Journal of the Royal Meteorological Society*, vol. 101, pp. 193–202, 1975.

29 The Brandt Commission Report foresees some cities of thirty million inhabitants by around the year 2000 in the Third World, so that the now well-known conditions of poverty, homelessness and overcrowding in Calcutta today should perhaps be seen as a first example of the pattern of growth. The density of some rural populations on marginal land may be no less serious.

30 For many details of the history of rainfall in Africa, see S. E. Nicholson, 'Saharan climates in historic times', in M. A. J. Williams and H. Faure (eds). *The Sahara and the Nile*, pp. 173–200, Rotterdam, A. A. Balkema, 1980; also, for earlier times, S. E. Nicholson and H. Flohn, 'African environmental and climatic changes and the general circulation in Late Pleistocene and Holocene', *Climatic Change*, vol. 2, pp. 313–48, Dordrecht, Reidel, 1980.

31 R. C. Rainey, 'Airborne pests and the atmospheric environment', *Weather*, vol. 28, no. 6, pp. 224–39, London, 1973. See also R. C. Rainey, E. Betts and A. Lumley, 'The decline of the desert locust plague in the 1960s: control operations or natural causes?', *Philosophical Transactions of the Royal Society of London*, series B, vol. 287, pp. 315–44, 1979.

32 For example, in 1438 we are told of these things happening around Smolensk. Among the accounts of similar disasters in earlier times, the chronicles of the monasteries tell us that in AD 1215 in the district of Novgorod the frost destroyed all the harvest, there was a great famine, people ate pine bark and gave up their children into slavery for bread. As in Scotland in the 1690s there was the horror of the common grave, but in the end so many died that they could not all be buried and the dogs could not eat all the bodies which lay in the towns, villages and fields. Again in 1230 frost destroyed the harvest

and severe famine struck Novgorod and Smolensk: not only were the human corpses eaten by the survivors but 'people killed and ate each other'. And plague followed the exposure of so many corpses. (These details are extracted from the excerpts from the chronicles of the time quoted by I. E. Buchinsky in *The Past Climate of the Russian Plain*, Leningrad (in Russian), Gidrometeoizdat, 1957.)

33 In a letter in *The Lancet*, dated 25 March 1961, Dr Moynahan estimated that, whereas most adults in the more advanced countries today have a two months reserve of fat, the obese may have enough to survive for eight months or more on very short rations.

34 The various references in the records of past climate in many very different parts of the world to instances of cannibalism in conditions of extremity may prompt the question whether cannibalism, wherever it is found, always and only began when communities found themselves extremely short of food.

16 THE CAUSES OF CLIMATE'S FLUCTUATIONS AND CHANGES

1 D. V. Hoyt, 'Variations in sunspot structure and climate', *Climatic Change*, vol. 2, no. 1, pp. 79–92, Dordrecht and Boston, USA, Reidel.

2 J. D. Hays, J. Imbrie and N. J. Shackleton, 'Variations in the Earth's orbit: pacemaker of the ice ages', *Science*, vol. 194, pp. 1121–32, 1976.

3 A. Berger, 'The Milankovitch astronomical theory of paleoclimates: a modern review', in A. Beer, K. Pounds and P. Beer (eds), *Vistas in Astronomy*, vol. 24, pp. 103–22, Oxford, Pergamon Press, 1980. N. Calder, 'Arithmetic of ice ages', *Nature*, vol. 252, pp. 216–18, London, 15 November 1974. G. J. Kukla, 'Missing link between Milankovitch and climate', *Nature*, vol. 253, pp. 600–2, London, 20 February 1975.

4 H. H. Lamb, 'Volcanic dust in the atmosphere; with a chronology and assessment of its meteorological significance', *Philosophical Trsnsactions of the Royal Society*, series A, vol. 266, no. 1178, pp. 425–533, 1970. Updated to 1976–7 in *Climate Monitor*, vol. 6, no. 2, pp. 54–67, Norwich, University of East Anglia, 1977.

5 The greatest loading of the atmosphere with volcanic dust in recent times seems to have been in 1815–16 and 1835–6. High figures are also indicated, at least for the northern hemisphere, in 1783–4 and perhaps in the 1690s.

6 R. A. Bryson and B. M. Goodman, 'Volcanic activity and climatic changes', *Science*, vol. 207, pp. 1041–4, 7 March 1980.

7 K. K. Hirschbroeck, 'A new world-wide chronology of volcanic eruptions', *Palaeogeography, Palaeoclimatology, Palaeoecology*, vol. 29, pp. 223–41, Amsterdam, Elsevier, 1979–80.

8 R. A. Bryson and T. B. Starr, 'Indications of Chandler compensation in the atmosphere', K. Takahashi and M. M. Yoshino (eds), *Climatic Change and Food Production*, Tokyo, University Press, 1978.

9 These results are quoted from a study by G. L. Potter, H. W. Ellsaesser, M. C. MacCracken and F. M. Luther, 'Possible climatic impact of tropical deforestation', *Nature*, vol. 258, pp. 697–8, London, 25 December 1975.

10 The production of cement contributes a small proportion, somewhat under 2 per cent, of the rate of increase.

11 S. Manabe and R. T. Wetherald, 'The effects of doubling the CO_2 concentration on the climate of a general circulation model', *Journal of Atmospheric Sciences*, vol. 32, pp. 3–15, 1975.

W. L. Gates in 'A review of modeled surface temperature changes due to increased atmospheric CO_2,' (*Report No. 17*, Climatic Research Institute, Oregon State University, Corvallis, 1980) summarizes the position in this way:

> there is a consensus among modelers that increasing the concentration of CO_2 in the atmosphere will lead to an overall warming. The more simplified models . . . generally yield a surface temperature rise of several degrees C for a doubling of atmospheric CO_2. . . . General circulation models . . . generally confirm the simpler models' warming and . . . emphasize the climate's sensitivity to the treatment of the oceans. Using . . . a motionless ocean . . . (and no indication of the ocean's heat capacity) GCMs yield . . . average warming from about 3 °C in the tropics to about 6 °C in high latitudes for a doubling of CO_2. . . . Calculations . . . with a realistic geography . . . with an ocean . . . to follow the normal seasonal variation . . . yield only about 0.2 °C overall warming for doubled CO_2. . . . When . . . a shallow mixed-layer ocean [is] used . . . a more realistic average warming of about 2 °C is inferred for doubled CO_2.

Despite the writer's undoubted authority it is hard to see that the word 'realistic' here means anything more than 'as expected'.

12 G. N. Plass, 'The carbon dioxide theory of climatic change', *Tellus*, vol. 8, pp. 140–54, Stockholm, 1956.

13 National Academy of Sciences, Washington, DC, Climate Research Board, Ad Hoc Study Group on Carbon Dioxide and Climate (Chairman. J. G. Charney). Report – A Scientific Assessment (22 pp.), 1979.

14 A recent study by G. Ohring of Tel Aviv University, Israel (*Pageoph.*, vol. 117, pp. 851–64, Basle, 1979) supports the earlier conclusion of work by S. I. Rasool and S. H. Schneider (*Science*, vol. 173, pp. 138–41, 1971) that the general effect of dust in the atmosphere is cooling near the Earth's surface.

15 I am indebted for this information to Dr P. Brimblecombe of the University of East Anglia, Norwich, from his work on the history of air pollution.

16 H. Flohn, 'Estimates of a combined greenhouse effect as background for a climate scenario during global warming', in J. Williams (ed.). *Carbon Dioxide, Climate and Society*, pp. 227–37, Proceedings of the International Institute for Applied Systems Analysis (IIASA) Oxford, Pergamon Press, 1978.

17 Coincidentally Europe was experiencing an exceptional heat wave at the time, in the second of the two great warm summers of the 1970s; and both Europe and much of North America had enjoyed an unbroken run of three to six mild winters.

18 Probably in Devonian times some 350–400 million years ago.

19 S. H. Schneider and C. Mass, 'Volcanic dust, sunspots and temperature trends', *Science*, vol. 190, pp. 741–6, 1975.

20 P. Brimblecombe, 'Attitudes and responses to air pollution in medieval England', *Journal of the Air Pollution Control Association*, vol. 26, no. 10, pp. 941–5, 1976. P. Brimblecombe and C. Ogden, 'Air pollution in art and literature', *Weather*, vol. 32, p. 285, 1977. P. Brimblecombe, 'London air pollution 1500–1900', *Atmospheric Environment*, vol. 11, pp. 1157–62, 1977.

21 The recorded frequency of fog, and of dense fog impeding traffic, at Greenwich in southeast London, decade by decade through the nineteenth century, was as shown in the following table.

These figures, originally given by R. C. Mossman in his work 'The noninstrumental meteorology of London 1713–1896', *Quarterly Journal of the Royal Meteorological Society*, vol. 23, p. 287, 1897, are repeated with many other details in *London Weather* by J. H. Brazell, published by Her Majesty's Stationery Office for the Meteorological Office in London, 1968, 270 pp.

	Average yearly number of days with fog	Average yearly number of days with dense fog
1811–20	19	2.4
1821–30	19	2.5
1831–40	26	5.2
1841–50	22	3.9
1851–60	33	7.6
1861–70	39	8.1
1871–80	49	9.0
1881–90	55	9.3

As the Clean Air Act took effect at Glasgow airport the average number of hours of fog in every hundred hours of light winds fell after 1961 from about forty-four to just over twenty between 1964 and 1969.

22 The word 'smog, (smoke + fog) came into use first in the United States to describe one of these disasters when many people died at Donora in Pennsylvania in 1948.

17 FORECASTING

1 J. Namias, 'Seasonal interactions between the North Pacific Ocean and the atmosphere during the 1960s', *Monthly Weather Review*, Washington, DC, vol. 97, no. 3, pp. 173–92, 1969.

2 A periodicity close to a hundred years in length has been suggested by analyses of Chinese rainfall and drought studies covering the last five hundred years, of tree rings in Japan, of the central England temperatures, and more tentatively in the shorter records for eastern North America, and the frequency of cyclones affecting the Atlantic seaboard.

A farm diary from Jaeren in southwest Norway reporting the severity of the frost (which struck 1.3 m into the ground even near the coast) and coastal ice in early 1838 drew attention to the fact that the last comparable occasion had been almost a hundred years earlier in 1740 (though similar severity was also known to have occurred in 1709).

Nevertheless, in all these cases the records are far too short for proof of the longer-term persistence, or reality, of a periodicity of this length.

3 Consideration of other European records, particularly the detailed Swiss records (especially snow) analysed by Pfister and the English temperature series, as well as the reports of snowy winters in Scotland and of ice in Danish waters, suggests that at least from the sixteenth century onwards the eighties are somewhat underrepresented in Easton's rating of the severe winters which we have tabulated here in table 8. (Compare the upper and lower lines of table 9.) At least 1684 and 1685 would be included if these other records were used, and 1584, 1586 and 1587 all have some claim to be considered.

4 There is some evidence that, if we had firmer reports, the year 1496 should be added to this sequence as it appears in table 8. The record of the years

ending in 84 or thereabouts is scarcely less remarkable, if the cases mentioned in footnote 3 are given their due weight. Incomplete data suggest that 1486 should perhaps be added to the sequence.

5 A. Berger, 'The Milankovitch astronomical theory of paleoclimates – a modern review', *Vistas in Astronomy*, vol. 24, pp. 103–22, Oxford, Pergamon Press, 1980. See also A. Berger. J. Guiot and G. Kukla, 'Milankovitch theory of climatic changes – the monthly insolation approach', and G. Kukla and A. Berger, 'The astronomic climate index'. Both papers were presented at the International Scientific Assembly, 'The Life and Work of Milutin Milankovič', held by the Serbian Academy of Sciences and Arts, Belgrade, 9–11 October 1979.

6 G. Woillard, 'Abrupt end of the last interglacial in north-east France', *Nature*, vol. 281, no. 5732, pp. 558–62. London, 18 October 1979.

7 One thousand pollen grains were identified, and the species represented were counted, in sections a quarter of a millimetre thick at intervals of one millimetre, which seems to correspond to six years. This time-scale depends on the dating of the main features of the core length as taken from the bog and assumes a constant deposition rate. It is certainly unrealistic to suppose that deposition was really constant at the average rate, but this gives a useful first estimate of the rapidity of the changes here reviewed. Some further confidence may be gained by study of the laminations – presumably year-layers in the deposit on a former lake bed – at this point and elsewhere in the same bog: deposition seems to have been continuous and undisturbed since.

8 There is indeed an implication that this assumption may be justified in the work of Schneider and Mass described on p. 339; fn 19, p. 405.

9 J. Williams, 'Global climatic disturbance due to large-scale energy conversion systems', in M. H. Glantz, H. Van Loon and E. Armstrong (eds), *Multidisciplinary Research Related to the Atmospheric Sciences*, Boulder, Colorado, National Center for Atmospheric Research (NCAR/ 3141–78/1), 1978.

10 My colleague at the University of East Anglia, Professor F. J. Vine, has given the following figures for the capacity of various alternative (renewable) energy resources, which should be seen in relation to present world energy consumption in the region of 8–10 million megawatts (i.e. $8–10 \times 10^{12}$ watts), or an overall average of around 2000 watts for each of the world's more than 4000 million people:

Energy source	Theoretical total capacity (watts)	Probably realizable total capacity (watts)	Realized so far (watts)
Tidal	10^{12}	3×10^{10}	3×10^8
Hydroelectricity	3×10^{12}	3×10^{12}	3×10^{11}
Geothermal (Earth's interior heat flow)	3×10^{13}	10^{11}	10^9
Ocean temperature gradients	10^{13}	10^{11}	0
Photosynthesis	10^{13}	10^{11}	0
Wind	3×10^{14}	10^{12}	insignificant
Solar	10^{16}	more than 10^{13}	insignificant

These figures were given in Professor Vine's inaugural lecture, 'Can you fuel all the people for all time?', 17 February 1976.

11 See particularly U. Siegenthaler and H. Oeschger, 'Prediction of future CO_2 concentrations in the atmosphere', in W. Bach *et al.* (eds), 'The carbon dioxide problem – an interdisciplinary survey', vol. 36, pp. 783–6 – fascicule 7, pp. 767–890 of Experientia, Basle, Birkhäuser Verlag, 1980. Also *Energy and Climate*, a monograph in the *Studies in Geophysics* series, published by the National Academy of Sciences, Washington, DC, in 1977.

12 This is among the useful points treated by Irene Smith in 'Carbon dioxide and the "greenhouse effect"', Report number ICTIS/ER 01 of the International Energy Agency's IEA Coal Research, London, 1978.

13 See, for example, C. Norman, 'Will world population double?' *Nature*, vol. 264, pp. 7–8, London, 4 November 1976.

14 C. D. Keeling and R. B. Bacastow, 'Impact of industrial gases on climate', *Energy and Climate*, published in the *Studies in Geophysics* series by the National Academy of Sciences, Washington, DC, 1977.

15 See, e.g., H. Flohn, 'A scenario of possible future climates – natural and man-made', *World Climate Conference, Geneva, February 1979*, pp. 243–66, Geneva, World Meteorologica Organization, 1979; also 'Die Zukunft unseres Klimas: Fakten und Probleme', *Promet*, pp. 1–21, 2 March 1978.

16 Among the most useful studies of this aspect of the CO_2 problem, reference should be made to W. Bach, 'Impact of increasing atmospheric CO_2 concentrations on global climate – potential consequences and corrective measures', *Environment International*, vol. 2, pp. 215–28, Oxford, Pergamon Press, 1979; also T. M. L. Wigley, P. D. Jones and P. M. Kelly, 'Scenario for a warm, high-CO_2 world', *Nature*, vol. 283, pp. 17–21, London, 3 January 1980.

17 See, for example, J. Williams, W. Häfele and W. Sassin, 'Energy and climate – a review with emphasis on global interactions', *World Climate Conference, Geneva, February 1979*, pp. 267–89, Geneva, World Meteorological Organization, 1979.

18 See G. D. Robinson, 'The effluents of energy production – particulates', *Energy and Climate*, pp. 61–71, published in the *Studies in Geophysics* series of monographs by the National Academy of Sciences, Washington, DC, 1977.

19 See, for instance, J. M. Mitchell, 'The natural breakdown of the present inter-glacial and its possible intervention by human activities', *Quaternary Research*, vol. 2, no. 3, pp. 436–45, London, Academic Press, 1972; also 'The changing climate', *Energy and Climate*, pp. 51–8 published in the *Studies in Geophysics* series of monographs by the National Academy of Sciences, Washington, DC, 1977.

20 'Comparing the United States temperature trends with the supposed global trend, a warning note' appears in G. Kukla, T. R. Karl and J. Gavin, 'U.S. versus hemispheric temperature trends', in *Proceedings of the Eleventh Annual Climate Diagnostics Workshop*, pp. 114–28, Champaign, Illinois, 1986.

21 R. Kee, *The Green Flag*, London, Weidenfeld & Nicolson, 1972; London, Penguin Books edition in three volumes; here quoted from vol. 1, *The Most Distressful Country*, 1989, pp. 171–8.

22 Chr. Pfister, *Klimageschichte der Schweiz 1525–1860*, Band I, pp. 118, 138, Bern and Stuttgart, Verlag Paul Haupt, for *Academica Helvetica*, 6.

23 See, for instance, T. D. Davies, P. M. Kelly, P. Brimblecombe, G. Farmer and R. J. Barthelmie, 'Acidity of Scottish rainfall influenced by climatic change', *Nature*, vol. 322, pp. 359–61, 1986.

24 M. Parry, *Climate Change and World Agriculture*, London, Earthscan Publications, 1990, 157 pp.

25 For example, S. D. Burt, 'A new North Atlantic low pressure record', *Weather*, vol. 42, no. 2, pp. 53–6, 1987; S. D. Burt, 'Another new North Atlantic low pressure record', *Weather*, vol. 42, no. 2, pp. 53–6, 1987. H. Flohn, A. Kapal, H. R. Knoche and H. Mächel, 'Recent changes of the tropical water and energy budget and of mid-latitude circulations', *Climate Dynamics*, vol. 4, pp. 237–52, Berlin, Heidelberg, 1990. H. H. Lamb, *Weather, Climate, and Human Affairs*, pp. 93–4, London, Routledge, 1988, 364 pp.

26 Cf. H. H. Lamb, *Historic Storms of the North Sea, British Isles and North-west Europe*, esp. pp. 17–24, Cambridge, Cambridge University Press, 1991, 204 pp.

27 I. Weiss and H. H. Lamb, 'Die Zunahme der Wellenhöhen in jüngster Zeit in den Operationsgebieten der Bundesmarine, ihre vermutliche Ursache und ihre voraussichtliche weitere Entwicklung', *Fachliche Mitteilungen*, vol. 4, 160, Porz-Wahn, Geophys. Beratungsdienst der Bundeswehr, 1970.

28 G. Farmer, 'What's happened to the southern hemisphere weather?' *Weather*, vol. 38, no. 12, pp. 387–8, 1983. G. R. Bigg, 'El Niño and the Southern Oscillation', *Weather*, vol. 45, no. 1, pp. 2–8, 1990.

29 P. Handler, 'The effect of volcanic aerosols on global climate', *J. Volcanology and Geothermal Processes*, 1989.

18 WHAT CAN WE DO ABOUT IT?

1 W. O. Roberts and H. Lansford, *The Climate Mandate*, p. 106, San Francisco, W. H. Freeman & Co., 1979, 197 pp.

2 M. H. Glantz, 'Value of a reliable long-range climate forecast for the Sahel – a preliminary assessment', Boulder, Colorado, National Center for Atmospheric Research, 1976; and *The Politics of Natural Disaster – the Case of the Sahelian Drought*, New York, Praeger, 1976.

3 'Climatic change: its potential effects on the United Kingdom and the implications for research', London, HMSO, 1980, 19 pp.

4 *WCP Newsletter* no. 1, 15 August 1980, World Climate Programme office, WMO Secretariat, case postale no. 5. Geneva.

5 C. Tickell, *Climatic Change and World Affairs*, originally published 1977 in the series of Harvard Studies of International Affairs (no 37), and published by Pergamon Press, 1978.

6 Information supplied by Dr S. H. Schneider from US Department of Agriculture statistics.

7 Such runs of extremes have occurred in a number of parts of the world since the 1950s and have caused serious losses to the insurance industry.

8 I am indebted to Dr A. Bourke of Dublin, formerly President of the World Meteorological Organization's Commission for Agricultural Meteorology, for much of the information in this paragraph.

9 Editorial (here abbreviated and partly paraphrased) in *Climatic Change*, vol. 2, pp. 203–5, Dordrecht, Holland, and Boston, USA, Reidel, 1980.

10 A policy of producing surpluses would be liable to attack and ridicule, branding the 'wheat mountains' as a scandalous waste and a missed opportunity to lower prices.

11 R. A. Bryson and T. J. Murray, '*Climates of Hunger*', Madison, University of Wisconsin Press, 1977, 171 pp.

12 Tickell, op. cit., p. 37

NOTES

13 In a recent article the Chairman of the Esso Petroleum Company in Britain, Dr A. W. Pearce (in the 47th Melchett Lecture to the Institute of Energy, 11 December 1979) argued that with the world's known oil reserves now sufficient for less than thirty years consumption at the present rate and the rate of new oil finds declining, oil is now too precious to use for heating. It should be progressively reserved as transport fuel and for the production of lubricants, fertilizers and other chemical products.

14 In Tickell, op. cit.

15 See J. M. Walker's Conference Report in *Weather*, vol. 35, no. 11, pp. 332–5, 1980.

411

SUGGESTIONS FOR FURTHER READING

GENERAL

Lamb, H. H. *Climate: Present, Past and Future – Volume 1: Fundamentals and Climate Now*, London, Methuen, 1972, 613 pp.

Lamb, H. H. *Climate: Present, Past and Future – Volume 2: Climatic History and the Future*, London, Methuen, 1977, 835 pp.

National Academy of Sciences *Understanding Climatic Change: A Program for Action*, Washington, DC, 1975, 239 pp.

Wigley, T. M. L., Ingram. M.J, and Farmer. G. (eds) *Climate and History: Studies in Past Climates and their Impact on Man*, Cambridge University Press, 1981, 530 pp.

SPECIFIC TOPICS

Braudel, F. *The Mediterranean and the Mediterranean World in the Age of Philip II*, trans. S. Reynolds, 2 vols, London, Fontana/Collins, 1972 and 1973, 1375 pp.

Butzer, K. W. *Studien zum vor- und frühgeschichtlichen Landschaftswandel im Sahara*, Mainz, Abhandlungen der Akademie der Wissenschaften und Literatur, math.-naturwissenschaftliche Klasse No. 1, 1958; 49 pp.

Godwin, H. *History of the British Flora*, 2nd edn, Cambridge University Press, 1975.

Iversen, J. *The Development of Denmark's Nature since the Last Glacial*, Copenhagen, Geological Survey of Denmark, V series, no. 7-C, 1973.

Le Roy Ladurie, E. *Times of Feast, Times of Famine: A History of Climate since the Year 1000*, trans. B. Bray, New York, Doubleday, 1971, 426 pp.

Parry, M. L. *Climatic Change and Agricultural Settlement*, Folkestone, England, Dawson, and Hamden, Conn., Archon Books, 1978, 214 pp.

Pfister, C. *Agrarkonjunktur und Witterungsverlauf im westlichen Schweizer Mittelland 1755–1797*, Bern, Geogr. Inst. der Universität, 1975, 279 pp.

Schneider, S. H. *The Genesis Strategy: Climate and Global Survival*, New York and London, Plenum Press, 1976, 419 pp.

Tickell, C. *Climatic Change and World Affairs*, Harvard Studies in International Affairs No. 37, Oxford, Pergamon, 1978, 75 pp.

412

ADDITIONAL FURTHER
READING FOR THE
SECOND EDITION

Other aspects and other treatments are introduced by:

Bradley, R. S. and Jones, P. D. *Climate since A.D. 1500*, London and New York, Routledge, 1992, 679 pp.

Briffa, K. R., Bartholin, T. S., Eckstein, D., Schweingruber, F. H., Karlén, W., Zetterberg, P. and Eronen, M. 'Fennoscandian summers from A.D. 500: temperature changes on long and short time scales', *Climate Dynamics*, vol. 7, pp. 111–19, 1992.

Folland, C. K., Owen, J. A., Ward, M. N. and Colman, A. W. 'Prediction of seasonal rainfall in the Sahel region of Africa using empirical and dynamical models', *Journal of Forecasting*, vol. 10, pp. 21–56, 1991.

Grove, J. M. *The Little Ice Age*, London and New York, Methuen, 1988, 498 pp.

Handler, P. 'Possible association between the climatic effects of stratospheric aerosols and the surface temperatures of the eastern tropical Pacific Ocean', *Journal of Climatology*, vol. 6, pp. 31–41, 1986.

Hansen, J., Fung, J., Lacis, A., Rind, D., Lebedeff, S., Ruedy, R. and Russell, G. 'Global climate changes as forecast by the Goddard Institute for Space Studies using a 3-dimensional model', *Journal of Geophysical Research*, vol. 93 (D8), pp. 9341–64, 1988.

Harington, C. R. (ed.) *The Year Without a Summer? World Climate in 1816*, Ottawa, Canadian Museum of Nature, 1992, 576 pp.

Idso, S. B. *Carbon Dioxide and Global Change: Earth in Transition*, Tempe, Arizona, Institute of Biospheric Research, 1989, 292 pp.

Lindgren, S. and Newmawn, J. 'The cold and wet year 1695: A contemporary German account', *Climatic Change*, vol. 3, no. 2, pp. 173–87, 1981.

Newhall, C. G. and Self, S. 'The volcanic explosivity index (VEI): an estimate of explosive magnitude for historical volcanism', *Journal of Geophysical Research*, vol. 87(C2), pp. 1231–8, 1982.

Stothers, R. and Rampino, M. R. 'Historic volcanism, European dry fogs, and Greenland acid precipitation, 1500 B.C.–A.D. 1500', *Science*, vol. 222, pp. 411–3, 1983.

INDEX

Bold page numbers refer to illustrations.

414